To Bill
with continuing
appreciation for my
teacher

Tom
4/15/13

Urban Assemblages

This book takes it as a given that the city is made of multiple partially localized assemblages built of heterogeneous networks, spaces and practices. The past century of urban studies has focused on various aspects – space, culture, politics, economy – but these too often address each domain and the city itself as a bounded and cohesive entity. The multiple and overlapping enactments that constitute urban life require a commensurate method of analysis that encompasses the human and non-human aspects of cities – from nature to socio-technical networks, to hybrid collectivities, physical artefacts and historical legacies, and the virtual or imagined city.

This book proposes – and its various chapters offer demonstrations – importing into urban studies a body of theories, concepts, and perspectives developed in the field of science and technology studies (STS) and, more specifically, Actor-Network Theory (ANT). The essays, research articles and interviews included in this volume examine artefacts, technical systems, architectures, places and eventful spaces, the persistence of history, imaginary and virtual elements of city life, and the politics and ethical challenges of a mode of analysis that incorporates multiple actors as hybrid chains of causation. The chapters are attentive to the multiple scales of both the object of analysis and the analysis itself. The aim is more ambitious than the mere transfer of a fashionable template. The authors embrace ANT critically, as much as a metaphor as a method of analysis, deploying it to think with, to ask new questions, to find the language to achieve more compelling descriptions of city life and of urban transformations. By greatly extending the chain or network of causation, proliferating heterogeneous agents, non-human as well as human, without limit as to their enrolment in urban assemblages, ANT offers a way of addressing the particular complexity and openness characteristic of cities.

By enabling an escape from the reification of the city so common in social theory, ANT's notion of hybrid assemblages offers richer framing of the reality of the city – of urban experience – that is responsive to contingency and complexity. Therefore *Urban Assemblages* is a pertinent book for students, practitioners and scholars as it aims to shift the parameters of urban studies and contribute a meaningful argument for the urban arena and government policies.

Ignacio Farías holds a PhD in European Anthropology of the Humboldt University of Berlin, is Senior Researcher of the Social Science Research Center Berlin and Associate Researcher at the Diego Portales University in Santiago de Chile. His main research topics include social and cultural theory, cultural urban studies, economic sociology and anthropology of tourism.

Thomas Bender is University Professor and Professor of History at New York University. He is a historian of the United States. His books include *Toward an Urban Vision*, *The Unfinished City* and he has co-edited *Urban Imaginaries*. He also writes on urban design and development issues for various publications, including *The New York Times* and the *Harvard Design Magazine*.

Questioning Cities
Edited by Gary Bridge, *University of Bristol*, UK and
Sophie Watson, *The Open University*, UK

The 'Questioning Cities' series brings together an unusual mix of urban scholars under the title. Rather than taking a broadly economic approach, planning approach or more socio-cultural approach, it aims to include titles from a multidisciplinary field of those interested in critical urban analysis. The series thus includes authors who draw on contemporary social, urban and critical theory to explore different aspects of the city. It is not therefore a series made up of books which are largely case studies of different cities and predominantly descriptive. It seeks instead to extend current debates through, in most cases, excellent empirical work, and to develop sophisticated understandings of the city from a number of disciplines including geography, sociology, politics, planning, cultural studies, philosophy and literature. The series also aims to be thoroughly international where possible, to be innovative, to surprise, and to challenge received wisdom in urban studies. Overall it will encourage a multidisciplinary and international dialogue always bearing in mind that simple description or empirical observation which is not located within a broader theoretical framework would not – for this series at least – be enough.

Global Metropolitan
John Rennie Short

Reason in the City of Difference
Gary Bridge

In the Nature of Cities
Urban political ecology and the
politics of urban metabolism
*Erik Swyngedouw, Maria Kaika and
Nik Heynen*

Ordinary Cities
Between modernity and
development
Jennifer Robinson

Urban Space and Cityscapes
Christoph Lindner

City Publics
The (dis)enchantments of urban
encounters
Sophie Watson

Small Cities
Urban experience beyond the
metropolis
David Bell and Mark Jayne

Cities and Race
America's new black ghetto
David Wilson

Cities in Globalization
Practices, policies and theories
*Peter J. Taylor, Ben Derudder,
Piet Saey and Frank Witlox*

**Cities, Nationalism, and
Democratization**
Scott A. Bollens

Life in the Megalopolis
Lucia Sa

Searching for the Just City
*Peter Marcuse, James Connelly,
Johannes Novy, Ingrid Olivio,
James Potter and Justin Steil*

**Globalization, Violence and the
Visual Culture of Cities**
Christoph Lindner

Urban Assemblages
How Actor-Network Theory
changes urban studies
*Edited by Ignacio Farías and Thomas
Bender*

Urban Assemblages

How Actor-Network Theory changes urban studies

Edited by Ignacio Farías and Thomas Bender

Routledge
Taylor & Francis Group

LONDON AND NEW YORK

First published 2010 by Routledge
2 Park Square, Milton Park, Abingdon, Oxon OX14 4RN

Simultaneously published in the USA and Canada
by Routledge
270 Madison Avenue, New York, NY 10016

Routledge is an imprint of the Taylor & Francis Group, an informa business

Typeset in Times New Roman by
Swales & Willis, Exeter, Devon
Printed by the MPG Books Group in the UK

British Library Cataloguing in Publication Data
A catalogue record for this book is available from the British Library

Library of Congress Cataloging-in-Publication Data
Farías, Ignacio, 1978–
    Urban assemblages : how actor-network theory changes urban
    studies / Ignacio Farías and Thomas Bender.
        p. cm.—(Questioning cities)
    Includes bibliographical references.
    1. Cities and towns—Growth. I. Bender, Thomas. II. Title.
    HT119.F37 2009
    307.76—dc22   2009010537

ISBN 10: 0-415-48662-9 (hbk)
ISBN 10: 0-203-87063-8 (ebk)

ISBN 13: 978-0-415-48662-0 (hbk)
ISBN 13: 978-0-203-87063-1 (ebk)

Contents

List of illustrations   xi
List of contributors   xiii
Acknowledgements   xvii

Introduction: decentring the object of urban studies   1
IGNACIO FARÍAS

PART 1
Towards a flat ontology?   **25**

1   Gelleable spaces, eventful geographies: the case of
    Santiago's experimental music scene   27
    MANUEL TIRONI

2   Globalizations big and small: notes on urban
    studies, Actor-Network Theory, and geographical scale   53
    ALAN LATHAM AND DEREK P. MCCORMACK

3   Urban studies without 'scale': localizing the
    global through Singapore   73
    RICHARD G. SMITH

4   Assembling Asturias: scaling devices and cultural
    leverage   91
    DON SLATER AND TOMAS ARIZTÍA

    Interview with Nigel Thrift   109
    IGNACIO FARÍAS

viii   *Contents*

PART 2
A non-human urban ecology   **121**

5   How do we co-produce urban transport systems
    and the city? The case of Transmilenio and Bogotá   123
    ANDRÉS VALDERRAMA PINEDA

6   Changing obdurate urban objects: the attempts to
    reconstruct the highway through Maastricht   139
    ANIQUE HOMMELS

7   Mutable immobiles: building conversion as a
    problem of quasi-technologies   161
    MICHAEL GUGGENHEIM

8   Conviction and commotion: on soundspheres,
    technopolitics and urban spaces   179
    ISRAEL RODRÍGUEZ GIRALT, DANIEL LÓPEZ GÓMEZ AND NOEL GARCÍA LÓPEZ

    Interview with Stephen Graham   197
    IGNACIO FARÍAS

PART 3
The multiple city   **207**

9   The reality of urban tourism: framed activity and
    virtual ontology   209
    IGNACIO FARÍAS

10 Assembling money and the senses: revisiting
    Georg Simmel and the city   229
    MICHAEL SCHILLMEIER

11 The city as value locus: markets, technologies, and
    the problem of worth   253
    CAITLIN ZALOOM

12 Second empire, second nature, secondary world:
    Verne and Baudelaire in the capital of the nineteenth century 269
    ROSALIND WILLIAMS

    Interview with Rob Shields   291
    IGNACIO FARÍAS

Postscript: reassembling the city: networks and
urban imaginaries 303
THOMAS BENDER

Index 325

Illustrations

Figures

1.1	Gigs and events of Santiago's EMS	34
1.2	Netlabels related to Santiago's EMS	35
1.3	The EMS's digital media	35
1.4	The EMS's project ecology	36
1.5	The spaces of Santiago's EMS: socialization, performance and rehearsal	39
1.6	The temporality of the EMS's performance spaces	41
2.1	Heuristic schemas of geographical scalar hierarchy	58
2.2	Alternative morphologies of spatial organization: networks, flows and atmosphere	66
3.1	Time-line of key dates, 1889–2002	77
3.2	Locations of Singaporean law firms' overseas offices prior to their involvement in JLVs	78
5.1	Transmilenio in Bogotá	124
5.2	Transmilenio articulated bus	127
5.3	Transmilenio bus station	129
5.4	The Avenida Caracas	130
6.1	Municipal Flat	143

Tables

2.1	The world's biggest marathons	56
3.1	Ten largest off-shore (UK and US) firms prior to JLVs	76
3.2	Ten largest on-shore (Singaporean) firms prior to JLVs	78
3.3	Joint law ventures, 1999–2002	81

Contributors

Tomas Ariztía is Assistant Professor in the Sociology Department at Universidad Diego Portales and Associate Researcher of ICSO at the same university. He holds a PhD in Sociology from the London School of Economics for his dissertation on the assembling of middle class cultures in Chile. He also holds an MA in Sociology from the Universidad Católica de Chile. His main areas of interest are consumption, cultural economy and cultural sociology. He is interested in bridging contemporary social theory and empirical ethnographic research.

Thomas Bender is University Professor of the Humanities and Professor of History at New York University. His work has focused on American intellectual history and the history of cities and city culture. Most recently, he has been exploring the nature of historical narrative and alternative framings for national history, which has resulted in two books: *Rethinking American History in a Global Age* (2002) and *A Nation among Nations: America's Place in World History* (2006). His works in the field of urban studies include *Community and Social Change in America* (1978); *Toward an Urban Vision* (1975); *New York Intellect: A History of Intellectual Life in New York City, from 1750 to the Beginnings of Our Own Time* (1987); *The Unfinished City: New York and the Metropolitan Idea* (2002); and (with Alev Çinar) *Urban Imaginaries: Locating the Modern City* (2006).

Ignacio Farías is Senior Researcher at the Social Science Research Center Berlin and Associate Researcher of ICSO at the Universidad Diego Portales in Chile. His research interests are urban studies, social and cultural theory, tourism studies, economic sociology and cultural production. He is currently researching creative production in artists' studios and architecture offices. He has published in *Zeithistorische Forschungen/ Studies in Contemporary History, EURE: Latin American Journal of Urban and Regional Studies* and *Berliner Blätter. Ethnographische und ethnologische Beiträge*. Together with José Ossandón he has also edited two volumes on new approaches to social systems theory.

Noel García López holds an MA in psychosocial research for the Universitat Autònoma de Barcelona (UAB) and works as a researcher in Spora Sinergies, a psychosocial research consultancy in Barcelona, Spain. He is the founder of Ciutat Sonora, an international research network on sound anthropology. He has published in different edited volumes on sound spaces, technopolitics, and sound practices of urban daily life.

Michael Guggenheim is a researcher at the Department of Anthropology at the University of Zürich. His research interests are science studies, sociological theory, sociology of expertise and sociology of architecture. He is currently working on a history and ethnography of change of use of buildings. He is the author of *Organisierte Umwelt* (transcript, 2005), an ethnography of environmental consulting. He recently edited (together with Ola Söderström) *Re-Shaping Cities: How Mobility Shapes Urban Form* (Architext, Routledge, 2009).

Anique Hommels works as an Assistant Professor at the Department of Technology and Society Studies of Maastricht University. She studied issues of urban sociotechnical change and the problems involved in urban redesign from an STS perspective. This research resulted in the publication of her book *Unbuilding Cities* in 2005 (MIT Press) and in articles in *Urban Affairs Review* and *Science, Technology and Human Values*. She currently works on large-scale technological infrastructures and their vulnerabilities.

Alan Latham teaches at the Department of Geography, University College London. His research focuses on issues around sociality, mobility and public space. His main research interests are everyday patterns of sociality through which urban dwellers go about 'making a world' for themselves in the city, globalization and the cultural economy of cities, and corporeal mobility, particularly in terms of how certain internationally mobile individuals and groups use globalization to create life projects that are strung across enormous distances. He has published his work in journals such as *Environment and Planning A, Environment and Planning D: Society and Space, European Urban and Regional Research*, and *Progress in Human Geography*.

Daniel López Gómez is a researcher and Assistant Professor in the Department of Psychology, Universitat Oberta de Catalunya (UOC). He is currently working from an STS perspective on the implementation of new technologies in caring settings, specifically home telecare. There are three main matters of concern in his work: 1) the emergence of new spatialities and temporalities of care; 2) the emergence of new practices of caring and security due to the increasing importance of the technologies of accountability in this context; and 3) the enactment of hybrid forms of autonomy and independence. He has recently published some papers on these topics in *Environment and Planning D: Society and Space*, the *Sociological Review*

Monographs, Space & Culture and *Alter: European Journal of Disability Research.*

Derek P. McCormack is University Lecturer in Geography in the School of Geography, and a fellow of Hertford College, Oxford. He has published on geographies of affectivity and materials, the spaces of moving bodies, and the relation between cinema and geopolitics. His research has been at the forefront of a number of agendas within geography, including non-representational theory, performance and affect. His work has appeared in major geography journals including *Transactions of the Institute of British Geographers, Progress in Human Geography*, and *Environment and Planning D: Society and Space.*

Israel Rodríguez Giralt is a Professor of Social Psychology in the Open University of Catalonia, Barcelona, Spain. As an STS researcher he has experience in social research and particularly in science, technology and democracy studies. His Ph.D. thesis (2008) concerned the implications that an STS approach has for the analysis of contemporary collective action. His current research interests are the role of the Internet in how expert and non-expert knowledge is combined by social movements in strategies developed to participate in public debates and to create mechanisms for more participatory and creative models of democracy. He is editor of *Malababa: Contrapublicidad, resistencias y subculturas* and associate editor of *Athenea: Revista de Pensamiento e Investigación social.*

Michael Schillmeier is an Associate Professor at the Department of Sociology at Munich University, Germany. He teaches mainly disability studies, science and technology studies (STS), sociology of the body/senses, social theory and methodology. He received his Ph.D. ('More than Seeing: The Materiality of Blindness') at Lancaster University, UK. He writes mainly on the dynamics of social ordering, collectivity and the material praxis of social knowledge. He has published his work in journals including *Human Affairs, History of Human Sciences, Mobilities, Environment and Planning D*, and *Disability & Society.*

Don Slater is Reader in Sociology at the London School of Economics. His main research involvements are in economic sociology and theories of consumption and market society; information technology, new media and development; and visual culture (particularly photography). Publications include *Consumer Culture and Modernity; The Internet: An Ethnographic Approach* (with Daniel Miller); and *The Technological Economy* (with Andrew Barry). He is currently working on a book called *New Media, Development and Globalization.*

Richard G. Smith is Senior Lecturer in Human Geography, and Co-Director of the Centre for Urban Theory, at Swansea University in the UK. His research on global and world cities, published over the past decade in such journals as *Cities, Urban Geography, IJURR, Annals of the Association of American Geographers, Environment and Planning A*, and *Progress in Human Geography*, has pioneered new network approaches. His current research concerns the poststructuralist city.

Manuel Tironi is an Assistant Professor in the Department of Sociology at Pontificia Universidad Católica de Chile. He holds a Master's degree in city and regional planning from Cornell University and a Ph.D. in urbanism from Universitat Politècnica de Catalunya. His research has focused on three main areas: 1) creative industries and the cultural economy of cities; 2) innovation and organizational studies; and 3) nature and society. His work has been published in *EURE: Latin American Journal of Urban and Regional Studies, European Planning Studies, Journal of Urban Regeneration and Renewal, Journal of Urban Technology* and several edited books.

Andrés Valderrama Pineda is currently a Ph.D. student at the Design and Innovation Research School, Technical University of Denmark and visiting scholar at the Universidad de los Andes in Colombia. He researches on the design of large urban technological systems; technology and society; engineering studies; and de-colonial theory. He has published in *Revista de Estudios Sociales de la Ciencia (REDES), Built Environment* and *Technology and Culture*.

Rosalind Williams is the Bern Dibner Professor of the History of Science and Technology in the Program in Science, Technology, and Society at the Massachusetts Institute of Technology. She served as Dean of Undergraduate Education and Student Affairs at MIT from 1995 to 2000. A cultural historian of technology, she has published books on the emergence of a 'dream world' of consumption in late-nineteenth-century France; on underground worlds, real and imagined, as models of a technological environment; and on the manifestations of the 'information age' at MIT at the turn of the millennium. In her current work she is examining the transformation of the relationship between human beings and the Earth in the later nineteenth century, using imaginative literature as a source of evidence and insight regarding the unprecedented human domination of the planet.

Caitlin Zaloom is a cultural anthropologist and an Associate Professor at New York University, where she teaches urban studies and American studies in the Department of Social and Cultural Analysis. Her book *Out of the Pits: Traders and Technology from Chicago to London* examines the cultural geographies of financial markets and the material forms of economic reason. Currently, she is researching Americans' contemporary practices of financial debt in relation to home, health and faith.

Acknowledgements

We wish to thank the Center for Metropolitan Studies, particularly Heinz Reif and Oliver Schmidt in Berlin and Volker Berghan in New York. We would like to express our gratitude to all those at Columbia University who made possible the occasion for the initial discussion of the themes of this book and those at the Technical University in Berlin who made possible the organization of an international workshop on the subject, and especially to the many participants of both events whose questions and arguments greatly advanced our thinking. We wish to thank New York University for appointing Ignacio Farías as a visiting scholar and for providing the funds for Thomas Bender to travel to Santiago de Chile. We are grateful to the Dean of Social Sciences Benjamin Vicuña at Diego Portales University for facilitating our meeting there. These two visits were crucial to the development of the project and book. Ignacio Farías would also like to thank the people at the School of Sociology of the Diego Portales University, and his colleagues at the Social Science Research Center Berlin.

Introduction

Decentring the object of urban studies

Ignacio Farías

We are perhaps confronting a Tardean moment in urban studies. Myriads of small, lateral and almost peripheral changes, petty movements and subtle displacements are occurring which could suddenly gain momentum and dissipate flows of imitative repetition, radically changing the field. This hasn't yet happened, and its likeliness is something we certainly cannot know, but exploring these micro and subtle displacements can unveil surprising tendencies. Indeed, scholars in urban studies have begun to explore relational, symmetrical and even flat perspectives to make sense of cities, urban phenomena and transformations, thus challenging conventional understandings of their object of study. Even though we certainly can't speak of critical numbers, there have been some remarkable works (particularly Amin and Thrift 2002). These have involved not just a change in the vocabulary, but also the discovery of new settings and new objects of research. The city and the urban do indeed look quite different when explored with symmetrical and radically relational eyes. But how different, something we still cannot fully foresee, is what this collective volume of fresh new works in urban studies attempts to explore.

This volume engages in a much needed exploration in urban studies beyond the strong structuralistic programme still informing the largest portions of the field. It is worrying, to say the least, to ascertain that the last significant theoretical quantum leap in urban studies occurred in the 1970s with the Marxist political economy. This strong programme of urban studies still underlies the most influential approaches and issues in urban research, from global city networks to scalar structuration over notions of a symbolic and a creative economy. Sticking to this rather dated paradigm entails indeed various risks: the risk of taking meta-narratives of structural change for an explanation of urban life; the risk of losing sight of the actual complex and multiple cities we live in; the risk of disconnection from contemporary theoretical developments in social sciences. This volume explores a way out of this 'urban impasse' diagnosed over 15 years ago by Nigel Thrift (1993). It is neither a call for a new form of urban studies nor a reader. It is rather an exploration of the new insights into the city that can be gained if one dares to engage in urban studies with the theoretical tools of contemporary social science. And a timely question in that context is certainly how Actor-Network Theory (and post-ANT developments and discussions) might change urban studies.

There is also something Heideggerian about this volume, for it is the question itself that is primarily at stake. Consequently, its very terms, the nouns, not the verb, are subject to careful evaluation. In one way or another, most contributors to this volume engage indeed in both an appraisal of the composition, strength and temperament of this actor, ANT, of which it is claimed that it could act upon and even change urban studies, and a less or more critical diagnostic of the current state, conventions and blind spots of urban studies. Thus, this volume is not just the first collective experiment in exploring the city, its urban life, spaces and collectives, with tools provided by ANT, but it is also an experiment in which the collectives compounded, ANT and urban studies, are under scrutiny, being tested and eventually redefined.

Now, before explaining how and why the three parts of this volume were assembled the way they were, I would like to introduce the reader to some of the tools ANT has to offer for urban studies and discuss how ANT can challenge urban studies. The overall challenge – and this should be clear from the outset – affects the stable and bounded way urban studies has mostly conceived the city. The notion of urban assemblages in the plural form offers a powerful foundation to grasp the city anew, as an object which is relentlessly being assembled at concrete sites of urban practice or, to put it differently, as a multiplicity of processes of becoming, affixing sociotechnical networks, hybrid collectives and alternative topologies. From this perspective, the city becomes a difficult and decentred object, which cannot any more be taken for granted as a bounded object, specific context or delimited site. The city is rather an improbable ontological achievement that necessitates an elucidation.

OPENING UP THE ACTOR-NETWORK TOOLBOX FOR URBAN STUDIES

Latour's shifting positions regarding the label Actor-Network Theory, first chopping it into pieces (1999) and then vindicating it (2005), suggest that we are not dealing here with a clearly defined object. Mostly only spelled out letter by letter, ANT is an acronym criticized by many and pragmatically used by many more to rapidly convey a sense of the particular kind of research they are involved in and sometimes even praising. In any case, as soon as the composite 'actor-network' and the notion of 'theory' are articulated together, this is done very carefully and complemented with further definitions of the kind of approach actually meant by that: the study of associations, a symmetrical perspective, a sociotechnical analysis, and so on.

This should make it evident that ANT is less a matter of precise definitions than one of an (allegedly) shared sense regarding the objects researchers investigate and are curious about, and the kind of studies and discussions they engage in. It would be inaccurate to define it as a theory, for it does not aim at providing explanatory theoretical constructs for any particular state of affairs. It involves rather a certain sensibility towards the active role of non-human actors in the assemblage of the world, towards the relational constitution of objects, and the sense that all this calls for symmetrical explanations. We could add little to what has been already written about ANT's mode of engagement with the world and research (Callon 2001; Latour 2005; Law and Hassard 1999), but let us recall these three central principles: radical relationality, generalized symmetry, and association.

It is well known that what ANT does is in the first place to extend the principle of relationality beyond language (as in de Saussure), beyond culture (as in Lévi-Strauss) and beyond communication (as in Luhmann) to all entities: hence its radicality. Objects, tools, technologies, texts, formulae, institutions and humans are not understood as pertaining to different and incommensurable (semiotic) realms, but as mutually constituting each other (Law 1992). The methodological principle sustaining such radical relationality is called by Michel Callon (1986) the principle of generalized symmetry, which pleads for the use of a common conceptual repertoire to describe and analyse the relations between humans and non-humans. Now, this unveiling of hybrid chains of actants partaking of the social does not aim at deconstructing the social, but at understanding the associations that make up the social. The social is thus not a thing, but a type of relation or, better, associations between things which are not social by themselves (Latour 2005). These three general principles also act as mediators to wider strands of theory and research providing for constant overflows. ANT emerges thus as an imbroglio of theories, concepts and studies constantly being framed, overflowed and reframed. As a sensibility towards research and the world, it shouldn't surprise anyone to discover that through and behind ANT multiple theoretical fibres act surreptitiously, pushing this reshaping of urban studies sometimes in quite different directions. One should then ask: which are the materials, workers, contractors, machines and engineers of this 'open building site' (Callon 2001: 65) called ANT that are particularly relevant for reshaping urban studies?

One should first state that the featuring of the active role of non-humans has allowed ANT to establish multiple connections and even to enrol and canalize different traditions within the field of science and technology studies (STS). Paying attention to these interconnections is particularly crucial, for indeed the understanding of cities as sociotechnical systems and enormous artefacts was initially developed by historians and sociologists of technology researching Large Technological Systems (Coutard 1999; Hughes 1983; Summerton 1994) and the Social Construction of Technology (Pinch and Bijker 1984). Two influential examples of these are Thomas Hughes's (1983) study of electrification in the Western metropolis and Aibar and Bijker's (1997) study of town planning, both as technologies differently framed and negotiated by different groups of actors, articulating material and social components of the city simultaneously.

The articulation of these perspectives with an ANT perspective on cities was indeed a major endeavour of Graham and Marvin's celebrated book *Splintering Urbanism* (2001). Indeed, their basic understanding of the city as a process of assembling together technical and social aspects remains being a major theoretical fibre informing ANT studies of the urban, even after Coutard and Guy's (2007) fair critique of a soft form of technological determinism expressed in its universal alarmism. The latest studies presented by Graham have shown the enormous potential of this approach to unveil urban processes otherwise overlooked in urban studies. His work on strategies of urban warfare has patently shown that urban infrastructures are a life-or-death matter for cities and citizens (Graham 2005, 2006a, 2006b). As highly political as the latter is the issue of urban maintenance and repair taken up by Graham together with Thrift (2007). Particularly the interconnection between a politics of maintenance and repair and economic cycles of acquisition and disposal has become today a major subject of geopolitical struggles and environmental politics.

Besides featuring the city as a 'mecanosphere' (Amin and Thrift 2002: 78), ANT encourages and offers tools to the urban scholar for studying and reflecting on the architectural and built environment of cities and its production. Indeed, a particularly noticeable development is the enormous growth of ANT-inspired studies of architectural practice. David Brain (1994) was probably the first to feature architects as engineer-sociologists making sense with things. Also Michel Callon's (1996) little-known reflections on architectural work of conception and projection have been crucial in originating a new strand of research on the work of 'heterogeneous engineering' (Law 1987, 1992) conducted in architects' studios and the role of physical models, perspective drawing, 3-D design computer programs and the like (Houdart 2006; Yaneva 2005a, 2005b, 2008). Gina Neff and colleagues' (2009) current research and findings on cognitive dissonance and uneasy collaboration among architects, builders and engineers triggered by virtual 3-D models of buildings shows from yet another perspective that buildings are not static objects, but 'moving projects' (Latour and Yaneva 2008), which certainly do not stop moving when built. The work of Simon Guy, Ola Söderström, Ralf Brand and others emphasizes also this latter point. The dynamic co-evolution of the built environment and society is at the core of their innovative works on issues such as sustainable architecture (Guy and Moore 2005), sustainable mobility (Brand 2008) and the mediatory role of the built environment.

Only one small step separates all these intellectual strands acting through ANT from a notion of 'cyborg urbanization' (Gandy 2005) which in recent years has often been heard in urban studies (an early example is Swyngedouw 1996). Despite some slightly technology-driven uses of it (Mitchell 1998, 2003), this notion introduces two further nuances on what it means to look at the city through the eyes of ANT. Indeed, at least since Donna Haraway's (1985) 'A Manifesto for Cyborgs' was published, the notion of this cyborg as a 'hybrid creature composed of organism and machine' (1985: 1) involves primarily a challenge for common notions of human 'nature'. Consequently, 'cyborg urbanization' displaces the focus of attention from the city as a sociotechnical system and enormous apparatus to these hybrid forms of being human, of shaping an urban public sphere and experiencing the city through technological networks and life-supporting systems. Rob Shields' (2006) discussion of 'Flânerie for Cyborgs' suggests that the cyborg is not simply defined as a body with prostheses, but, like the flâneur, it emerges relationally at concrete sites of action, such as home, market, workplaces, states, schools, hospitals, churches and electronic spaces like the web. Cities thus constitute particularly intensive places of cyborgization, in which multiple cyborgian existences and practices are bundled together. Regarding a cyborgian urban public realm, Girard and Stark's (2006) research on the socio-technologies of public assembly in New York City after 9/11 is quite suggestive, for it shows that urban publics are enabled by the multiple sociotechnical set-ups necessary for sensing and experiencing, sense-making and imagining, and taking action and demonstrating, rather than a public dialogical urban sphere.

The second avenue of study and reflection opened by this notion of 'cyborg urbanization' involves the role of urban natures and the political ecologies of cities. ANT's stress on the agency of natures, ecologies and even ethologies, which can classically be brought back to Latour's (1988) and Callon's (1986) studies of microbes and scallops, respectively, has not just inspired scholars to look at urban natures, but also allowed for multiple connections and allies inside and outside the urban studies world. A pioneer in this field is certainly Erik Swyngedouw (1997; Swyngedouw and Heynen 2003), who has devoted years to investigating the political ecology of cities and its complex relations with urbanization, modernization, politics of scale, social justice and other central issues in urban studies. Another interesting forebear is William Cronon's study (1991) of the natural causation of Chicago's path of development, which has become a praised masterwork also by ANT scholars for explaining the progressive composition of a metropolis without invoking the agency of social contexts and other phantasmagorical agents. Yet another case in point is Manuel DeLanda's (2000) history and philosophy of the integral part played by geological and biological processes in the history of the world and particularly of cities. Now, and coming back to the issue of the public assemblies mentioned above, all these elements make particularly urgent the question of whether and how a parliament of things and life forms is possible (Latour 2004). This hasn't been discussed in urban studies, but a good starting point could be found in some of the newest discussions in legal studies about the rights of non-humans (Teubner 2006).

All this said, one should note that ANT involves more than an opening towards dimensions and agents otherwise not considered in urban studies. Its fervent anti-structuralist position, as well as its strong empirical commitment to actual sites and lines of activity, converts it into a major challenge for otherwise widely used and accepted distinctions, neo-Marxist structuralist narratives and explanatory models at work here. This affects sensible issues in urban studies, such as the production of space or the dynamics of urban economies. Indeed, ANT provides a radical account of space and time as consequences, effects or, even, dependent variables of the relations and associations making up actor-networks. From this perspective, space is not an underlying structure produced by capital relations (Smith 1992) or state strategies (Brenner 2004) or whatsoever. Thinking space and scale as a product which somehow becomes independent from the set of practices that produced it (what structuration ultimately means) would involve falling into the trap of fetishism, in the Marxian sense of taking for real and ontologically autonomous what is rather an attribute of particular actor-networks and urban sites. Space, scale and time are rather multiply enacted and assembled at concrete local sites, where concrete actors shape time–space dynamics in various ways, producing thereby different geographies of associations.

Probably the most notable example of this empirical commitment with actual and concrete urban sites is Latour and Emilie Hermant's mosaic *Paris ville invisible* (1998). This book shows that Paris exists in no one space or scale, but is differently enacted at multiple sites. Space, time and the city itself are produced or, better, emerge thus in ways conditioned by the types and extension of the actor-networks operating at these local sites. In this manner ANT destabilizes the autonomy and explanatory priority attributed to space in urban studies, substituting the key notion of sites in plural for it. Sites are defined not by spatial boundaries or scales, but by types and lines of activity, and spaces emerge through the networks connecting different sites (Latour 2005). Thus, while ANT becomes a loosely coupled medium through which post-structuralist (especially Foucault) and non-representationalist (Thrift 2007) notions of space affect urban studies, all these challenge and object to commonly invoked spatial formations in urban studies, be these a globally operating neoliberal economy or locally integrated urban clusters of production.

As this suggests, ANT also involves providing new insights into the dynamics of urban economies. The move in ANT towards the study of economy occurred mainly through Callon's (1998) pioneering examination of the performativity of economics and the resulting challenge to the classical 'new economic sociology' (Swedberg 1997). Economic action, argued Callon (1998), is not embedded in social networks (which the urban sociology and economic geography gladly understood as settled in urban spatial clusters), but it is framed activity which in order to occur requires rather its disentangling from the social. Thus, if economic action is embedded in anything, it is in economics which through knowledge and devices performs market activity. This approach transforms economic anthropology and sociology in such fundamental ways that it necessitates a new approach on the economies of cities beyond the embeddedness paradigm (Granovetter 1985; Zukin and DiMaggio 1990) informing discussions on global cities (Sassen 2005), the creative economy (Florida 2002) and the like. Such a new take has been notoriously attempted by Ash Amin and Nigel Thrift. In *Cities* (2002) they proposed a major shift in the consideration of the economic role of cities. The idea that production clusters obtain competitive advantages from their local environment is questioned and with it the notion of cities as units of economic production. The economic role of cities would rather reside in their demand effects. A new 'post-institutionalist' take on urban institutions is thus implemented that emphasizes their role as instances for demand agglomeration, not regulatory machines.

ANT's reshaping of our view of urban infrastructures, built environments, ecologies, urbanites, practices, spaces, economies and other central issues of urban studies ultimately involves an empirical and philosophical investigation into the ontological status of cities. Indeed, even though ANT started out studying the fabrication of knowledge in laboratories, and was thus at the crux of the 'science wars', ANT wasn't simply about recasting the epistemological status of scientific knowledge, but about understanding the production of such knowledge as a praxis-grounded ontological achievement. Seen in these terms, to argue that it is the ontological status of cities that is at stake here doesn't imply subscribing to the idea that only in recent days would a turn to ontology in STS be taking place (Woolgar *et al.* 2008). At least Bruno Latour's intellectual project has always been about 'the systematic comparison of the modes of existence' (2008: 9), such as science, technology, religion, law or politics, that make up the 'infraphysics of Europe, or regional ontology (depending on how we want to call it)' (2008: 9). And this is not new. Ten years ago Latour (1999) famously referred to Michael Lynch's suggestion that ANT should be really called 'actant-rhizome ontology', a horrible but accurate term, which makes evident the Deleuzian framework in which ANT operates. Indeed, ANT shares some central features of Deleuzian metaphysics, such as the Bergsonian notion (Deleuze 1988) that reality is qualitative multiplicity, to which I will come back in the next section. Beyond that, ANT makes out of Deleuze's philosophy of creation an empirical project focused on the generative capacities of actor-networks and the new entities (objects, technologies, truths, economic actors) and dimensions (times, spaces) brought into being.

One could go on and on and point to further intellectual strands and academic traditions acting and passing through ANT, noting how each of them decentres urban theory and research in different ways. But for the moment it might suffice to make clear that ANT entails an imbroglio of theories, authors, questions and sensibilities, which can't be strictly used as a theory of anything. This ANT-inian imbroglio we just started to open up provides rather a useful and plastic toolbox for urban studies, a broad pool of perspectives for questioning cities in new ways. What all these perspectives have to offer is a rich theoretical ground to develop radically relational and symmetrical understandings of the city: challenging distinctions between global and local, close and far, inside and outside, notions of place, propinquity and boundedness, ceasing to attribute priority to socio-cultural symbols, structures or practices over urban natures and technological infrastructures, radically rethinking the basis of urban power and knowledge and the notion of urban regimes and so on. Thus, opening the actor-network toolbox for urban studies entails not only a new way of posing research questions, but also new ways of doing research in the city. And it provides too a new interdisciplinary space for the interplay of the social sciences and the humanities.

But this certainly isn't an easy task, not least because there are no templates that one could mechanically transport into urban studies. Indeed, many of the rich resources ANT places at the disposal of the urbanist should be first disentangled from the sophisticated sociotechnical settings in which they were originally developed and translated into the urban realm: a broader context, less spectacular than a lab or a trading floor, but much more or very differently complex. The translation as creative and transformative process is indeed the key; for transformed gets not just ANT, but also the very field of urban studies. Indeed, the advancement this volume involves, namely, to explore how ANT might change urban studies and how the city is to be thought after accepting the challenge of ANT, comprises not only attesting the composition and strength of ANT, but engaging also in a critical appraisal of urban studies.

URBAN ASSEMBLAGES; OR DECENTRING THE OBJECT OF URBAN STUDIES

If there is an overall challenge ANT poses to urban studies, it does not consist simply of the partial displacements and subtle changes we identified above. There is indeed no necessity to think that highlighting the sociotechnical composition of cities should be more challenging than flat conceptions of space, or that cyborgian understandings of urbanites are more groundbreaking than questioning the urban embeddedness of economic production. All these emphases involve indeed suggestive avenues of research introducing new discussions and dimensions into the field, but they can't by themselves challenge urban studies as a whole. Rather we need to identify fundamental displacements transforming the very ground of urban studies, affecting the very way cities as objects of research are conceived. And, indeed, if the ontological stance of ANT should be credited for any one particular thing, this should be the immense ethnographic accuracy and analytical sophistication to follow and conceptualize objects. This started, indeed, very early with Latour and Woolgar's (1986) ascertainment of immutable mobiles: objects that despite their displacement in Euclidean space maintain a stable position and remain immutable in network space (see also Law 1986). And the latest developments in (post-)ANT are also associated with an examination of challenging objects, be they fluid technologies (de Laet and Mol 2000) or multiple bodies (Mol 2002).

Such 'object lessons' (Law and Singleton 2005) might be of high relevance for urban studies, for even when conceptualizations of 'the city' might vary a great deal in the details, they remain rather monolithic in the assumption that cities (should) involve some stability of shape, that cities are spatial formations, socioeconomic entities or sites for action or culture that can (should) be positively identified and distinctly delimited. Indeed, the obduracy of conceptions of the city in the field is striking. Since the early attempts of sociologists, historians, economists and urban planners to develop a field of urban studies, the reality of the city has been understood in highly stable and bounded ways. While much of contemporary research has revealed and explored the relational or processual aspects of city life, in most of the cases the city is understood as 'one' entity that can be identified, observed and investigated across multiple contexts of representation and practice. Indeed, what often seem to be radically new ways of imagining cities and their contemporary transformations, highlighting for example connections overcoming urban boundaries or critically assessing socioeconomic divisions, are in most cases pretty much framed in the same old schemes. The result is thus a binary: cities are bounded or fragmented, singular or dual, one or many.

There are at least three major understandings of the city that can be traced back to the early days and that still inform urban research today without much change. In a nutshell, these involve understanding cities as spatial forms, as economic units and as cultural formations. These are, of course, not alternative views, and indeed it is difficult to imagine any interesting urban research not building upon multiple connections between these three and other dimensions. For analytical purposes it is however useful to disentangle these basic understandings to explore the extent to which they converge in the assumption that cities constitute bounded or stable entities, be these spatial, economic or cultural.

So let's briefly rehearse some well-known arguments. The ecological approach to the city brought forward by the Chicago School provided a quite sophisticated spatial understanding of the city, urban life and dynamics of urban growth. The basic premise in the work of Burgess (1925), McKenzie (1926) and, to a lesser extent, Park (1952) was that cities constitute a definite spatial environment within which the human urban community settles down in discernible sociospatial patterns as a result of ecological processes, such as competition for location, invasion, successions and so on. This perspective contributed with crucial concepts and insights to the relationships between neighbourhoods, socioeconomic structure and segregation, the dynamics of real-state markets and other urban phenomena. It also provided the ground for the trade-off approach, social area analysis and other perspectives aimed at understanding the internal spatial dynamics and differentiation of cities as a result of social and economic forces. But barring the arrays of very sophisticated models developed in this line, what this perspective brings to the fore is a pervading understanding of cities as spatial units, as spatially delimited places. One could argue that it was precisely the primacy of such a common-sensical notion of cities that especially in the 1980s made so interesting and attractive the study of urban formations that challenged the idea of bounded spaces (Fishman 1987; Garreau 1991). Edward Soja's notion of *Postmetropolis* (1997, 2000) is a good case in point, for when it gets to its spatial features he speaks of a spread of 'exopolis':

> Some have called these amorphous implosions of archaic suburbia 'Outer Cities' or 'Edge Cities'; others dub them 'Technopoles', 'Technoburbs', 'Silicon Landscapes', 'Postsuburbia', 'Metroplex'. I will name them, collectively, *Exopolis*, the city without, to stress their oxymoronic ambiguity, their city-full non-cityness. These are not only exocities, orbiting outside, they are ex-cities as well, no longer what the city used to be.
>
> (Soja 1996: 238–9)

Interestingly, this direct contraposition of exopolis to the notion of a bounded city is possible because both, city and exopolis, metropolis and postmetropolis, are being primarily imagined as objects in Euclidean and geometrical space.

Another tradition of urban studies, which can be traced back to the classical works of Max Weber (1986) on the role of cities for capitalism, understands the city as an economic unit, sometimes even as an economic actor, which involves particular forms of government and urban regimes. From early on, this notion became highly influential, especially as it was complemented with larger geographical perspectives that understood cities as economic entities competing within larger systems of cities. Walter Christaller's (1933) central place theory, for example, explained size, location and economic development and specialization of cities with a pioneering network model that weighed up the quality of the node ('threshold') and of its ties ('range'). Indeed, a great part of the literature on the contemporary in inter-urban competition, urban systems, global cities and more recently also creative cities is keyed upon this understanding of cities as economic actors, entities that act. Peter Marcuse (2006) has recently pointed to the hidden contents of such notions:

> a city does not compete for the Olympics, certain groups within it do, others often object mightily. This idea of the city as an actor is perhaps the most politically loaded . . . [of] usages, for it implies a harmony of interests within the city; what's good for one (generally the business community) is good for all.
>
> (Marcuse 2006: 3)

As Marcuse notes, this usage is not just common in political talk. Saskia Sassen and other authors discussing globalization and cities sometimes slip too into this usage: 'cities will become the major actors in the new global economy' (Sassen 1991: 14). The city emerges here as a different type of object, not a spatial one, but a network object, which holds its shape and position as a consequence of its relations with other network objects.

The problem with describing cities as global, informational or creative entities is that of synecdoche, that is, taking a part of an object to be the object (Amin and Graham 1997). Now, interestingly, the right avoidance of synecdoches and the politically correct ascertainment of differences between the parts often lead to the collapse of the object. The notions of 'dual city' (Mollenkopf and Castells 1991) and 'divided city' (Fainstein *et al.* 1992) have been used since the 1980s to highlight the intensification of social polariza-tion in cities. Even though these notions were intensively discussed and criticized for being a simplistic analytical construct – the rich and the poor, the global and the local, etc. – the alternatives suggested, such as the frag-mented city or the quartered city, even though they involve a more specific analysis of sociospatial polarization, share the basic idea of a divided city: here, again, urban theory runs into the same Hamletian dilemma we pointed out above: if the city is not to be conceived as one political-economic actor, it is imagined as not holding together. Marcuse is in this regard quite clear: 'you don't get one thing, a "city", when you put all [interest] groups together' (2006: 5). The city, it seems, can be one or simply not be.

A third major line of reflection on cities goes back to Georg Simmel's (1903) groundbreaking essay on the mental life of big-city dwellers, Robert Park's (1925, 1929) understanding of the city as a state of mind and Louis Wirth's (1938) definition of urbanity as a way of life. In the foreground here is a conception of the city as culture – anonymous, vertiginous, public – and a claim for an ethnographic study of its various manifestations. Urban culture is seen thus as what constitutes the city:

> Much of what we ordinarily regard as the city – its charters, formal organisation, buildings, street railways, and so forth ... become[s] part of the living city only when, and in so far as, through use and wont they connect themselves, like a tool in the hand of man, with the vital forces resident in individuals and in the community.
>
> (Park 1925: 3)

Understanding the city as a product of human nature and, particularly, of human culture necessitates thus the investigation of the customs, beliefs, social practices and general conceptions of life prevalent among its inhabit-ants. In the tradition of the Chicago School, neighbourhoods provided the key analytical unit to follow different manifestations of urban culture. Urban culture appears here as the underlying common ground connecting different urban niches, ethnic groups and moral areas. In the tradition of everyday urbanism associated with such authors as Henri Lefebvre (1991), Michel de Certeau (1988) and Manuel Delgado (2007), urban culture is looked for and studied in the ubiquitous spaces of the urban everyday. In this context, culture is not invoked to understand how the city holds together despite its internal differentiation, but to understand the urban as a process of dissemination, transformation and even revolution. Urban culture is not just a 'state' of mind or 'a' way of life manifesting itself in slightly different ways in different urban areas, but involves a creative movement and uncontrollable spaces in between urban areas. Urban culture is recognized here in residual and transient spaces which are seen as even opposed to 'the' city made of bounded places, fixed meaning and big history.

Looking at cities as culture raises the same type of dilemmas we observe above for spatial and economic understandings of cities. The problem lies in the stability of the object. As cultural formations, cities are imagined as fluid objects that remain identical despite slight variations from case to case. They are fluid, for their manifestations vary, but keep their shape and identity. Thus, while it is possible to look at many lively Euro-American cities as resembling expressions of a fluid urban culture, as soon as differences go too far between and within cities, one is confronted with the paradox of cities to which an urban culture cannot be ascribed. And this is precisely the dilemma: understood as fluid cultural objects cities are either the locus of a fluid urban culture or they are not. Seen in this way, it is not surprising that authors writing in the tradition of everyday urbanism point to a profound gap between the city and the urban life. Building on de Certeau and Lefebvre, Delgado (2007) retrieves a generative notion of non-city in order to grasp the relentless urban dissemination of creative practices and transient spaces. The non-city, not to be confused with the negativity of Marc Augé's (1995) account of 'non-places', is about pedestrian tactics, street choreographies, indifference, anonymity and other features that define urban culture:

> What constitutes the city is . . . a non-city which is not the opposite to the city . . . but a perpetual unmaking of what is already made, and a ceaseless remaking of what we saw disintegrating in front of our eyes.
>
> (Delgado 2007: 63, transl. IF)

From this perspective, it is not just that urban culture cannot be predicated of all cities, but that it constitutes a vital force that cannot be fixed and stabilized in any notion of city.

This rough summary is certainly very basic and incomplete and one could fairly object to the gross separation of spatial, economic and cultural dimensions and to some of the implied generalizations. The field is more complex than this and we know it. But, at the same time, one should concede that it catches fundamental notions of important strands of urban thinking enabling thus a fair critique, namely, that the city has been mostly thought of as a bounded unit and a stable object: a spatial form, an economic-political entity, a cultural formation. Moreover, insights and concepts critically developed to question this idea of 'a' city haven't effected any fundamental transformations, for they too rapidly move to the other side of the binary: the city exists or not; it is one or many. While notions of exopolis, divided cities and non-cityness do the work of rejecting the idea of 'a' singular and integrative city, they make the object implode. The mere ascertainment of profound sociospatial polarization and insurmountable gaps between urban culture and the city does not lead to a renovated understanding of the city as an object, but often to a renouncing.

The main contribution that the ANT approach has to offer lies ultimately in the delineation of an alternative ontology for the city, an alternative understating of this messy and elusive object. Such an alternative ontology can here only commence to be imagined. However, three central principles or notions strike as impregnable. Firstly, the city's mode of existence resembles less a notion of 'out-thereness' than one of 'in-hereness'. This principle is derived from John Law's (2004) discussion of the realistic assumptions of social science methods, which have involved understanding objects as entities independent and even prior to our actions, with definite boundaries and constituted within a singular shared reality. Law shows that from early on ANT has involved countering this metaphysics of presence by looking at how objects are being made and unmade at particular sites of practices. Even 'out-thereness' is crafted and produced 'in here', as enormous effort is put into making objects achieve independency and anteriority, and representational practices are involved in producing definiteness. It follows from that discussion that cities are a variable product of concrete practices, constructed *in situ*, in here. A second major point is that the city is not socially constructed, but enacted into being in networks of bodies, materialities, technologies, objects, natures and humans. Bruno Latour (2003) has pointed out that underlying the notion of construction is the assumption of a powerful creator, such as society, structure, culture, discourse, constructing and creating *ex nihilo*. Moreover, since the construction of reality is mostly understood in epistemological terms, the materials and intermediaries involved in the construction are deprived of any active role. This is certainly not how we should imagine the city being locally constituted. Annemarie Mol's (2002) notion of enactment is much more accurate in understanding how objects are brought into being. Similar to the notion of performance (of subjects), the enactment of objects, such as the city, is not just social, but also material, and involves the heterogeneous ecologies of entities acting at sites and contexts of practice.

The principle that follows from this is that the city is a multiple object. This suggestion is, of course, based on Mol's (2002) groundbreaking book *The Body Multiple*, where she suggests that differences in the ways an object (such as a body) is enacted at different moments and sites are not to be understood epistemologically as different perspectives on the object, but ontologically, acknowledging that different realities are being enacted here and there, now and then. This understanding touches upon the Bergsonian assertion of the multiplicity of reality. For Bergson (Deleuze 1988) multiplicity results from the temporal or better durational nature of the real, for it is this temporal dimension that makes potential possibilities and virtual tendencies into a constitutive aspect of the object. Through time and space the same object proves to be multiple. And if this applies to a body moving within a hospital, it certainly applies to cities being enacted in multiple different ways at different sites and times. This feature of the city is also stressed by Amin and Thrift, when they suggest that

> the city is made up of potential and actual entities/associations/ togetherness ... The accumulation of these entities can produce new becomings – because they encounter each other in so many ways, because they can be apprehended in so many ways, and because they exhibit 'concrescence'.
>
> (Amin and Thrift 2002: 27)

Concrescence means here that the encounters and associations of urban entities produce emergent urban realities and that becoming is an emergent process. Now, what is interesting is that these multiple enactments or multiple becomings are not understood as fluidly following from each other, but as discontinuous, even contradictory and mutually exclusive. As Mol (2002) suggests, they collide with each other, overlap, interfere, and form thereby a multiplicity that has to be managed, coordinated or even held apart. Understanding the city as a multiple object involves thus a major challenge for urban research: identifying, describing and analysing these multiple enactments of the city and understanding how they are articulated, concealed, exposed, and made present or absent (cf. Law and Singleton 2005).

The notion of urban assemblages in the plural form provides an adequate conceptual tool to grasp the city as a multiple object, to convey a sense of its multiple enactments. There are many reasons for using this notion. Firstly, it is a term that provides a concrete and graspable image of how the city is brought into being and made present in ensembles of heterogeneous actors, material and social aspects. This idea of a sociomaterial and sociotechnical ensemble is the most literal meaning of assemblage. It is, indeed, a rather imprecise translation of Deleuze and Guattari's (1981) notion of *agencement*, a quite common term in French for the arrangement or fitting together of different elements (Phillips 2006). Correspondingly, it allows and encourages the study of the heterogeneous connections between objects, spaces, materials, machines, bodies, subjectivities, symbols, formulas and so on that 'assemble' the city in multiple ways: as a tourist city, as a transport system, as a playground for skateboarders and free-runners ('parkour'), as a landscape of power, as a public stage for political action and demonstration, as a no-go area, as a festival, as a surveillance area, as a socialization space, as a private memory, as a creative milieu, as a huge surface for graffiti and street-artists, as a consumer market, as a jurisdiction etc. As Marcus and Saka have pointed out, the notion of assemblage becomes thus 'the major thrust' of social and cultural theory to focus on the 'always-emergent conditions of the present' (Marcus and Saka 2006: 101–2). It makes possible a double emphasis: on the material, actual and assembled, but also on the emergent, the processual and the multiple.

The emergent constitution of urban assemblages suggests indeed that they are not to be understood as results of the mere encounter or sum of multiple elements. Attending to the shared etymological root of *agencement* and agency, one should note that urban assemblages enable new types of activity and agency (Muniesa *et al.* 2007). Agency is thus an emergent capacity of assemblages. We move thus towards a second notion of urban assemblages. As DeLanda (2006) points out, *agencements* or assemblages are to be thought of in terms of relations of exteriority. This means that the relations between heterogeneous elements, out of which an urban assemblage is made, do not necessarily alter the identity of each of the particular elements. A tourist urban assemblage might well require the concurrence of political buildings, art galleries or public bus routes and not necessarily alter any of these particular entities. Assemblages do not form wholes or totalities, in which every part is defined by the whole, but rather emergent events or becomings. Urban assemblages designate thus the processes through which the city becomes a real-state market, a filmic scene, a place of memory; it is the action or the force that leads to one particular enactment of the city. It is interesting to notice that Bruno Latour (2005) does not use the noun 'assemblage', but the verb 'assembling' to understand how the social comes together through associations between human and non-human elements. The social, the assembling process, is external to these actants, which do not become social, but remain human, material, technological and biological. Similarly, the notion of urban assemblages understands that the urban is an emergent quality of the multiple assemblage process, which is not pre-existent in the streets, the buildings, the people, the maps etc. The city is thus not an out-there reality, but is literally made of urban assemblages, through which it can come into being in multiple ways.

As with the notion of 'global assemblages' introduced by Collier and Ong (2004), the plural is also crucial. The global, they observe, involves abstract forms, such as a market calculation, which are actualized in complex infrastructural conditions. The global is always thus an actual global. And this actual global is plural. It does not produce similar effects everywhere and its functioning and meaning might greatly vary depending on the multiple determinations of its assemblages. In our terms, the global is multiple, and this multiplicity results from the tension between the virtual and the actual. The actual city, as the global, exists only in concrete assemblages and provides no encompassing form for its multiple enactments. The city is thus a contingent, situated, partial and heterogeneous achievement: an ontological achievement, indeed, as it involves the enactment of an object otherwise inexistent.

THE ASSEMBLAGE OF *URBAN ASSEMBLAGES*

Every book, every object, every project has multiple points of origin. One has to choose some of them to tell stories. In the case of this book an important one was 5 October 2006 at the International Affairs Buildings of Columbia University, where the Center for Metropolitan Studies of Berlin (CMS), to which I was affiliated as a DFG doctoral candidate, was hosting its first conference in New York City. One of the CMS's New York-based board members, Thomas Bender, Professor of History at New York University, provided the opening keynote lecture. To my surprise, in his talk he discussed in detail the different kinds of challenges Actor-Network Theory poses to urban history and urban studies. After that conference Tom and I decided to organize an international workshop on what we called 'Urban Ontologies: Importing ANT into Urban Studies', which took place in May 2007 at the offices of the CMS in Berlin. Even though only a handful of authors contributing to this volume participated in that workshop, the presentations, discussions and exchange among participants made evident that we were not just experimenting with highly interesting perspectives and vocabularies, but entering a whole new field. Later that year, during my three-month stay at NYU, we confronted the real proof of fire: problematizing the lack of connections between ANT and urban studies. We looked for and then enrolled other colleagues interested in joining our enterprise, and we identified a series, series editors, an editor in Andrew Mould, and a press in Routledge that was supportive of our effort. It was a difficult and paradoxical process, for it involved a non-existing object about which our friends and colleagues, six anonymous reviewers (whom we want to sincerely thank) and many other actors seemed to have a lot to say. Perhaps the most decisive moment, before the project became an immutable mobile, occurred one year later with both Tom and me seated at a table on the loud and sunny terrace of the Social Science Building of the Universidad Diego Portales in Santiago de Chile. There we finally began to figure out what this whole book was to be about, from the issues and the structure to the right tone and style.

This story is not just an ingratiating way to thank all the institutions that supported us; it aims rather to highlight the unique geographical trajectory through which this book came into existence. Indeed, Berlin, New York and Santiago de Chile do not count among the major sites for the discussion, enhancement and transformation of ANT. This is a book mostly written at the borderlands of ANT with authors working in eight countries from three continents and writing on case studies from many more places. And this is something clearly reflected in the articles of the book, which do not represent any internal debate within ANT, but seek to open up fresh approaches and new avenues of interdisciplinary research more generally. Written at the boundary between urban studies and ANT, it is also aimed at a broad audience.

The book is structured in three main parts. Each of these seeks to illuminate a different kind of challenge for urban studies and provide new research questions, new subject matters and new vocabularies. The contributions selected for each of these parts have been chosen strategically to illuminate different topics particularly relevant in contemporary urban studies. The chapters are mostly research oriented and exemplary in character. They do not seek to elaborate ultimate positions on any of these issues or represent theoretical discussions. The research chapters of each part are followed by one interview with a leading scholar, whose work partly frames the orientation of that part. Since the 'translation' of actor-network tools into the field of urban studies is being developed in an almost experimental mode, this book has a prospective rather than a retrospective nature. It takes the risk of reflecting on new strands influencing urban studies and it seeks, thereby, to set a new research agenda.

The first part, 'Towards a flat ontology?', offers rich empirical materials and incisive discussions on the spatial and scalar configuration of urban assemblages. Distinctions between the local and the global, small and large, as well as forms of spatial agglomeration, are rendered here into emergent qualities of urban assemblages. Scalar structuration and clustering are rejected as underlying structural processes and consequently as analytical categories for the study of cities. Instead this part moves towards a relational understanding of spatial formations, which doesn't take away space from actors, as though it would exist out there independently, but imagines space (and eventually scale) as fields of activity or better as attributes of certain urban assemblages. It is at this point that the question 'What does it mean for urban studies to take a flat perspective?' becomes crucial. In Chapter 1, 'Gelleable spaces, eventful geographies: the case of Santiago's experimental music scene', Manuel Tironi provides major insights for reimagining the stability of the sociospatial formations of creative scenes, beyond the bounded notions of space informing most studies of creative production. Neither spatial proximity nor spatial fixity hold together the artistic scene under Tironi's scrutiny, but its capacity to spatially and organizationally gel, couple and decouple through its emergent eventuality. In Chapter 2, 'Globalizations big and small: notes on urban studies, Actor-Network Theory, and geographical scale', Alan Latham and Derek P. McCormack explore the usefulness of a notion of spatial flatness by contrasting the kind of account it provides of globalization with an account based on the idea that the world is intrinsically hierarchical and scalar. By looking at the recent history of mass urban sporting events they argue that flat perspectives do not involve getting rid of scales, but redefining them as the assemblage of extensive attributes. In Chapter 3, though, entitled 'Urban studies without "scale": localizing the global through Singapore', Richard G. Smith draws on his research on the restructuring of Singapore's legal services to argue that, through being attentive to how actors define situations and networks (or work-nets) in their terms, using their own dimensions and touchstones (rather than those of the social scientist), it is clear that scalar processes do not exist and consequently should not be invoked to try to explain, for example, how Singapore is participating in the global economy. But what if actors themselves engage in scaling practices based on notions of the global and the local? In Chapter 4, 'Assembling Asturias: scaling devices and cultural leverage', Don Slater and Tomas Ariztía present a rich case study on the assembling of social scales. Closely following the scaling practices and devices put in action by governmental agencies, an international cultural centre, local youth, and themselves as researchers, they unveil how 'fudging' regarding what is meant by 'local', 'global', 'culture' and so on is crucial for attempts at 'leveraging' Asturias into the global circuit of culture. This part closes with an interview with Professor Nigel Thrift, Vice-Chancellor of the University of Warwick. Thrift is one of the most influential scholars in the field of human geography and has led the way for the incorporation of non-representational theories, especially ANT, into the field of urban studies.

A number of explorations into 'A non-human urban ecology' build the core of the second part. While the ascription of an active role to non-human aspects of the city is a well-accepted point in urban studies, it is still rather difficult to find good examples and analyses that successfully both avoid asymmetrical explanations, whether technological determinism or social constructivism, and also engage with urban issues and processes. This second part gathers exceptional works that vividly show that cities and urban life involve a hybrid ecology of natural, material, mechanical and technological elements. They assess the architectures, infrastructures and even sound technologies as key pieces ensuring the stability and change of urban assemblages. In this way, this part opens up the study of the non-human ecology of the city to otherwise overlooked objects and actants. In Chapter 5, 'How do we co-produce urban transport systems and the city? The case of Transmilenio and Bogotá', Andrés Valderrama Pineda takes on a rather classical object for sociotechnical studies of the city, the design and implementation of a transport system, and carefully shows to what extent design decisions regarding particular elements of new transport systems, such as buses or bus stations, were and still are entangled with the physical reorganization of the city and thus in the lived experience of citizens. Anique Hommels explores in Chapter 6, 'Changing obdurate urban objects: the attempts to reconstruct the highway through Maastricht', the far-reaching consequences of such co-productions and the stubbornness and immutability of urban infrastructures once in place, anchored in their own history. Her chapter studies the attempts to reconstruct the highway that cuts through the Dutch city of Maastricht, which from the late 1950s until today (2009) remain unfruitful, and explores the strengths and limitations of three models of obduracy: dominant frames, embeddedness and persistent traditions. But the non-human ecology of the city, particularly its built environment, changes much more rapidly and in much more contingent, flexible and plastic ways than often thought. In Chapter 7, 'Mutable immobiles: building conversion as a problem of quasi-technologies', Michael Guggenheim explores precisely the obduracy and plasticity of buildings and the extent to which they can be considered technologies. His rich fieldwork focusing on legal controversies on change of use suggests that buildings are neither technologies nor objects, but quasi-technological mutable immobiles. In Chapter 8, 'Conviction and commotion: on soundspheres, technopolitics and urban spaces', Israel Rodríguez Giralt, Daniel López Gómez and Noel García López study the effects of sound technologies in the production of urban spaces as a particular sort of technopolitics, looking especially at how collectivities get ordered and organized by two strategies or logics of sound space modulation: a logic of conviction and a logic of commotion. This second part concludes with an interview with Stephen Graham, Professor of Geography at Durham University. His work on urban technological infrastructures has had a major impact in the field of urban studies, contributing powerfully to the move from humanistic understandings of cities towards sociotechnical perspectives.

The third part, entitled 'The multiple city', stresses the plurality of urban collectives populating, circulating in and constituting the urban public sphere, questioning thereby the possibility of an all-encompassing urban assembly. When looked at in detail, as the chapters in this part do, the public spaces and spheres of the city reveal themselves indeed to be plural and to put into play different cities. The city is literally multiple different things, has multiple different forms, gathers multiple different publics, fulfils multiple different functions, triggers multiple different practices, and so on. The public urban sphere is not a singular realm for citizen negotiation of access to spaces, identities, urban representations or values, but is made of multiple orders of value and groups of people often running parallel to each other. Tourists, blind people, traders and writers, to name just those characters mentioned in this part, but certainly many other city publics, are assembled together in multiple ways in the streets and stages of the city, enacting thereby multiple cities: in literary imagination, in touristic worlds, in the economy, in blind everyday life. In Chapter 9, 'The reality of urban tourism: framed activity and virtual ontology', I pose the question 'How can tourism become an urban reality?' and begin by looking at how bus tours, and their complex sociotechnical arrangement, make tourist activity in urban settings possible. The emergence of an urban destination, though, involves the constitution of a virtual urban ontology which enables and is enabled by actual tourist activity. In Chapter 10, 'Assembling money and the senses: revisiting Georg Simmel and the city', Michael Schillmeier shows that, for Simmel, urban spaces already constituted intensive spaces of intermittent circulations and translations of human and non-human configurations. By looking at everyday practices of blind people in the city such as 'going shopping' or 'dealing with money', Schillmeier reassembles Simmel's work on the senses with his insights on the money economy and urban life and opens up thereby the conventional use of Simmel in urban studies, which often suffices in highlighting the psycho-cultural effects of big-city public culture. In Chapter 11, 'The city as value locus: markets, technologies, and the problem of worth', Caitlin Zaloom discusses how urban assemblages enable not just price-setting for the market economy but the production of value in the economy. Not just at the Chicago Board of Trade, a skyscraper enacting the whole history of the city in its relation to the global financial economy, but also at the virtual markets of the financial economy, Zaloom unveils the economic quality and necessity of urban assemblages. In Chapter 12, 'Second empire, second nature, secondary world: Verne and Baudelaire in the capital of the nineteenth century', Rosalind Williams asks whether a different picture of mid-nineteenth-century urbanization and literature emerges if, instead of Charles Baudelaire, one looks at the character and work of Jules Verne. Stressing Verne's continuing connections with Nantes and reading his recently uncovered novel *Paris in the Twentieth Century*, Williams discusses his literary efforts to build a 'human world' in the midst of the rapidly emerging 'human empire' through art, social ties and technologies. This part concludes with an interview with Rob Shields, Henry Marshall Tory Chair and Professor of Sociology, Arts and Design at the University of Alberta. Shields pioneered in the early 1990s the exploration of space from non-representational perspectives and recently again with his explorations of the urban as a virtuality.

This volume, which is something of a prospectus based on rigorous analysis of highly specific urban issues, invites reflection on the ways the contributions have demonstrated what urban studies under the aegis of ANT might be good at doing and what limitations might be revealed. Can such work capture the quotidian experience of the city as well as the way it works? Is there an urban imaginary that could frame urban experiences and articulate multiple urban assemblages? What is the relation of networks to that imaginary? Thomas Bender's 'Postscript: Reassembling the city: networks and urban imaginaries' stresses the usefulness and fascinations of working with the idea, insights and sensibility of ANT, explores the work and agency of both networks and imaginaries in reassembling the city, and finally examines the politics of urban studies built on ANT.

REFERENCES

Aibar, E. and Bijker, W. (1997) 'Constructing a City: The Cerda Plan for the Extension of Barcelona', *Science, Technology & Human Values*, 22(1): 3.

Amin, A. and Graham, S. (1997) 'The Ordinary City', *Transactions of the Institute of British Geographers*, 22: 411–29.

Amin, A. and Thrift, N. (2002) *Cities: Reimagining the Urban*, Cambridge, Oxford: Polity.

Augé, M. (1995) *Non Places: Introduction to an Anthropology of Hypermodernity*, New York: Verso.

Brain, D. (1994) 'Cultural Production as "Society in the Making": Architecture as an Exemplar of the Social Construction of Cultural Artifacts', in D. Crane (ed.), *The Sociology of Culture*, Oxford: Blackwell, pp. 191–220.

Brand, R. (2008) 'People plus Technology: New Approaches to Sustainable Mobility', *Built Environment*, 34(2): 133–9.

Brenner, N. (2004) *New State Spaces: Urban Governance and the Rescaling of Statehood*, New York, Oxford: Oxford University Press.

Burgess, E.W. (1925) 'The Growth of the City: An Introduction to a Research Project', in R. Park and E.W. Burgess (eds), *The City: Suggestions for Investigation of Human Behaviour in the Urban Environment*, Chicago: University of Chicago Press, pp. 47–62.

Callon, M. (1986) 'Some Elements of a Sociology of Translation: Domestication of the Scallops and the Fishermen of St-Brieuc Bay', in J. Law (ed.), *Power, Action, and Belief: A New Sociology of Knowledge?*, London: Routledge & Kegan Paul, pp. 196–233.

Callon, M. (1996) 'Le travail de la conception en architecture', *Situations: Les Cahiers de la recherche architecturale*, 37(1er trimestre): 25–35.

Callon, M. (1998) 'The Embeddedness of Economic Markets in Economics: Introduction', in M. Callon (ed.), *The Laws of the Market*, Oxford, Malden, MA: Blackwell Publishers.

Callon, M. (2001) 'Actor-Network Theory', in Neil Smelser and Paul Baltes (eds), *International Encyclopedia of the Social and Behavioral Sciences*, Amsterdam: Pergamon Press, pp. 62–6.

Christaller, W. (1933) *Die zentralen Orte in Süddeutschland*, Jena: Gustav Fischer.

Collier, S.J. and Ong, A. (2004) 'Global Assemblages, Anthropological Problems', in S.J. Collier and A. Ong (eds), *Global Assemblages: Technology, Politics, and Ethics as Anthropological Problems*, London: Blackwell.

Coutard, O. (1999) *The Governance of Large Technical Systems*, London, New York: Routledge.

Coutard, O. and Guy, S. (2007) 'STS and the City: Politics and Practices of Hope', *Science, Technology & Human Values*, 32(6): 713.

Cronon, W. (1991) *Nature's Metropolis: Chicago and the Great West*, New York: W.W. Norton.

de Certeau, M. (1988) *The Practice of Everyday Life*, Berkeley, London: University of California Press.

de Laet, M. and Mol, A. (2000) 'The Zimbabwe Pump: Mechanics of a Fluid Technology', *Social Studies of Science*, 30: 225–63.

DeLanda, M. (2000) *A Thousand Years of Nonlinear History*, New York: Swerve Editions.

DeLanda, M. (2006) *A New Philosophy of Society: Assemblage Theory and Social Complexity*, London, New York: Continuum.

Deleuze, G. (1988) *Bergsonism*, New York: Zone Books.

Deleuze, G. and Guattari, F. (1981) 'Rhizome', *Ideology and Consciousness*, 8: 49–71.

Delgado, M. (2007) *Sociedades movedizas: Pasos hacia una antropología de las calles*, Barcelona: Anagrama.

Fainstein, S.S., Gordon, I. and Harloe, M. (1992) *Divided Cities: New York and London in the Contemporary World*, Oxford: Blackwell.

Fishman, R. (1987) *Bourgeois Utopias: The Rise and Fall of Suburbia*, New York: Basic Books.

Florida, R. (2002) *The Rise of the Creative Class: And How It's Transforming Work, Leisure and Everyday Life*, New York: Basic Books.

Gandy, M. (2005) 'Cyborg Urbanization: Complexity and Monstrosity in the

Contemporary City', *International Journal of Urban and Regional Research*, 29(1): 26–49.

Garreau, J. (1991) *Edge City: Life on the New Frontier*, New York: Doubleday.

Girard, M. and Stark, D. (2006) 'Socio-Technical Assemblies: Sense-Making and Demonstration in Rebuilding Lower Manhattan', in D. Lazer and V. Mayer-Schoenberger (eds), *Governance and Information: The Rewiring of Governing and Deliberation in the 21st Century*, New York, Oxford: Oxford University Press.

Graham, S. (2005) 'Switching Cities Off', *City*, 9(2): 169–94.

Graham, S. (2006a) 'Cities and the "War on Terror" ', *International Journal of Urban and Regional Research*, 30(2): 255–76.

Graham, S. (2006b) 'Cities, War, and Terrorism: Towards an Urban Geopolitics', *Annals of the Association of American Geographers*, 96(1): 216–18.

Graham, S. and Marvin, S. (2001) *Splintering Urbanism: Networked Infrastructures, Technological Mobilities, and the Urban Condition*, London, New York: Routledge.

Graham, S. and Thrift, N. (2007) 'Out of Order: Understanding Repair and Maintenance', *Theory, Culture & Society*, 24(3): 1.

Granovetter, M. (1985) 'Economic Action and Social Structure: The Problem of Embeddedness', *American Journal of Sociology*, 91(3): 481–510.

Guy, S. and Moore, S.A. (2005) *Sustainable Architectures: Cultures and Natures in Europe and North America*, New York: Spon Press.

Haraway, D. (1985) 'A Manifesto for Cyborgs: Science, Technology, and Socialist Feminism in the 1980s', *Socialist Review*, 80: 65–107.

Houdart, S. (2006) 'Des multiples manières d'être réel', *Terrain*, 46: 107–22.

Hughes, T.P. (1983) *Networks of Power: Electrification in Western Society, 1880–1930*, Baltimore, MD: Johns Hopkins University Press.

Latour, B. (1988) *The Pasteurization of France*, Cambridge, MA: Harvard University Press.

Latour, B. (1999) 'On Recalling ANT', in J. Law and J. Hassard (eds), *Actor Network Theory and After*, Oxford: Blackwell, pp. 15–25.

Latour, B. (2003) 'The Promises of Constructivism', in D. Ihde and E. Selinger (eds), *Chasing Technoscience: Matrix for Materiality*, Bloomington: Indiana University Press, pp. 27–46.

Latour, B. (2004) *Politics of Nature: How to Bring the Sciences into Democracy*, Cambridge, MA: Harvard University Press.

Latour, B. (2005) *Reassembling the Social: An Introduction to Actor-Network Theory*, Clarendon Lectures in Management Studies, Oxford, New York: Oxford University Press.

Latour, B. (2008) 'Coming Out as a Philosopher', in Acceptance speech for the third Siegfried Unseld Preis, Frankfurt.

Latour, B. and Hermant, E. (1998) *Paris ville invisible*, Paris: Les empêcheurs de penser en rond, La Découverte.

Latour, B. and Woolgar, S. (1986) *Laboratory Life: The Construction of Scientific Facts*, Princeton, NJ: Princeton University Press.

Latour, B. and Yaneva, A. (2008) 'Give Me a Gun and I Will Make All Buildings Move': An ANT's View of Architecture', in R. Geiser (ed.), *Explorations in Architecture: Teaching, Design, Research*, Basel: Birkhäuser, pp. 80–9.

Law, J. (1986) 'On the Methods of Long-Distance Control: Vessels, Navigation and the Portuguese Route to India', in J. Law (ed.), *Power, Action and Belief: A New Sociology of Knowledge*, London: Routledge & Kegan Paul, pp. 234–63.

Law, J. (1987) 'Technology and Heterogeneous Engineering: The Case of Portuguese Expansion', in W.E. Bijker, T.P. Hughes and T.J. Pinch (eds), *The Social Construction of Technological Systems: New Directions in the Sociology and History of Technology*, Cambridge, MA: MIT Press, pp. 111–34.

Law, J. (1992) 'Notes on the Theory of the Actor-Network: Ordering, Strategy, and Heterogeneity', *Systemic Practice and Action Research*, 5(4): 379–93.

Law, J. (2004) *After Method: Mess in Social Science Research*, London: Routledge.

Law, J. and Hassard, J. (1999) *Actor Network Theory and After*, Oxford: Blackwell, Sociological Review.

Law, J. and Singleton, V. (2005) 'Object Lessons', *Organization*, 12(3): 331–55.

Lefebvre, H. (1991) *The Production of Space*, London: Basil Blackwell.

Marcus, G. and Saka, E. (2006) 'Assemblage', *Theory, Culture & Society*, 23(2–3): 101–6.

Marcuse, P. (2006) 'What is "the" City?', Unpublished text, Center for Metropolitan Studies, Technical University of Berlin.

McKenzie, R.D. (1926) 'The Scope of Human Ecology', *American Journal of Sociology*, 32: 141–54.

Mitchell, W.J. (1998) 'Cyborg Civics', *Harvard Architecture Review*, 10: 164–75.

Mitchell, W.J. (2003) *Me++: The Cyborg Self and the Networked City*, Cambridge, MA, London: MIT Press.

Mol, A. (2002) *The Body Multiple: Ontology in Medical Practice*, Durham, NC: Duke University Press.

Mollenkopf, J. and Castells, M. (1991) *Dual City: Restructuring New York*, New York: Sage.

Muniesa, F., Millo, Y. and Callon, M. (2007) 'An Introduction to Market Devices', in M. Callon, Y. Millo and F. Muniesa (eds), *Market Devices*, Malden, MA, Oxford: Blackwell, Sociological Review, pp. 1–12.

Neff, G., Fiore-Silvast, B. and Dossick, C. (2009) 'Model Failure: Assemblages, Performances, and Uneasy Collaborations in Commercial Construction', Paper presented to American Sociological Meetings, San Francisco, August.

Park, R.E. (1925) 'The City: Suggestions for the Investigation of Human Behavior in the Urban Environment', in R.E. Park and E.W. Burgess (eds), *The City: Suggestions for Investigation of Human Behavior in the Urban Environment*, Chicago: University of Chicago Press, pp. 1–46.

Park, R.E. (1929) 'The City as a Social Laboratory', in T.V. Smith and L.D. White (eds), *Chicago: An Experiment in Social Science Research*, Chicago: Chicago University Press, pp. 1–19.

Park, R.E. (1952) *Human Communities: The City and Human Ecology*, Glencoe, IL: Free Press.

Phillips, J. (2006) 'Agencement/Assemblage', *Theory, Culture & Society*, 23(2/3): 108–9.

Pinch, T. and Bijker, W. (1984) 'The Social Construction of Facts and Artefacts: Or How the Sociology of Science and the Sociology of Technology might Benefit Each Other', *Social Studies of Science*, 14(3): 399.

Sassen, S. (1991) *The Global City: New York, London, Tokyo*, New York: Princeton University Press.

Sassen, S. (2005) 'The Embeddedness of Electronic Markets: The Case of Global Capital Markets', in K. Knorr-Cetina and A. Preda (eds), *The Sociology of Financial Markets*, New York: Oxford University Press, pp. 17–37.

Shields, R. (2006) 'Flânerie for Cyborgs', *Theory, Culture & Society*, 23(7–8): 209–20.

Simmel, G. (1903) 'Die Großstädte und das Geistesleben', in T. Petermann (ed.), *Die Großstadt: Vorträge und Aufsätze zur Städteausstellung*, Dresden: Jahrbuch der Gehe-Stiftung Dresden, pp. 185–206.

Smith, N. (1992) 'Geography, Difference and the Politics of Scale', in J. Doherty, E. Graham and M. Malek (eds), *Postmodernism and the Social Sciences*, London: Macmillan, pp. 57–79.

Soja, E. (1996) *Thirdspace: Journeys to Los Angeles and Other Real-and-Imagined Places*, London: Blackwell.

Soja, E. (1997) 'Six Discourses on the Post-Metropolis', in S. Westwood and J. Williams (eds), *Imagining Cities: Scripts, Signs, Memory*, London: Routledge.

Soja, E. (2000) *Postmetropolis: Critical Studies of Cities and Regions*, Oxford: Blackwell.

Summerton, J. (1994) *Changing Large Technical Systems*, Boulder, CO, Oxford: Westview Press.

Swedberg, R. (1997) 'New Economic Sociology: What Has Been Accomplished, What Is Ahead?', *Acta Sociologica*, 40(2): 161.

Swyngedouw, E. (1996) 'The City as a Hybrid: On Nature, Society and Cyborg Urbanization', *Capitalism, Nature, Socialism*, 7: 65–80.

Swyngedouw, E. (1997) 'Power, Nature, and the City: The Conquest of Water and the Political Ecology of Urbanization in Guayaquil, Ecuador: 1880–1990', *Environment and Planning A*, 29(2): 311–32.

Swyngedouw, E. and Heynen, N. (2003) 'Urban Political Ecology, Justice and the Politics of Scale', *Antipode*, 35(5): 898–918.

Teubner, G. (2006) 'Rights of Non-Humans? Electronic Agents and Animals as New Actors in Politics and Law', *Journal of Law and Society*, 33(4): 497.

Thrift, N. (1993) 'An Urban Impasse?', *Theory, Culture & Society*, 10(2): 229–38.

Thrift, N. (2007) *Non-Representational Theory: Space, Politics, Affect*, London: Routledge.

Weber, M. (1986) *The City*, Glencoe, IL: Free Press.

Wirth, L. (1938) 'Urbanism as a Way of Life', *American Journal of Sociology*, 44(1): 1–24.

Woolgar, S., Neyland, D., Lezaun, J., Cheniti, T. and Sugden, C. (2008) 'A Turn to Ontology in STS? Ambivalence, Multiplicity and Deferral', Session in 'Acting with Science, Technology and Medicine', Four-Yearly Joint Conference of the Society for Social Studies of Science (4S) and European Association for the Study of Science and Technology (EASST), Rotterdam.

Yaneva, A. (2005a) 'A Building is a Multiverse', in B. Latour and P. Weibel (eds), *Making Things Public: Atmospheres of Democracy*, Karlsruhe, Cambridge, MA: KFZ, MIT Press, pp. 530–5.

Yaneva, A. (2005b) 'Scaling Up and Down: Extraction Trials in Architectural Design', *Social Studies of Science*, 35(6): 867.

Yaneva, A. (2008) 'How Buildings "Surprise": The Renovation of the Alte Aula in Vienna', *Science Studies*, 21: 8–29.

Zukin, S. and DiMaggio, P. (1990) *Structures of Capital: The Social Organization of the Economy*, New York: Cambridge University Press.

Part 1

Towards a flat ontology?

1 Gelleable spaces, eventful geographies

The case of Santiago's experimental music scene

Manuel Tironi

INTRODUCTION: CLUSTERS, FIXITIES AND CULTURAL PRODUCTION

This chapter is about a relatively cohesive group of young individuals – college students, post office clerks, school teachers, unemployed freelancers – who are involved in doing (performing, promoting) avant-garde music in Santiago, Chile. It deals, in brief, with Santiago's experimental music scene. And while music is at stake, the focus will not zoom in to music itself but to the urban spaces and the knowledge economies this scene performs. Or, more straightforwardly, this study focuses on the *stability* of the scene (Bijker 1997). The main question is: how can Santiago's experimental music scene *exist* and, in addition, be productive and innovative?

The link between stability, urban spaces and knowledge hinges around the concept of cluster, an analytical construct that is at the eye of the storm in this study. So, another way to put it is that this chapter deals with the enactment of a creative cluster in Santiago. But the notion of 'cluster' needs to be quotation-marked, because this chapter, rather than utilizing this concept, *deconstructs* it, so to speak. Indeed, if one follows the conventional scholarship on economic agglomeration and localized economies, the applicability of the concept of 'cluster' to Santiago's experimental music scene is not fully clear. The problem does not reside in the scene's material precariousness (although critical), nor in its institutional weakness (certainly an obstacle), nor even in its market marginalization (indeed problematic): against all odds, as will be explained later on, Santiago's experimental music scene *behaves* as a cluster. The scene is reported to have the key defining features of a cluster, such as value-added creation, economic spillovers and horizontal/vertical linkages.

The problem resides in the scene's *spatiality*. Santiago's experimental music scene defies a critical – perhaps the most important – assumption of cluster theory: the supposition of a *bounded spatial ontology*. A robust scholarship on the clustering of economic activity indicates that particular spatial orderings (of people, knowledge, firms, institutions, cultures and objects) create the best conditions for value-added and innovation-oriented economic production.

These spatial orderings may stretch along regions or be circumscribed to neighborhoods, but they always refer to a definite and ontologically closed territory where the 'being there' (Gertler 2003), among people or firms, is enacted: the embeddedness of economic activity in a fixed space, both physical and social, is the condition of possibility for any *localized* economy, out of which emerges a vast array of Marshall-inspired analytical devices (district, milieu, quarter, cluster, etc.). But Santiago's experimental music scene, in spite of its cluster-like functionality, does not conform to the expected cluster spatiality, as will be demonstrated in this chapter. Not only does the physical place of the scene not fit conventional cluster-like territories, but neither does the nature of the agents that create – and are shaped by – these spatialities.

Now we return to stability: if Santiago's experimental music scene lacks the necessary spatiality to reach cluster status and if this spatiality is, in turn, the main source of permanence in cluster creation, then how can the scene reach stability? How can it *exist* without the proper spatial medium? Here we face a dilemma: should we (a) reject Santiago's experimental music scene as a cluster case because it does not conform to its conventional socio-spatial syntax, or should we (b) reject the *definition of space* within conventional cluster theory for being unable to explain emergent and heterogeneous localized economies, such as Santiago's experimental music scene?

Based on ethnographic data and drawing on science and technology studies, especially on actor-network theorists, I explore the second option. In doing so, I understand spatiality as the process by which both the object/ agent and the space in which the former is embedded are mutually enacted. Thus the core challenge regarding Santiago's experimental music scene is not simply to identify the physical place of the scene, but to reveal how this scene is ordered and organized, assuming that in this ordering and organizing spaces and agents are co-constitutive in a topological field (Law 2000; Law and Mol 1994, 2000). The main task, then, is to contest the univocal urbanism underlying cluster theory. Or, from a more politically charged perspective, the main goal is to question the contours of an urban ontology that reproduces, far too dangerously, the elements of Euro-American, modern urbanism of developed countries.

This chapter won't tackle the above enterprise in its full complexity, but it will shed some light on one specific question: are there other ways to understand the spatiality of localized economic activities, particularly cultural industries in developing countries? More a descriptive exercise than an explanatory endeavor, this chapter will try to unveil the contradictions at work in the enactment of Santiago's experimental music scene and to find an analytical device to make sense of the latter without reproducing a circumscribed, linear ontology of the urban.

One could be tempted here – as I indeed am – to let go of the term 'cluster': perhaps, in light of its distinctive features, Santiago's experimental music scene is not a cluster but something *else*. Maybe we need a different analytical/semantic device to indicate – and define – the complexities of an urban production network in which its agglomeration pattern is more liquid/nomadic than solid/fixed. But one reason keeps me from taking that step: paraphrasing Gieryn (2002: 45), if the rigidities of the 'cluster' are to become more friendly for empirical analysis, then perhaps we need to look more closely at the 'middle range'. More straightforwardly, to keep – for now – the notion of 'cluster' as the analytical reference enables us to fully scrutinize the ontological assumptions and empirical limitations of the term and therefore to better understand how exactly the notion of 'cluster' needs to be revisited.

The rest of the chapter, then, falls into four sections. The next section discusses the concept of cluster and revises the debate on localized cultural production, highlighting the problematic notion of 'locality' mobilized by this debate. I then describe Santiago's experimental music scene as a cluster, arguing that *functionally* – its outputs and contours – the scene operates as a de facto cluster. In the fourth section, and based on ethnographic fieldwork, I demonstrate that the spatiality of Santiago's experimental music scene contests conventional cluster theory. I argue that the nature of the geography of the actors that constitute – and are constituted by it – have to be understood in the light of a heterogeneous and non-linear definition of locality. To this end I outline, drawing on an ANT framework, two controversies that define the complex nature of the scene's topology. Finally, I propose the concept of 'gelleable mobile' as the suited analytical tool for understanding Santiago's experimental music scene.

LOCALITY, CULTURAL PRODUCTION AND THE CITY

Within the fields of economic geography, urban sociology and innovation studies, it has become a commonplace to situate the features of the local – shared worlds-of-life and co-presence – at the heart of contemporary economic development (Gertler 2003; Howells 2000). After a century of apocalyptic prognoses predicting the disappearance of locality, first by the hands of modernity and then by the homogenizing forces of globalization, locality proved to be alive and playing a key role in the new global economy (Amin and Thrift 2002; Savage *et al.* 2005; Smith 2000). The *global city* theory (Friedmann 1986; Sassen 1991; Taylor 2004) was one of the results of such a reviving.

Moreover, the so-called 'knowledge economy' put a premium on the 'production, acquisition, absorption reproduction, and dissemination' of tacit knowledge (Gertler 2003: 76). Indeed, economic geographers realized that the force of agglomeration remained strong even though transportation and communication costs continued to decline (Storper and Venables 2004). The answer was, they discovered, in the 'buzz' – that special something 'in the air' – produced by informal, dialogic, temporary and non-articulated interactions. After the seminal insight of Michael Polanyi ('we can know more than we can tell', 1966: 4), economic geography and management studies then recognized that the competitive base of firms did not rely solely on codified, formal and explicit knowledge, but also on tacit, experienced and practical knowledge:

> The idea is that, in a competitive era in which success depends increasingly upon the ability to produce new or improved products and services, tacit knowledge constitutes the most important basis for innovation-based value creation ... when everyone has relatively easy access to explicit/codified knowledge, the creation of unique capabilities and products depends on the production and use of tacit knowledge.
>
> (Gertler 2003: 78–79)

Tacit knowledge, moreover, is intrinsically spatial: it has strong agglomerating effects. The scholarship on the subject has elaborated three interrelated arguments in this respect. First, since tacit knowledge defies codification and is acquired and produced in *practice* by interacting or doing (Howells 2000; Maskell and Malmberg 1999), it is difficult to exchange over a long distance: tacit knowledge is produced in co-presence. Second, tacit knowledge is spatially *sticky*, 'since two parties can only exchange such knowledge effectively if they share a common social context ... [and] important elements of this social context are defined locally' (Gertler 2003: 78). And, third, innovation itself is increasingly supported by *socially organized* learning (Camagni and Maillat 2006; Gertler 2003; Lundvall and Johnson 1994), that is, on a network of interactions between economic entities, research institutions and public agencies operating locally or regionally (Morgan 1997). Locality – propinquity, interpersonal interactions and bounded spaces – became, in sum, the locus of innovation and economic development in the new global era.

This line of research grew in popularity. Alfred Marshall's ideas about industrial districts and the 'industrial atmosphere' were revived, originating a new breed of concepts – clusters, milieus, quarters, districts – thought to capture the benefits of 'being there' for innovation-driven firms and competitive regions. The success of place-specific knowledge-intensive economies such as Silicon Valley reinforced this 'new localism' (Amin and Thrift 2002).

Experts on localized economies broadened their research spectrum to include non-conventional economic sectors, such as cultural and creative industries (Castillo and Haarich 2004). Cultural production became a fundamental sector in the knowledge economy for its economic importance (Kong 2000; Pratt 1997; Scott 2000), but also for its symbolic relevance as an image-making catalyst (Evans 2003). Accordingly, a number of scholars contested the Marxist political economy perspective of urban geographers who saw in creative districts mere gentrified enclaves proper to late capitalism's mode of (urban) accumulation (Deutsche and Ryan 1984; Mele 2000; Podmore 1998; Shaw 2006; Zukin 1989). On the contrary, they understood urban cultural agglomerations as key innovation incubators in the knowledge economy.

A fundamental argument in this direction is that creative and cultural industries, like all knowledge-intensive sectors, have a notable tendency towards clustering (Kloosterman and Stegmeijer 2004; Scott 2000, 2004), because

> [i]n the cultural industries, we typically find relatively small companies which are very dependent on extremely specific high-quality knowledge and which, in addition, have to deal with rapidly fluctuating demand [and on] the development of dedicated suppliers and the creation of an 'atmosphere'.
>
> (Kloosterman and Stegmeijer 2004: 2)

Moreover, it was argued, the locational logic of creative industries is highly sensible to urban features. For example, Florida (2002a, 2002b, 2005) suggests that the 'creative class' – the engine of the new creative economy – is attracted to bohemian, authentic and culturally dense places (Clark *et al.* 2002; Florida 2002b; Lloyd 2004, 2006; Markusen and Schrock 2006; Sabaté and Tironi 2008). Thus economic development thrives in cities gifted with artistic milieus and creative clusters. Indeed, Florida (2005) indicates that the locations of artists and high-tech industries are correlated in cities, while Markusen and King (2003) assert that artistic production expands the region's export capacity, supports regional industries through its services in marketing, architecture and web design, and helps retain current business and residents by enhancing a place's 'lovability'.

The scholarship on localized cultural production, then, has reproduced (without contestation) an ontology of the local as a static, bounded and representational entity. The local appears as the site of informal and primary relations, use values and community; the local is the place of propinquity and parochial, face-to-face interactions. Not surprisingly, field research on creative industries has heavily focused on the (inner-city) 'neighborhood' as the primary object of study (see for example Crewe 1996; Crewe and Beaverstock 1998; Hutton 2006; Indergaard 2003; Sabaté and Tironi 2008). Paradoxically, then, artistic milieus, creative clusters and bohemian districts are simultaneously the epitomes of the new knowledge-driven, global-oriented economy and the last resorts of traditional localism.

It is necessary, then, to approach cultural production and its urban clustering recognizing the power of place, but acknowledging, as well, its heterogeneous, networked and liquid composition, especially when analyzing cities – and practices – that do not conform to conventional urban spatialities.

SANTIAGO'S EXPERIMENTAL MUSIC SCENE [1]

Santiago's experimental music scene (EMS) is marked by a paradox: on the one hand, it's a highly innovative and productive cultural industry. On the other hand, its spatial organizing – including both the space and the actor-networks that create and populate it – does not conform to the conventional definitions of locality utilized by the mainstream research on urban cultural clusters, for, when analyzing the geography of the EMS and the ontological nature of its agents, there is nothing that can be nearly called a 'cluster' or a 'district'. Analytically, there are two possible ways out. One solution is to reject the *case* for not fitting into the model (the EMS is either not innovative or it has – although it is hard to see – a district-like spatiality). The other solution is to reject the *model* for its incapacity to explain the case under scrutiny. The evidence gathered during one year of ethnographic fieldwork supports the second option.

Between January 2007 and March 2008, I conducted ethnographically based research to identify the organizing principles of the EMS. Santiago's alternative music scene has grown significantly since the 1990s, but it was only in the early 2000s that an innovative and independent sub-scene emerged and expanded beyond Chile's national borders. In contrast with the 'mainstream' alternative scene in Santiago, this sub-scene embraced more avant-garde paths of musical exploration (see Kruse 2003 and Thornton 1996 for the entanglements and differences between 'alternative', 'independent', 'popular' and 'mainstream' scenes). The scene gathers a variety of musical projects – from electronica to folk, from *musique concrète* to hiphop – but they all share at least three main principles:

- *Hybridity*: all projects mix and remix different types of musical categories, making it impossible to associate a project with one, established musical identity.
- *Non-conventional procedures*: all projects engage in the exploration for new possibilities of creation (field recordings, circuit bending, plunderphonics, instrument recycling) and diffusion (netlabels, art performances, concept installations).
- *Commercial marginalization*: for the above characteristics, the EMS has little (or no) access to mainstream, commercial markets and audiences. This marginalization is, often, self-inflicted: artistic production is done 'for the love of art' (Bourdieu 1993; Ley 2003).

The last point reverberates with the institutional precariousness of Santiago's EMS. Cultural policy in Chile is still weak and partial. The main policy instrument for the promotion of cultural production is the National Fund for the Arts (Fondart in Spanish). This fund subsidizes cultural and art projects in several categories (including music). However, the priority is still focused on relatively consolidated artists belonging to the 'art scene', especially when it comes to experimental projects. Unknown and unconnected artists outside the established art circuit doing avant-garde music (or theater, installations or audiovisuals) are not likely to benefit from Fondart subsidies. In addition, municipal cultural institutions – perhaps the most fitted for the promotion of small-scale off-market cultural projects – are scarce, manage extremely reduced budgets and orient their limited funds to projects with wider popular impact.

Although small, unpublicized and precarious, Santiago's EMS is highly dynamic. As a matter of fact, it is possible to suggest that the scene function-ally operates as a de facto cluster, for it complies with at least three features of Marshall-like districts. First, the scene is productive and innovative. In other words, the scene produces value added. In spite of its material precariousness, the scene is highly creative. It gathers around 40 projects ranging from one-member sonic projects to more conventional rock-like bands that pivot around a (semi)continuous circuit of live gigs, performances and festivals. The vitality of this circuit has been praised by the international media for its quality and creativity. The scene has been featured by newspapers and maga-zines in New York, Los Angeles and Buenos Aires. Cumshot Records, a collective project of noise music, has been invited to perform at São Paulo Art Biennial, and Pueblo Nuevo, a netlabel focused on avant-garde electronic music, has recently won an important French award. An Australian news-paper commented, in its music section, that one of the 'nicest surprises' of 2006 was 'discovering an incredibly exciting, self-contained scene in Santiago – Gepe, Javiera Mena, Prissa, Julia Rose, World Music – that may just make Chile the "New Sweden" ' (Carew 2006: n.p.).[2] Thus Santiago's EMS, despite its marginal economic position, has entered the global circuit of cultural production, a highly sophisticated and valued niche-oriented sector that is key in the new knowledge economy (Scott 2004).

Second, the effervescence of the scene has produced economic and indus-try spillovers. The capacity of a localized economy to generate multiplier effects on lateral industries is a key indicator of cluster performance (Feldman 2000). Santiago's EMS has triggered the emergence of a quasi-commercial, semi-informal music industry that organizes events, designs flyers and posters, and deals with promotion. In addition, and perhaps more important, the scene is sustained by a number of music labels, most of them on-line. Today there are at least ten netlabels, housed in Santiago, dedicated to the promo-tion of avant-garde music. Some of them – Jacobino Discos, Pueblo Nuevo and Quemasucabeza – have even enlisted international projects.

These labels have little – if any – economic return and they are all run by the same members of the scene. Nonetheless these labels, together with spe-cialized media devices, are instrumental for the publicity of the scene's work. Indeed, Santiago's EMS is supported by several Internet-based music maga-zines and information resources. These weblogs not only promote live gigs and other events of the scene, but also connect the Chilean experimental music community with connoisseur international information. In terms of Bathelt, Malmberg and Maskell (2004), these on-line magazines are the scene's 'global pipelines'.

Third, Santiago's EMS has developed – and is constituted by the inter-action between – vertical and horizontal linkages, a requirement for cluster formation (Bathelt *et al.* 2004; Richardson 1972). The vertical dimension

Figure 1.1 Gigs and events of Santiago's EMS (source: the author)

Figure 1.2 Netlabels related to Santiago's EMS (source: the author)

refers to 'nodes that are functionally dissimilar, but that carry out complementary activities – a situation often described as a production system of input/output relations' (van Heur 2007: 15). The relation established by these differentiated and complementary nodes tends to be based more on cooperation and less on competition, while 'it is the interaction between these nodes that leads to an efficient and economically effective cluster' (van Heur 2007: 15). In Santiago's EMS, vertical linkages are developed around the interactions between music projects and at least three economic nodes: netlabels, venues and media devices. With the three of them, the scene has established relatively cohesive productive links in which music projects benefit from support given by these nodes, while these nodes depend on the 'success' of the scene and its artistic production.

The horizontal dimension, in contrast, refers to nodes undertaking similar activities. Thus the relation between these nodes is based on competition, for the success of one node will be at the expense of others (van Heur 2007: 15). Monitoring, copying and adapting are, therefore, the key actions for horizontally positioned nodes. In Santiago's EMS, horizontal linkages are deployed around the relation between musical projects. However, owing to the reduced scale of the scene and its material precariousness, horizontal linkages are more oriented towards cooperation and less to competition. As will be depicted later, many members of the scene perform in more than one project, and there

Figure 1.3 The EMS's digital media (source: the author)

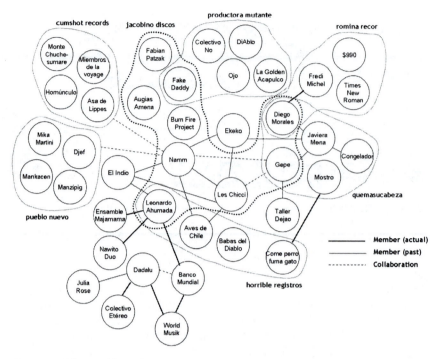

Figure 1.4 The EMS's project ecology (source: the author)

is an extended interchanging among projects of instruments, industry tips, technical services and promotional support.

In sum, Santiago's EMS performs as a cluster; it is productive and innovative, it has produced economic spillovers and it has developed vertical and horizontal linkages. But the EMS lacks a fundamental cluster element: its spatiality. In the next section I turn towards two controversies that illuminate this conflict. Together, they demonstrate that neither the geography underpinning the scene nor the nature of the actor-networks performing it comply with the conventional wisdom on cluster formation and local economic development.

TOPOLOGIES, SPACES AND PERFORMATIVITY

Space is not given in the order of things. Rather the opposite, space, as actor-network theorists remind us, is performed. Topology gives us some clues to deal with this situation. For Law,

topology is a mathematical game which explores the possibilities and properties of different forms of continuity – and the different spaces which express or allow those continuities. And there is, at least in principle, an indefinite number of ways of defining what will count as (spatial) continuity.

(Law 2000: 4)

Why are there an indefinite number of ways to define spatial continuity? Because objects themselves – the stuff from which continuity is made – are heterogeneous and networked assemblages. So, '[t]he ANT argument, then, is that when a (network) object is performed, so, too, a (network) world is being enacted. *But a network world is a topos*' (Law 2000: 6, emphasis in the original). This is to say, objects do not move *in* space; they *create* it. But, since objects are networked entities whose elements include spatiality, spaces create, too, what an object *is*. Spaces, then, 'are being made along with the objects it contains' (Law 2000: 6). The co-construction of spaces and objects allows for indefinite possible continuities or topoi.

If we apply a topological sensibility to cluster theory, we would have to accept at least three facts. First is that a cluster node (commonly a *firm*) and the territoriality in which this actor establishes its relational network are *simultaneously* enacted: there are not spatially deprived 'firms' on the one hand and objectless 'spaces' on the other, but heterogeneously engineered networks of firms and spaces performatively enacted. Second, we would also have to accept that the possible outcomes from this co-construction are undefined or, more strongly, that there isn't *one* actor–space assemblage, but multiple: the shape, content and extension of a cluster – indeed the cluster itself – depend on the performative continuity established by firms and their (co-created) spatiality. And, third, we would have to accept that the definition of 'firm' – together with the one of 'space' – is also obliquous: if objects and spaces are together at the same time in a topological field that multiplies and redefines objectiveness and spatiality, then objects (firms) and spaces are themselves a fluid constellation of elements in constant redelineation.

This conceptualization is at hand when analyzing Santiago's EMS, for neither its actors nor its spaces comply with the conventional wisdom on creative clusters and cultural economic agglomeration – yet the scene performs as one. In this section I explore the topology of Santiago's EMS, that is, I describe (a) the ways in which the actors perform a 'scene' by a number of organizing *spatial* and *knowledge* practices that are (b) co-constitutive, forming (c) an emergent actor–space network that does not comply with traditional cluster features.

Dispersion, mobility and multiplicity

The activities and agents of Santiago's EMS are not concentrated in space. There is nothing like a 'neighborhood', 'quarter', 'district' or 'milieu' that might characterize the physical relation between the different nodes making up the scene. On the contrary, the scene is distributed throughout the city.

Figure 1.5 depicts three key spaces for the scene: socialization spaces (pubs, discotheques, restaurants and live music venues where the scene meets – usually over drinks – to informally share information, talk and watch each other); performance spaces (where the scene musically displays itself and enacts its publicness); and rehearsal spaces (where the bands and projects design and manufacture their products). These three (networked) sites compose the scene's physical spatiality, and each one of them plays a key role in the enactment of the scene. They are what Camagni and Maillat (2006) call the 'support spaces' of a milieu, and cluster theory assumes that these should create an integrated and geographically tight relational meso-space (Giuliani and Bell 2004). The central hypothesis is that the clustering of these support spaces facilitates the flow of information and cooperation, creates an intangible 'industrial atmosphere' that propels creativity and enhances (symbolic and material) economies of scale (O'Connor 2004).

And, indeed, Santiago's EMS shows signs of agglomeration around – a broadly defined – downtown Santiago, but *only* for socialization and performance spaces. Rehearsal spaces, far from being concentrated in downtown Santiago, are distributed throughout the city. So another way to put it is that *production* is dislocated from *consumption*: the spaces where the 'stuff' of the scene is made does not conform to a coherent and relational geography with the spaces where the scene 'shows' and 'recreates' itself.

The reason for this asynchrony has to be found in two fundamental and interrelated facts. First, most of the scene's members are still living – and rehearsing – in their parents' places. Here, then, we face a paradox: an avant-garde movement that, instead of breaking with the 'petit bourgeois way of life' by colonizing and recreating an alternative cultural realm in marginalized urban spaces (as usually artists and cultural entrepreneurs do (Deutsche and Ryan 1984; Ley 2003; Lloyd 2006)), *embeds* itself and its cultural production in the most conventional social and physical space available: the space of the family. The everyday spaces of the practices of middle-class families are entangled with those making experimental sonic explorations. Dadalu, a member of Colectivo Etéreo, makes this situation clear:

> Now we rehearse at [Colectivo Etéreo's] CO2's place, in Las Condes,[3] because his turntables are there, and we rehearse in the dining room. He used to live only with his dad and he had an independent room with his turntables to rehearse. But now a brother arrived to live with them, so now we have to rehearse in the dining room.
>
> (September 2007)

And it is not that the members of the scene *want* to stay or rehearse at their parents', but that they can't afford otherwise. So, secondly, we confront a highly precarious scene that doesn't have the purchasing power or the institutional support to create a creative milieu. The scene hasn't spatialized in a

REHEARSAL SPACES   SOCIALIZATION SPACES   PERFORMANCE SPACES

Figure 1.5 The spaces of Santiago's EMS: socialization, performance and rehearsal (source: the author)

site – like Hackney and Shoreditch in London; the Quartier Latin in Paris; Prenzlauer Berg and Wedding in Berlin; Poblenou in Barcelona; Williamsburg (and Greenwich Village and SoHo) in New York; and Palermo 'Hollywood' in Buenos Aires – in which the aesthetical disposition of artists is performed and their work/life, production/consumption spaces are integrated.

Of course, the scene *occurs* in Santiago. One could argue, therefore, that what is at stake is not agglomeration itself, but the scales of spatial concentration (i.e. that the scale of Santiago's EMS is metropolitan, not sectorial). But the main point is not scalar, but gravitational. The specificity of Santiago's EMS is its multipivotality, its decenteredness: the physical spatiality of the scene is dichotomized, with one spatial layer centered on Santiago's official – and relatively clusterized – cultural space, and with the other dispersed throughout the city, without any focal center.

Yet the spaces of Santiago's EMS are not only multiple, dislocated and decentered: they are also *mobile*. Indeed, the scene is deployed by and actualized in a network of sites, places and venues that are in constant movement. So even when some activities of the scene – those related to consumption, publicity and socialization – may agglomerate in a specific city location, within this (extra) boundedness the scene flows through a (intra) temporary and contingent physical spatiality.

As shown in Figure 1.6, the scene hasn't created 'spatial fixes' to balance the inherent volatility of cultural economies. The scene reports having utilized 45 live music venues in the last three years. Of those, 19 are no longer functioning, and 12 are only occasionally used (less than twice over a 12-month period). And out of the 14 venues in actual use at the time of the research, only two have been functioning for more than one year. Of those, only Bar Uno – a small and unequipped bar in Bellavista – has being symbolically and functionally appropriated as the 'scene's place', a site that is exclusively devoted to the scene and that has entered into the scene's imaginary as a heterotopy, as an abnormal spatial development where the scene can *be*.

However, Bar Uno, at the time of the fieldwork, had lost its prominent position within the scene's network. On the one hand, the venue was going through management restructuring and live gigs were being restricted. But more significantly, Ervo Perez, the head of Productora Mutante, a key actor in the organization of the scene and the 'contact person' between Bar Uno and the bands, was out of the country. In other words, the enactment of the scene hinges on the entrepreneurial abilities of one single individual. Or as a member of DiAblo puts it:

> Ervo left and the pace [of live gigs] slowed down a lot, even though he left [the contact with] Bar Uno and Taller Sol opened. This year we had significantly less activity than last year. Dizzlekzico aren't playing anymore, Innombrable neither, Neurotransmisor hasn't played for a long time, Cumshot is organizing gigs in their own places, so it's like people

PERFORMANCE SPACES – NORMAL
PERFORMANCE SPACES – INTERMITTENT
PERFORMANCE SPACES – NOT IN FUNCTION

Figure 1.6 The temporality of the EMS's performance spaces (source: the author)

> aren't really motivated to organize things on their own . . . I've talked to
> some other bands which have been around the same time as us, and they
> don't want to organize shit.
>
> (Daniel, August 2007)

This brings us to another dimension of the scene's mobile nature: it is not
only that the places of the scene are constantly 'on the move', but that the
organizational logics of the scene are characterized by contingency and temporality. Or, put differently, the geography of the scene is inherently *eventful*:
without any sort of financial, spatial or institutional support, Santiago's
EMS depends on the ever-changing and always unpredictable flow of events
that, even for the members of the scene, seem always beyond control. Rodrigo,
from electronic project Olaüs Romer, makes explicit the contingent nature of
the scene: 'I've tried to organize [events], but I don't know why I have bad
luck. For example, I've organized gigs in which everything is set; I have the
promotional poster printed and then the venue stepped down' (July 2007).

There are no substantial (organizing) pivots to which to recur in case of diversion; thus the enactment of the scene is the result of the conjunction of a variety of contingencies that are 'glued' together for that specific moment only, and everything can always be, until the very last minute, different. Nelson, from Neurotransmisor, links this contingency to the practice – common in the scene – of looking for a venue. Recalling Neurotransmisor's live appearances, he says that one gig took place in Cerrillos, in a 'very well-known venue that was usually lent without cost. Then we played in Casa Usher, just because that was our rehearsal place. The other gig was because a friend saw a venue; he liked it and talked to the owner to do something there' (September 2007). It can therefore be extracted from Nelson's words that Neurotransmisor's live presentations have almost been anomalies, impossibilities, fortuitous exceptions that have somehow materialized.

The result is a liquid spatiality, a network of spaces that moves and emerges or disappears without a bounded framing. The scene doesn't have, in other words, a spatial 'obligatory point of passage' (Callon 1986; Law and Hetherington 2000), i.e. a central node which stabilizes the network, aligning other nodes in the same network, while becoming a mandatory means of access for all actors in the network. Obligatory points of passage permit accumulation, or what Law (2000: 9) calls 'a logic of strategic aggrandisement'. Indeed, Law – reflecting on Latour (1988) – argues that Pasteur's laboratory became an obligatory point of passage for French agricultural production. 'As a result the laboratory *accumulated* resources – a surplus – which might then be re-deployed to increase its location as an obligatory point of passage, a location of *capitalization*' (Law 2000: 9, emphasis added).

Santiago's EMS doesn't have a strategic point of accumulation, a single location which – by its capacity to amass and redistribute the surpluses of the scene – orders the network both spatially and organizationally. This deficiency contradicts, moreover, the evidence gathered in global cities: cultural innovation is always anchored in and propelled by specific spatial and knowledge hubs – whether the Factory in Warhol's Greenwich Village, Can Felipa in Poblenou, Barcelona, or Café Orbis Mundi in Wicker Park, Chicago – that order the network around a centripetal focal point and an institutional, political or social agenda.

In sum, the geography of Santiago's EMS is decentered, mobile, contingent and transient; it is, in brief, spatially multiple. However, this doesn't translate into dysfunctionality. In her study on the diagnosis and treatment of atherosclerosis, Mol (2002) shows that different 'atheroscleroses' are enacted depending on whether the disease is being looked at in a surgery room, in radiography, in the ultrasound department or in the operating theatre (Mol 2002; see also Law 2007). Mol's main argument is that the body, rather than being one unitary volume, is *multiple* – yet this multiplicity is made to cohere through a variety of practices. It could also be said, therefore, that Mol's bodies are not ontologically *centered* on one (biological, medical, expert, moral) definition but that they nonetheless reach stability. The same can be argued for the spatiality of Santiago's EMS: like Mol's multiple bodies, the spaces of the scene perform different spatial syntaxes which seem, from the outside, to be disconnected fragments. But, on the contrary, these decentered spatial networks 'find their way out'.

Obliquitous actants, porous identities

How can the spatiality of Santiago's EMS elude fragmentation and find coherence? A topological answer would have to look at the nature of actors themselves. If objects (actors) and spaces are co-constitutive, then there has to be *something* about the actors of the scene in the making of such multiple – although functionally coherent – spatiality, for the inexistence of a spatial 'obligatory point of passage' hasn't resulted in a network (or cluster) failure.

Analyzing Santiago's EMS, it is possible to recognize that, attuned with the multiplicity of the scene's spatiality, the *units* of this localized economy aren't, in fact, *unitary*. As a matter of fact, the agents of the scene are not firms properly, but *projects*: the unit of economic action in Santiago's EMS, rather than being a self-contained, enduring and institutionally rooted unity, as usually assumed by cluster theory (cf. Moulaert and Sekia 2003), is a task-oriented, market-responsive, transient and flexible actor-network (Boltanski and Chiapello 1999; DeFillippi and Arthur 1998; Grabher 2001, 2002a, 2002b). Moreover, the scene posits two additional challenges to the ontology of the 'firm': the scene's projects are characterized by a *porous* identity and *computer-mediated* intra-cluster 'buzz'.

Cluster theory assumes that firms have fixed and distinct economic identities. One of the key conditions of possibility for cluster formation is that firms have to be complementary and range from large to medium and small units (Marshall [1881] 2005; Mouleart and Sekia 2003). Firms, in other words, establish clear-cut identity frontiers differentiating each unit from the others, in both functional and dimensional terms. This assumption, however, doesn't apply to Santiago's EMS. On the contrary, the (economic and aesthetical) identity of the scene's projects is always being enacted, discussed and challenged.

Indeed, projects are *porous*: they are permeable entities that admit more than one functional and aesthetical identity. First, the members of the scene participate in more than one project; thus the borders from one project to the other are highly permeable. Reflecting on the effervescence of the scene, Walter Roblero – member of Congelador, one of the oldest projects of the scene – states:

> One thing that has helped [the emergence of new projects] is . . . this capacity [of younger people] to split themselves up so many times . . . it's like 'with this guy the project's name is that, and my solo project is that other', and they finally end up with, like, five different bands.
>
> (August 2007)

An important incentive for such rapid project incubation has been the increasing access to cheap music and recording technologies, with the concomitant shift from rock and roll to electroacoustics. 'We don't have the money to buy a drum set and to rent a rehearsal room to play it', says Fakuta (October 2007) from electronic duo Banco Mundial. The arrival of the computer, in this context, was liberatory and allowed for the free-flow development of sonic projects. Talking about Namm's origins in 2003, Pablo says:

> At that time [2003] we began exploring different stuff because Sebastián [Namm's partner] had a computer and he knew a bit about music software ... Then I got a computer and we started recording stuff at my place ... we began improvising, looping, mixing and making tracks. We realized that we could make music without having a band, that sometimes a computer was way more interesting.
>
> (November 2007)

In addition, web-based social networking platforms with streaming applications like MySpace, widely used by the scene, minimized diffusion costs (I'll come back to that later). Then, without major material or technological barriers to entry, the possibilities of forming a 'band' multiplied. 'We made like a pajamas party [with band mate Daniela] and smoked a joint, and we were high and we began recording stuff. *That's* World Musik. And, I don't know, people liked it,' remembers Fakuta (October 2007) about the origins of World Musik, her side-act together with Daniela, also a member of Julia Rose, Colectivo Etéreo and Iris. The result, then, is a highly complex and interlinked ecology of mergers, alliances and temporary collaborations in which the boundaries of each emergent project are constantly redefined. What counts as a 'project' or as a 'band' is the performative effect of a momentary association that has 'gelled' (White 1992; see also Sheller 2004) into a unitary agent.

The porosity of the scene is also evidenced in the multifunctionality of its agents. Rather than performing a specialized division of labor, each actor of the scene has internalized the functions and operations needed for the system's reproduction. There are not 'bands' on the one hand, and 'promoters', 'technicians' and 'designers' on the other: in order to exist, the members of the scene have dislocated their identity locus to permit, without losing their practical coherence, multiple tasking. Namm, besides its musical act, runs netlabel Jacobino Discos and functions as an event organizer; Ervo Perez, member of DiAblo, Colectivo No, Fake Daddy and La Golden Acapulco, is also the head of Productora Mutante, an organization that promotes noise projects and organizes gigs and festivals; Hector, aka Asa de Lippes and member of Mega Toy, runs Cumshot Records, both a netlabel and a semi-formal office for audiovisual services; Rodrigo from Olaüs Romer and Montaña Extendida designs posters and flyers, as does Diego from La Banda's; Mika Martini runs netlabel Pueblo Nuevo; Carlos from Mostro and

Come Perro Fuma Gato does the art for several projects and runs netlabel Horrible Registros; Daniel from DiAblo, Colectivo No and La Golden Acapulco, and Nicolás from Innombrable are sound engineers and, together, have recorded several bands.

The entrepreneurial capability of cultural agents has been studied elsewhere (Lloyd 2006). Moreover, it reverberates with the 'do it yourself' ideology inaugurated by punk and post-punk cultural vanguards in the 1970s. However, these approaches are usually framed in the context of a creative milieu in which collaboration and cooperation are part of a larger aesthetical disposition and refer to the artist's relative position in the (cultural) field (Lloyd 2006). In Santiago's EMS, on the contrary, this impulse is strategic: it's a means of survival in the face of a highly precarious environment. Pablo, member of Namm and director of Jacobino Discos, addresses this issue directly: 'to organize a gig requires sending emails, talking to people, talking to the media, distributing flyers, designing a nice poster, cutting tickets. It's hard for me to be cutting tickets the night of the gig, but I do it anyway' (November 2007).

In sum, projects are not stable entities. On the contrary, they form an ameba-like identity whose limits are being constantly redefined. In addition, the scene builds its communication and collaborative network not through face-to-face interactions but mostly over the Internet. This is, then, the second assault on the 'firm' ontology: the 'buzz', that intangible and informal industrial atmosphere that's 'in the air', is not supported on *human* but on technological mediation.

Santiago's EMS lacks a defined territory and stable economic actors in which to center its activities and practices. But it has managed, nonetheless, to create a local 'buzz' (Storper and Venables 2004), the intangible, spontaneous and informal information and communication ecology produced by spatial propinquity of industries: 'the idea that a certain milieu can be vibrant in the sense that there are lots of piquant and useful things going on simultaneously and therefore lots of inspiration and information' (Bathelt *et al.* 2004: 38). Yet in the case of Santiago's EMS, this 'buzz' is not 'created by face-to-face contacts, co-presence and co-location of people and firms within the same industry and place or region' (Bathelt *et al.* 2004: 38). On the contrary, it is the effect of a web of distantiated relations catalyzed by the Internet or, more specifically, by MySpace.

MySpace[4] is a social networking platform that allows for audio reproduction, streaming and downloading. These capabilities have made MySpace the preferred on-line communication medium for local music scenes in Chile and elsewhere (Noble 2008). For Santiago's EMS, MySpace is entangled in so many diverse ways with the organizing and productive practices of the scene that it is not possible to separate both elements. First, MySpace is where the scene's horizontal linkages are enacted: it brings into being the possibilities of competition and imitation within the scene. For example, Nicolás from Innombrable says:

> Out of the eight tracks of our record, six are there [in MySpace] to be downloaded. In that sense, our entire work is at everybody's disposition . . . you know, sometimes I also walk around[5] MySpace to listen to new music, to see what people are doing and what project people are involved in.
>
> (December 2007)

MySpace has become the scene's *place* of publicness. In the absence of a geographical realm in which competing agents can map out the industry's innovations, MySpace has become the site where the members of the scene can watch each other, check their innovations and hear their new products – to defy, emulate or transubstantiate them.

To be sure, MySpace is not just a promotional platform, a social network on which the scene can observe itself. MySpace, more radically, is a condition of possibility for the scene. 'There is no second opinion', says songwriter Calostro, reflecting on the projects of the scene that don't have an account in MySpace, 'and *that*'s what generates a band status' (July 2007). In other words, in order to be a band you have to be available in public space, and that place, for Santiago's EMS, is MySpace.

We then have a non-human actor (MySpace) that operates not only as an agglutinating device, a distributed and relational panopticon, but also as the realm in which the 'being there' is enacted and performed. So Santiago's EMS contests one of the most basic assumptions of clusters: that the 'buzz' is the only competitive advantage that cannot be formally traded by firms because it is the result of human – therefore unpredictable and elusive – interactions in localized settings. The scene's 'buzz' is, indeed, unpredictable and elusive, but not because of its human and spatially bounded nature.

FINAL DISCUSSION: FROM MUTABLE TO GELLEABLE MOBILES?

A topological description of Santiago's EMS would be like this. First, the scene has – at least – two spatialities, one being its Euclidean space (defined by a set of three-dimensional coordinates) and the other its network-space (defined by the heterogeneous assemblages that constitute every actor-network). Second, the shape (i.e. the scene) reaches continuity by being unstable *both* in network and in Euclidean space: the scene is an actor-network that is 'on the move' in Euclidean space (a mobile and episodic geography that has no center) and whose network-elements (the bands, the projects) are constantly changing (multilayered identities, multitasking, virtual non-human mediators). Law and Mol (2000) call 'fluid spatiality' this topological possibility (see also Law 2000, 2007; Law and Mol 1994) and 'mutable mobiles' the shapes enacted by and the entanglement of these topoi. Reflecting on the case of the Zimbabwe bush pump studied by de Laet and Mol (2000), Law and Mol argue:

the pump is a success which (to do the ethnography very quickly) spreads far and wide in Zimbabwe, into many of the villages that need a new water pump. So why is this? The answer is: *because it changes shape*. Of this pump and everything that allows it to work, nothing in particular necessarily holds in place. Bits break off the device and are replaced with bits which don't seem to fit ... Within Euclidean and network space alike, the bush pump is an object that changes shape. It looks different from one village to the next, and it works differently from one set-up to the next ... It is a *mutable mobile*.

(2000: 5, emphasis in the original)

It could be said that Santiago's EMS configures, too, a fluid spatiality in which the malleability of the scene in both its network and its Euclidean spaces enacts a mutable mobile, and it is precisely because the scene can change in network-space that it can move successfully in regional or Euclidean space. But in Law and Mol's account, malleability is a form of strategic adjustment; immutable mobiles are made mutable to cope with an adverse and unstable environment. The story of the bush pump, for example, is one of successful attuning: like Luhmann's structural coupling (1996), the pump regulates itself to adjust to the (changing) necessities of its environment.

But the story of Santiago's EMS is not one – primarily – of mutation. The scene has to struggle, like the pump in Zimbabwe, against a precarious and inhospitable environment. Yet its solution is not adjustable mutation, but *temporalization*: the network spatiality of the scene doesn't *change* in each event; its network elements stay put and there is no need to alter pieces and bits to make it stable and successful. But, on the contrary, the scene is enacted *temporally* for each – and only for each – specific occasion, dissolving right after; confronted with indeterminacy, the scene has to eventually actualize and assemble itself from scratch every time.

But how should this temporality be conceptualized? Harrison White's concepts of gels and publics (1992) may be of use. Equivalent to Law and Mol's fluid spatiality, White's notion of gel is an attempt to contest the rigidity and 'managerialist' nature of the often reified network concept (Law 2000). Or as Sheller puts it:

Gel is one alternative way of thinking about the interconnected social structures beyond the idea of network ... Whereas a network implies clean nodes and ties, then, a gel is suggestive of the softer, more blurred boundaries of social interaction. It also challenges our notions of scale, boundary, and structuration. Rather than a clean break between the micro and the macro, the private and the public, or the local and the global, we can think in terms of this messy gel of sociality occurring at different scales and scopes.

(Sheller 2004: 27)

Gels, too, are a way of coping with 'inhomogeneous environments', always full of 'lumpy contingencies' (Sheller 2004: 47). But, while mutable mobiles deal with this indeterminacy by changing their network-spatiality, gelling is the process by which socialities couple and decouple from network-domains. For White, the capacity to decouple from networks 'makes it possible for levels of social organization, such as cities and organizations and families, to mix and blur into an inhomogeneous gel' (1995: 12; in Sheller 2004: 47–48). In practice, then, we don't distinguish separated and independent relational realms, but we interact with our ever-changing environment in a gellified manner. The urban and productive practices of Santiago's EMS, for example, are enacted, blurring dichotomies (artist/technician, performance/production, fixity/mobility, clustering/dissemination) and gelling – coupling and decoupling – momentarily, eventually and messily every time it is required.

Here, then, *time* is at stake: how can actors negotiate across gelling socialities and different social times? White (1995: 14; in Sheller 2004: 48) proposes the idea of publics as 'special moments or spaces of social opening that allow actors to switch from one setting to another, and slip from one kind of temporal focus to another'. This possibility is given by the trade-off between ambage and ambiguity inherent in contingent environments. Whereas ambiguity refers to fuzzy meanings and interpretations (to facilitate communication), ambage is a kind of fuzziness related to 'the concrete world of social ties' (White 1992: 107; in Sheller 2004: 48). The main point, as highlighted by Sheller (2004: 48), is that this trade-off 'creates a built-in tendency toward enabling switching from one set of relations to another. It suggests the idea that social actors are never simply one thing, but always carry with them multiple identifications and capacities to "play" different parts at once.'

It could be argued, therefore, that fluid spatiality comprises two types of network 'mutability' or, better said, of survival strategies in the face of environmental hostility: one referred to *space* (i.e. network malleability, the pump) and the other referred to *time* (i.e. network temporalization, Santiago's EMS). Indeed, Santiago's EMS could be conceptualized as a gelleable mobile, as an actor-network that, in order to achieve stability in a fluctuant environment, gels itself episodically and contingently to switch/ assemble different social spaces and times. The fuzziness of the scene's topoi – how to define the 'project', how to define the 'local' – permits this gelling capacity that enables the scene, in turn, to perform heterogeneous roles and to stabilize in spite of external contingencies.

Finally, the notion of a fluid, temporary productive activity – a gelleable mobile – is fundamental for the design and implementation of urban and cultural development policies in Santiago. If policy makers continue to assume an 'immutable immobile' definition of creative clusters, expecting the realization of a geographically and ontologically fixed 'quarter', 'milieu' or 'district', the chances of promoting Santiago's experimental music scene and tapping into its innovations and spillovers will be severely limited. This issue is critical not only for cultural industries in Santiago but in cities of the 'global South': marked by material and institutional precariousness, inorganic growth and hybrid urban structures, the Eurocentric 'cluster' model may not match Santiago's EMS, nor Dakar's 'world music' industry (Pratt 2004), nor Popayan's (Colombia) gastronomic agglomeration (Unesco 2007). Thus the definition of the scene as a gelleable mobile questions, too, the unquestioned transferal of economic development models that don't take into account the specificities of contemporary cities and organizational logics, especially in the developing world.

NOTES

1 The term 'scene' refers to a network of producers related to a particular music style or aesthetic disposition. The consumption side is not explicitly analyzed, mainly because, as in most niche-specific artistic scenes, it's difficult – not to say unnecessary – to separate between producers and consumers (Lloyd 2006).
2 The independent scene in Sweden generated international attention in the early 2000s – with bands such as José González, The Knife, Peter, Bjorn and John, and the Radio Dept – for the mix between a peripheral music market and a highly innovative scene that it represents.
3 A municipality in the western end of Santiago.
4 MySpace launched in 2003 and Rupert Murdoch's News Corp bought the site in 2005 for $580 million. A July 2006 estimate noted 'membership approaching 100 million in total' (Olson, Stefanie, 'MySpace blurs line between friends and flacks', on CNET's News.com, http://news.com.com/MySpace+blurs+line+between+friends+and+flacks/2009–1025_3–6100176.html).
5 *Doy vueltas* in Spanish, literally 'I spin around', an expression that indicates to walk around (or to browse, depending on the context) without a definite objective.

REFERENCES

Amin, A. and Thrift, N. (2002) *Cities: Reimagining the Urban*, London: Polity.
Bathelt, H., Malmberg, A. and Maskell, P. (2004) 'Clusters and Knowledge: Local Buzz, Global Pipelines and the Process of Knowledge Creation', *Progress in Human Geography*, 28: 31–56.
Bijker, W. (1997) *Of Bicycles, Bakelites, and Bulbs: Toward a Theory of Sociotechnical Change*, Cambridge, MA: MIT Press.
Boltanski, L. and Chiapello, E. (1999) *The New Spirit of Capitalism*, New York: Verso.
Bourdieu, P. (1993) *The Field of Cultural Production*, New York: Columbia University Press.
Callon, M. (1986) 'Some Elements of a Sociology of Translation: Domestication of the Scallops and the Fishermen of Saint Brieuc Bay', in J. Law (ed.), *Power, Action and Belief: A New Sociology of Knowledge?*, London: Routledge & Kegan Paul.

Camagni, R. and Maillat, D. (2006) *Milieux innovateurs: Théorie et politiques*, Paris: Economica-Anthropos.

Carew, A. (2006) 'Sounds of '06: Nicest Surprise', *The Age*, 29 December, viewed 17 January 2009, Online, available at: http://www.theage.com.au/news/music/the-sounds-of-06/2006/12/28/1166895393620.html?page=fullpage.

Castillo, J. and Haarich, S. (2004) 'Urban Renaissance, Arts and Culture: The Bilbao Region as an Innovative Milieu', in R. Camagni, D. Maillat and A. Matteaccioli (eds), *Ressources naturelles et culturelles, milieux et développement local (GREMI VI)*, Neuchâtel: Institut de recherches économiques et regionales (IRE).

Clark, T.N., Lloyd, R., Wong, K. and Jain, P. (2002) 'Amenities Drive Urban Growth', *Journal of Urban Affairs*, 24: 493–515.

Crewe, L. (1996) 'Material Culture: Embedded Firms, Organizational Networks and the Local Economic Development of a Fashion Quarter', *Regional Studies*, 30(3).

Crewe, L. and Beaverstock, J. (1998) 'Fashioning the City: Cultures of Consumption in Contemporary Urban Spaces', *Geoforum*, 29: 287–308.

DeFillippi, R. and Arthur, M. (1998) 'Paradox in Project Based Enterprise: The Case of Filmmaking', *California Management Review*, 40(2): 125–39.

de Laet, M. and Mol, A. (2000) 'The Zimbabwe Bush Pump: Mechanics of a Fluid Technology', *Social Studies of Science*, 30: 225–63.

Deutsche, R. and Ryan, C. (1984) 'The Fine Art of Gentrification', *October*, 31: 91–111.

Evans, G. (2003) 'Hard-Branding the Cultural City: From Prado to Prada', *International Journal of Urban and Regional Research*, 27: 2.

Feldman, M.P. (2000) 'Location and Innovation: The New Economic Geography of Innovation, Spillovers, and Agglomeration', in G. Clark, M. Feldman and M. Gertler (eds), *Oxford Handbook of Economic Geography*, Oxford: Oxford University Press.

Florida, R. (2002a) 'Bohemia and Economic Geography', *Journal of Economic Geography*, 2: 55–71.

Florida, R. (2002b) *The Rise of the Creative Class: And How It's Transforming Work, Leisure, Community and Everyday Life*, New York: Basic Books.

Florida, R. (2005) *Cities and the Creative Class*, New York: Routledge.

Friedmann, J. (1986) 'The World City Hypothesis', *Development and Change*, 17: 69–83.

Gertler, M. (2003) 'Tacit Knowledge and the Economic Geography of Context, or the Undefinable Tacitness of Being (There)', *Journal of Economic Geography*, 3: 75–99.

Gieryn, T. (2002) 'What Buildings Do', *Theory and Society*, 31: 35–74.

Giuliani, E. and Bell, M. (2004) 'When Micro Shapes the Meso: Learning Networks in a Chilean Wine Cluster', *SPRU Electronic Working Paper Series*, University of Sussex.

Grabher, G. (2001) 'Ecologies of Creativity: The Village, the Group, and the Heterarchic Organisation of the British Advertising Industry', *Environment and Planning A*, 33: 351–374.

Grabher, G. (2002a) 'The Project Ecology of Advertising: Tasks, Talents and Teams', *Regional Studies*, 36: 245–63.

Grabher, G. (2002b) 'Cool Projects, Boring Institutions: Temporary Collaboration in Social Context', *Regional Studies*, 36: 205–14.

Howells, J. (2000) 'Knowledge, Innovation and Location', in J.R. Bryson, P.W. Daniels, N. Henry and J. Pollard (eds), *Knowledge, Space, Economy*, London: Routledge.

Hutton, T. (2006) 'Spatiality, Built Form, and Creative Industry Development in the Inner City', *Environment and Planning A*, 38: 1819–41.

Indergaard, M. (2003) 'The Webs They Weave: Malaysia's Multimedia Super-Corridor and New York City's Silicon Alley', *Urban Studies*, 40: 379–401.

Kloosterman, R. and Stegmeijer, E. (2004) 'Cultural Industries in the Netherlands – Path-Dependent Patterns and Institutional Contexts: The Case of Architecture in Rotterdam', *Pettermanns Geographische Mitteilungen*, 148: 66–73.

Kong, L. (2000) 'Culture, Economy, Policy: Trends and Developments', *Geoforum*, 31: 385–90.

Kruse, H. (2003) *Site and Sound: Understanding Independent Music Scenes*, New York: Peter Lang Publishing.

Latour, B. (1988) *The Pasteurization of France*, Cambridge, MA: Harvard University Press.

Law, J. (2000) *Objects, Spaces and Others*, Lancaster: Centre for Science Studies, Lancaster University, Online, available at: http://www.comp.lancs.ac.uk/sociology/papers/Law-Objects-Spaces-Others.pdf.

Law, J. (2007) *Actor Network Theory and Material Semiotics*, Lancaster: Centre for Science Studies, Lancaster University, Online, available at: http://www.heterogeneities.net/publications/Law-ANTandMaterialSemiotics.pdf.

Law, J. and Hetherington, K. (2000) 'Materialities, Spatialities, Globalities', in J.R. Bryson, P.W. Daniels, N. Henry and J. Pollard (eds), *Knowledge, Space, Economy*, London: Routledge.

Law, J. and Mol, A. (1994) 'Regions, Networks and Fluids: Anaemia and Social Topology', *Social Studies of Science*, 24(4): 641–71.

Law, J. and Mol, A. (2000) 'Situating Technoscience: An Inquiry into Spatialities', Lancaster: Centre for Science Studies, Lancaster University, Online, available at: http://www.comp.lancs.ac.uk/sociology/papers/Law-Mol-Situating-Technoscience.pdf.

Ley, D. (2003) 'Artists, Aestheticisation and the Field of Gentrification', *Urban Studies*, 40: 2527–44.

Lloyd, R. (2004) 'The Neighborhood in Cultural Production: Material and Symbolic Resources in the New Bohemia', *City and Community*, 3: 343–71.

Lloyd, R. (2006) *Neo-Bohemia: Art and Commerce in the Postindustrial City*, New York: Routledge.

Luhmann, N. (1996) *Social Systems*, Palo Alto, CA: Stanford University Press.

Lundvall, B.-A. and Johnson, B. (1994) 'The Learning Economy', *Journal of Industry*, 1: 23–42.

Markusen, A. and King, D. (2003) 'The Artistic Dividend: The Arts' Hidden Contributions to Regional Development', Project on Regional and Industrial Economics, Humphrey Institute of Public Affairs, University of Minnesota.

Markusen, A. and Schrock, G. (2006) 'The Artistic Dividend: Urban Artistic Specialization and Economic Development Implications', *Urban Studies*, 43: 1661–86.

Marshall, A. ([1881] 2005) *Principios de Economía*, Madrid: Editorial Síntesis.

Maskell, P. and Malmberg, A. (1999) 'Localized Learning and Industrial Competitiveness', *Cambridge Journal of Economics*, 23: 167–86.

Mele, C. (2000) *Selling the Lower East Side: Culture, Real Estate, and Resistance in New York City*, Minneapolis: University of Minnesota Press.

Mol, A. (2002) *The Body Multiple: Ontology in Medical Practice*, Durham, NC, London: Duke University Press.

Morgan, K. (1997) 'The Learning Region: Institutions, Innovation and Regional Renewal', *Regional Studies*, 31: 491–504.

Moulaert, F. and Sekia, F. (2003) 'Territorial Innovation Models: A Critical Survey', *Regional Studies*, 37: 289–302.

Noble, J. (2008) 'The Power of Mobile Texting and the Connecting of Place in Creating a Cultural Scene', ANZCA08 Conference, Power and Place, Wellington, July.

O'Connor, J. (2004) ' "A Special Kind of City Knowledge": Innovative Clusters, Tacit Knowledge and the "Creative City" ', *Media International Australia*, 112: 131–49.

Podmore, J. (1998) '(Re)reading the "Loft Living": *Habitus* in Montréal's Inner City', *International Journal of Urban and Regional Research*, 22: 283–302.

Polanyi, M. (1966) *The Tacit Dimension*, New York: Doubleday.

Pratt, A. (1997) 'Employment in the Cultural Industries Sector: A Case Study of Britain, 1984–91', *Environment and Planning*, 29: 1953–76.

Pratt, A. (2004) 'The Music Industry in Senegal: The Potential for Economic Development', Report prepared for UNCTAD.

Richardson, G.B. (1972) 'The Organisation of Industry', *Economic Journal*, 82: 883–96.

Sabaté, J. and Tironi, M. (2008) 'Rankings, creatividad y urbanismo', *Eure*, 34: 5–23.

Sassen, S. (1991) *The Global City: New York, London, Tokyo*, Princeton, NJ: Princeton University Press.

Savage, M., Bagnall, G. and Longhurst, B. (2005) *Globalization and Belonging*, London: Sage.

Scott, A.J. (2000) *The Cultural Economy of Cities: Essays on the Geography of Image-Producing Industries*, London: Sage.

Scott, A.J. (2004) 'Cultural-Products Industries and Urban Economic Development: Prospects for Growth and Market Contestation in Global Context', *Urban Affairs Review*, 34: 461–90.

Shaw, W. (2006) 'Sydney's SoHo Syndrome? Loft Living in the Urbane City', *Cultural Geographies*, 13: 182–206.

Sheller, M. (2004) 'Mobile Publics: Beyond the Network Perspective', *Environment and Planning D: Society and Space*, 22: 39–52.

Smith, M.P. (2000) *Transnational Urbanism: Locating Globalization*, Boston, MA, Hoboken, NJ: Wiley-Blackwell.

Storper, M. and Venables, A. (2004) 'Buzz: Face-to-Face Contact and the Urban Economy', *Journal of Economic Geography*, 4(4): 351–70.

Taylor, P. (2004) *World City Network: A Global Urban Analysis*, London: Routledge.

Thornton, S. (1996) *Club Cultures: Music, Media, and Subcultural Capital*, Middletown, CT: Wesleyan University Press.

Unesco (2007) *Popayan: UNESCO City of Gastronomy*, http://unesdoc.unesco.org/images/0015/001592/159284e.pdf

van Heur, B. (2007) *The Clustering of Networked Creativity: Between Myth and Reality*, CMS Working Paper Series, Berlin: Center for Metropolitan Studies.

White, H. (1992) *Identity and Control: A Structural Theory of Social Action*, Princeton, NJ: Princeton University Press.

White, H. (1995) 'Where Do Languages Come From? I. Switching between Networks II. Times from Reflexive Talk', Center for the Social Sciences at Columbia University Pre-Print Series, Columbia University, New York.

Zukin, S. (1989) *Loft Living: Culture and Capital in Urban Change*, Piscataway, NJ: Rutgers University Press.

2 Globalizations big and small

Notes on urban studies, Actor-Network Theory, and geographical scale

Alan Latham and Derek P. McCormack

INTRODUCTION

In this chapter we want to explore two ways of thinking about social space: that of space as 'flat' and that of space as 'scalar' – that is to say that spatial is fundamentally defined by properties of scale. We are geographers so the origins of our interest in scale should be obvious. Geography is – or at least appears to be – all about scale, that of the region, the local, the nation and, more recently, the globe. Traditional geography concerned itself with a particular scale. Contemporary geographers are – or at least claim to be – more ambitious. They seek to understand how different spatial scales are 'articulated' with each other: how different social actors, and different social groups, have variable access to different scales, and how some actors can 'jump', or 'shift', with relatively little effort from scale to scale, and place to place, whilst others remain, in the powerful words of Doreen Massey, imprisoned by the local. We want to meditate on this use of the notion of scale. We want to consider why it has become so popular amongst certain parts of the discipline of geography and, through that, within the closely related field of urban studies. And we want to examine some of the limitations inherent in this way of thinking about the spatial. We want also to explore an alternative notion of space – that of space as 'flat'. If the notion of the world as consisting of a range of different scales seems intuitively reasonable, and intrinsically geographical, the idea that we should begin to think of the world as flat, as not defined a priori by spatial hierarchy, sounds unreasonable, perverse even. Yet this is the claim made by Actor-Network Theory (ANT) and those within human geography and urban studies who have been influenced by its arguments. We do not want to give away our punch line right at the start, but what we aim to demonstrate is the usefulness of this notion of spatial flatness as both a theoretical and a methodological heuristic.

We also have a second target, globalization. 'Globalization' is, of course, a faddy term and one which any number of writers have critiqued and placed into question. Nonetheless, if following David Held (in Dickens 2004: 9) we accept a working definition of globalization as being simply the 'stretching and deepening of social relationships and institutions across time and space' then we think it is hard to disagree that in all sorts of ways a range of globalization processes does characterize (and even to a certain extent define) the world that we inhabit. The question of globalization then becomes an issue of how we approach and work to understand the processes of 'stretching and deepening'. Globalization also becomes a question of our stance towards thinking about any underlying driver of globalization. Do we see globalization as a process, or collection of processes, that is in the last instance given momentum by the dynamics of capitalist expansion? Or do we see globalization as essentially under-determined – the outcome of a great range of different phenomena that sometimes seem to push in the same direction, but more often do not? This is not simply a question of taste. Which side we come down on in this issue profoundly shapes how globalization is understood. Returning to the question of scale and flatness, what we want to argue is that, if the central methodological precepts of ANT are followed and we start by assuming in the first instance that the world is flat, then we will end up with some very different accounts of what globalization is and how it is unfolding than if we start from the notion that the world is fundamentally and intrinsically hierarchical and scalar in nature. That – in a sentence – is the purpose of this chapter.

So, where to start? We want to begin with a short empirical case study. What is so intriguing and indeed puzzling about globalization is the strange imbroglios of objects, people, practices and materials that it brings together. There is something magical and wondrous about the mixture of things through which globalization has been (and is being) built. Since making sense of these often confusing and contradictory-appearing imbroglios is the principal challenge facing any social scientist interested in globalization it seems most straightforward to start with an example. Starting with an empirical case study also provides us with a way of judging the usefulness of the two different approaches to spatiality that this chapter seeks to examine.

BERLIN, NEW YORK, LONDON AND THE BIRTH OF THE BIG-CITY MARATHON

On 13 October 1974, 274 runners assembled for the first Berlin Marathon. Looping around a mostly forested route through the city's Grunewald forest, the race was won in a far-from-world-beating time of 2 hours and 44 minutes by the local runner Günter Hallas. Four years earlier Fred Lebow had organized the first New York City Marathon. Like the Berlin Marathon, the New York Marathon was a small affair. Run in Central Park, the event barely registered on the city's consciousness. Only 55 of the 127 starters finished, and the event received only a short report in the *New York Times*. Six years later Lebow inaugurated what was to become a revolution in marathon running. Considering how to widen the appeal of the marathon as a mass-participation event, Lebow and his co-organizers decided to run the marathon through the streets of New York's five boroughs. Profiting from the emerging enthusiasm for jogging and running, and a surprising interest in the idea of running *through* New York (it is not at all clear where this interest came from – it simply seemed to have a sense to people), hundreds of entry forms flooded into the offices of the New York Joggers Club. And, in November 1976, 2,090 people lined up to compete in the first all-city, road-based New York Marathon. In 1977 it was 4,823, in 1978 9,875 and, in 1979 when the British marathon runner Chris Brasher flew in from London to take part in the marathon, 11,532 other runners joined him at the start. 'Last Sunday, in one of the most trouble-stricken cities in the world,' a euphoric Brasher wrote in the British *Observer* newspaper the following week, 'men and women from 40 countries in the world, assisted by over a million black, white and yellow people, laughed, cheered and suffered during the greatest folk festival the world has seen.'[1]

Inspired by the New York Marathon, Brasher set about organizing a London Marathon. Starting in the south-east of the city, passing through the badlands of the city's East End and ending along the Mall, in its first staging in April of 1981 Brasher's marathon attracted over 20,000 entrants, of whom 7,055 made it to the starting line (policing restrictions having limited the maximum number of runners). In September of the same year, when the organizers of the Berlin Marathon, Sport Club Charlottenburg, moved their marathon on to the streets of central West Berlin (increasing their entries almost tenfold in the process), they were no longer part of a barely noticeable athletic sub-culture of long-distance running competitions. Along with New York and London, cities as diverse as Chicago, Honolulu, Amsterdam, Paris, Madrid, Stockholm, Copenhagen, Dublin, Barcelona, Valencia, Rotterdam, Frankfurt, Nottingham and Sheffield had become hosts of mass marathon events.

The question that you might well be asking at this point is: why should you care about any of this? Why should a mass of sweaty joggers once a year pushing through a city's streets be of any great interest to urban theorists? There are at least three answers to this question. Firstly, the emergence of the mass-participation urban marathon, and the fun runs, half-marathons and charity runs and walks to which they are closely related, speaks to the ways that cities are constantly generating new forms of collective life, novel ways of being together. Contrary to so many accounts of life in contemporary cities that stress the ways that our cities are defined by disconnection and the erosion of public life (see Davis 1990; Mitchell 2003; Sennett 1977, 1990; Sorkin 1992; Zukin 1995), if we look carefully it is possible to observe all sorts of emergent practices and events that are invigorating the collective life of our cities (see Amin 2007, 2008; Ryan 2006; Schwartz 1999; Thrift 2008). Secondly, the kinds of physical practices with which marathons are enfolded, such as running, jogging, fitness walking and in-line skating, speak to the multiple ways that the human body accommodates itself to the urban environment. Writers as diverse as Georg Simmel (1950), Lewis Mumford (1938), Walter Benjamin (1977), Margaret Crawford (1991), Wolfgang

Schivelbusch (1979) and Richard Sennett (1977, 1995) have stressed the passivity of the human body within the urban environment. Yet what we can see with events like the marathon is how urban environments also foster all sorts of styles of bodily practice that resituate the body's capacity for physical movement in a whole range of ways (see Borden 2001; Ehrenreich 2007; Latham and McCormack 2008; McCormack and Latham forthcoming; Schwartz 1992). And, thirdly, mass-participation urban marathons and the practices of jogging and running[2] of which they are a part point to the ways in which our world is populated by all sorts of strangely globalizing practices and events, the origins of which are often remarkably difficult to locate.

It is this third point that is most relevant to the concerns of this chapter. The emergence of mass-participation urban marathons is synchronous with the post-Bretton Woods, neo-liberal phase of globalization that has been the focus of so much recent urban research. And, indeed, the map of the world's largest marathons takes in many of those cities which global- and world-city researchers such as Peter Taylor, Saskia Sassen and John Rennie Short place at the top of the global urban hierarchy (see Table 2.1). How then should we understand the spread of the urban marathon from New York out into the rest of the world? And how should we understand this synchroneity? But it is also worth asking, should we be asking these kinds of questions? Or is to ask them to misunderstand what an urban marathon is? Is the interesting thing about the spread of the urban marathon over the past 30 years or so really

Table 2.1 The world's biggest marathons

World's Ten Biggest Marathons (no. of finishers, 2007)	Alpha World Cities, 2004	
New York (38,557)	Alpha++	London
London (35,694)		New York
Berlin (32,638)	Alpha+	Hong Kong
Paris (26,880)		Paris
Chicago (25,532)		Tokyo
Tokyo (25,139)		Singapore
Honolulu (20,692)		
Washington (20,679)	Alpha	Toronto
Boston (20,332)		Chicago
Los Angeles (18,013)		Madrid
		Frankfurt
		Milan
		Amsterdam
		Brussels
		Sao Paulo
		Los Angeles
		Zurich
		Sydney

best understood as a kind of globalization? To try to answer these questions we want to explore what it would involve to frame our understanding of spatiality of the urban marathon through a scalar analysis. This is more than just a prosaic exercise. Since what initially appears most striking about mass urban marathons is that they are big and that they have 'gone global', organizing an account of them that focuses on their scale (i.e. their size and geographical reach) would seem like a logical place to start.

THINKING THROUGH SCALE AND THE MASS-PARTICIPATION URBAN MARATHON

What then would it involve to try to understand the urban marathon through the lens of scale? Well, in the first instance it would be to insist that it is possible and necessary to consider the range of different geographical scales through which the marathon is organized. And, secondly, it is necessary to understand how actions within those different spatial scales are articulated together.

Of course the use of the concept of geographical scale is not straightforward. Introduced as a term that allowed a bridging between the fact of a city as an internal space and the fact that the city itself existed within a web of external relationships, initial conceptions of scale were simplistic to say the least. Peter Taylor (1981, 1982, 1984; see also 2004), who could make a reasonable claim to be the founder of a scale-based urban geography, suggested that cities could be understood as existing within a tri-scalar system of urban, nation state and world economy. Taylor's argument was that it was not possible to make sense of the kinds of relationships that defined a particular urban environment without explicitly understanding the position of that environment within the world economy. The fact that all cities were not alike was not a result of their own unique histories or dynamics of growth. They were different as a result of the different ways they had been 'inserted into the world economy'.

This, frankly, does not give us much analytical leverage. However, it does highlight three of the defining features of scalar-informed analyses of the urban: firstly, that it is possible to define an urban scale that is in some analytical sense distinctive from other scales of action; secondly, that this urban scale is structured (or in some cases determined) through its relationship to a range of other scales (the regional, the national, the global and so on); and, thirdly, that this scalar system was in some sense historically specific. Taylor's account of this scalar organization left relatively little space to consider the role this historical specificity played in the emergence of the world economy – the scale that he saw as being that which 'really mattered'. Other theorists, however, have been much more attuned to the possibility that individual scales, and the scalar hierarchies of which they are a part, are historically contingent. Thus, contemporary processes of globalization represent 'a reterritorialization of both socioeconomic and political-institutional spaces that unfolds simultaneously upon multiple, superimposed geographical scales' (Brenner 1999a: 432).

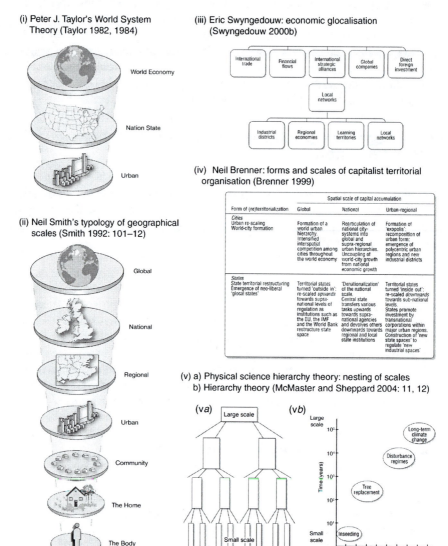

Figure 2.1 Heuristic schemas of geographical scalar hierarchy: Taylor, Smith, Swyngedouw, Brenner, natural sciences (source: the authors, and as shown)

It is this style of thinking about scale, popularized through the writing of Neil Smith (1984, 1992, 1993, 1996, 1997, 2002, 2004; Smith and Dennis 1987), Erik Swyngedouw (1992, 1996, 1997a, 1997b, 2000a, 2000b, 2004a, 2004b) and Neil Brenner (1997a, 1997b, 1998, 1999a, 1999b, 2000, 2001, 2003), informed to varying degrees by an engagement with the writings of the

French philosopher Henri Lefebvre and the Marxist geographer David Harvey, that has come to define much of urban geography and urban studies (see also Delaney and Leitner 1997; Herod 1997; Herod and Wright 2002; McDowell 2001; Marston 2000, 2004). For these writers scale is an indispensable aid to understanding the kinds of social, economic and political transformations that have been reshaping cities over the past quarter of a century. Put simply their argument (like Taylor's) is that the world is organized through a hierarchy of different spatial scales, each of which defines in important ways the capacity for certain actors to act (or indeed to be unable to act). This, of course, sounds intuitively logical. And, indeed, if we look out into the world we do in fact encounter much that seems to adhere to this intuition. Systems of government are organized through nesting scales of responsibility from the local ward and borough to the city, state and national level. When we think of businesses we often categorize them in terms of their scale of operation, from mom-and-pop stores through to national and international corporations like Starbucks or ICI or Exxon-Mobil.

What is clear, however, as Smith, Swyngedouw and Brenner take great pains to explain, is that the ordering of these scales is not an inherent element of the social world – even if it is often felt or understood by many to be so. So, while scale *an sich* might well be 'intrinsic to nearly all geographical inquiry', as Robert McMaster and Eric Sheppard (2004: 1) have claimed, the construction, ordering and maintenance of a particular scalar system is a complex historical-geographical achievement. Different societal systems have quite different ways of organizing space, and the hierarchy of scales that exists within any social formation is the product of complicated and on-going conflicts over the distribution of social power. In the words of Neil Smith:

> the geographical scales of human activity are not neutral 'givens', not fixed universals of social experience, nor are they an arbitrary methodological or conceptual choice . . . Geographical scale is socially as simultaneously a platform and a container of certain kinds of social activity. Far from neutral and fixed, therefore, geographical scales are the product of economic, political and social activities and relationships; as such they are as changeable as those relationships themselves . . . Scale is the geographical organizer and expression of collective social action.
>
> (in Brenner 2000: 367)

Geographical scale then defines a series of lines of force that structure the world in all sorts of ways from the scale of the individual body, through to that of the supra-regional, and the globe (see Marston 2000, 2004; Smith 1992). What is more, and of just as much interest, geographical scale thus emerges as a key terrain for political struggle and contestation. According to writers like Smith, Brenner and Swyngedouw, the dynamics of many of the central political issues that are confronting contemporary societies – and, within them, contemporary cities – are to a fundamental degree about scale.

If we want to understand the ways in which our world is be reorganized – or restructured, in the terminology of Smith, Brenner and Swyngedouw – then we need to be attuned to the ways in which key actors such as corporations, governments, trade unions and so on are attempting to both rearticulate the relationships between scales and define novel scales of action. Over the past 30 years or so the emerging notion of the 'global' is perhaps the most striking example of this – although we could also highlight how institutions such as the European Union and NAFTA have involved the construction of a range of novel supra-regional scales 'between' the national and the global.

Drawing on this framework of analysis, we could start to make some interesting observations about the emergence of the mass-participation urban marathon, and the kinds of spatialities through which it is constituted. First, we could explore the ways in which the organizers of early city marathons sought to project themselves on to the city-wide stage: working to move themselves away from being a minor, and largely invisible, subculture to a highly visible and mainstream (if still distinctive) group. We can see different patterns to this movement. In Germany, for example, by the early 1970s marathon running had already begun to take on a mass character as marathon events were opened to all comers (not just those formally aligned to an athletics club), with the biggest marathons attracting fields of over 2,000 runners. However, these marathons took place outside cities, often on quite tough terrain. (The Hornisgrinde Marathon, for example, founded a year before the Berlin Marathon, follows forest trails in a circuit around the tallest peak in the Black Forest. At its largest in 1974 and 1975, just under 1,000 competitors ran the Hornisgrinde Marathon.) The challenge that the organizers of the Berlin Marathon faced was reconfiguring the marathon as an 'urban' event, and indeed running itself as an appropriately urban activity (as opposed to something that rightly took place within a 'natural' landscape such as forest or countryside). In America, by contrast, the making 'mass' of the marathon was a through-and-through urban affair. Marathons had been held in American cities since the end of the nineteenth century. The Boston Marathon, America's most prestigious and longest-running marathon, finished in the centre of Boston – even if the majority of its course wound through the countryside outside the city. Boston excepted, however, these marathons were small-scale, minority events. The precursor to the New York Marathon, the Cherry Tree Marathon, held in the Bronx and organized by the New York Road Runners Club, involved just 20 runners when first run in 1961. The primary issue that faced race organizers like Fred Lebow was how to get people to take notice and get involved in the marathon. Taking the marathon out on to the streets and placing it against the backdrop of a city's more iconic places was an essential part of this projection.

But there is also a more subtle geography to this attempted projection. Fred Lebow, who transformed the New York Road Runners Club from an amateur club with membership of 270 in 1972 when he became its president into the largest running club in the world with a membership of over 30,000 and full-time professional staffing, did so by suggesting a series of affinities between the practice of running and the ideals of the successful professional. At the very moment that Lebow was taking his marathon out into the outer boroughs of New York, he was also trying to reconfigure the practice of running in general, and the marathon in particular, as a mass elite practice (that is to say a practice of and for the professional middle class and upper-middle class) or, to put things more geographically, as a practice for those who populated – either as employees or as residents – mid-town and down-town Manhattan. What is more, he brought all the skills, techniques and – perhaps most crucially – imperatives (such as growth, brand development and profit) of the business organization to the previously determinedly ama-teur world of American road racing. In doing this he constructed – and did so with a remarkable single-mindedness – a profound synergy between the worlds of downtown and mid-town corporate capitalism and the New York Marathon.

This brings us to a second – and closely related – way we might consider the scale of the urban marathon: how the organization of the marathon is articu-lated with various institutions of government and economy, and at what geographical scale this takes place. One of the biggest challenges faced by those seeking to establish urban marathons was building relationships with various different organizations involved in administering a city – the police, sanitation, parks and traffic departments, and business groups, to name some of the most central. This involves: 1) establishing a sense that the marathon was – or could be – a significant institution in the collective life of a city and hence worth talking to; and parallel to this 2) the creation of a professional and administratively competent organization for the marathon that under-stood formally *and* informally the way a city was governed and which pro-vided a consistent point of contact for the institutions of government. More than that, the kind of entrepreneurialism that drives the organizers of mass urban marathons, with their need to constantly expand their event, fits neatly with the kind of local state entrepreneurialism that from the mid-1970s onwards came to define the imperatives of municipal government politicians and officials. Just as it is possible to see the emergence of the mass-participation urban marathon as involving the development of a close syn-ergy between the urban marathon and corporate capitalism, it is also possible to see the marathon as related in a similarly synergistic fashion with the entrepreneurial local state.

So, third, we could also ask about the ways big-city marathons work to project themselves beyond the immediate horizon of the city's boundaries. Events like the New York, London and Berlin marathons are – quite obvi-ously – city-based events. But they also function through projecting a sense of global connectedness and significance. In their publicity they speak repeat-edly of the hundreds of different nationalities taking part in the event, the uniqueness of the size of the event, and the speed of the elite runners partici-pating. Indeed, this kind of event boosterism dovetails neatly into the kinds of global-city boosterist urban entrepreneurialism already mentioned. As a kind of 'global talk' these rhetorical tropes do two things. Firstly, they work to conjure a sense of the world *as* global (the whole world is part of the London, Berlin and New York marathons). Secondly, they locate the city where the marathon is taking place as a significant part of that globality. Through the fact that the marathon is defined by factors that spill out past the merely local and, indeed, the regional or national – through the quality and origins of the elite runners who are participating, the speed with which the fastest runners complete the course, the distance many of the non-elite runners have come just to compete in the event, and so on – the city where the marathon takes place defines itself as an important node within the wider global ecumene. The London Marathon is not just a large mass running event, it is 'The Greatest Race on Earth'. The New York Marathon is (unsurprisingly enough) 'The World's Greatest Road Race' and 'The World's Greatest Party'. The Berlin Marathon makes do with the rather more modest claim that it is the 'schnellste City-Marathon der Welt'.

WORKING TOWARDS AN ONTOLOGICALLY FLAT ANALYSIS, OR HOW TO APPROACH SCALE IN A DIFFERENT KEY

Looking at this sketch of an analysis of the emergence of the mass urban marathon, it is not hard to see why the concept of scale has been so attractive to urban theorists. Scale provides a powerful – in the sense of generating intelligible and engaging narratives – account of the relationship between the geography of 'the local' and that which goes beyond it: the national, the international, the global. It also provides a convenient frame for considering the way different elements within an urban area are articulated within a wider frame of action. And it provides a vehicle through which the patterns of geographical organization can be understood as historical achievements. Drawing on notions of scale, we can see how the marathon's emergence and its continued existence are enmeshed in a complicated web of institutional relationships, and we can gain a sense of the historicity of that enmeshing.

And yet, there is also something disappointingly *stale* about the account of the mass-participation urban marathon sketched above. We have ended up with an account, or rather we are in the process of ending up with an account, of the urban marathon that simply subsumes the marathon into the imperatives of an existing web of relationships. The success of the marathon is a – admittedly complicated – product of the ways in which it resonates with and amplifies the existing trajectories of certain key institutions and key groups within New York (or London, or Berlin, or any other city with a mass-participation marathon). Nothing really new or novel has been generated. Rather, when we look closely we see a set of familiar stories about urban spectacle, individual, corporate and governmental entrepreneurialism, and – if we come down to focus on the individual body – disciplining and symbolic one-up-manship. We are also starting to get a story about some sort of ersatz global ecumene, an ecumene underpinned by the hard realities of global capitalism. Now, undoubtedly there is some accuracy in such an account. But what might happen if we were to approach the marathon more naively? What if, instead of thinking that we have a basic sense of shape, size and scale of the marathon, we were to assume in the first instance that we have no sense of the size or scale of the marathon?

What we want to argue is that if, precisely by assuming – in the first instance – that the world is 'flat', we have no sense of the scales that might define it, we might actually be in a much better position to think about the scalar than if we assume from the start that the world is scaled and that we have some basic sense of what that scaling involves. In doing this we want to try to address three weaknesses in the kind of scalar analysis practised by writers like Brenner, Smith and Swyngedouw. In our view these weaknesses are:

1 *Scalar analysis is not very good at making sense of patterns of organization that fit outside its hierarchical spatial organization.* Scalar analysis, inspired as it is by a broadly neo-Marxist theoretical tradition, has been fundamentally concerned with ways in which various different state apparatuses generate spatial orderings within capitalism. Unsurprisingly, therefore, it is quite good at making sense of the great many patternings of political organization that are explicitly defined by scale. But scalar analysis struggles to make sense of spatial patternings that fall outside, or confound, the kinds of scalar orderings that define much of non-political life. This does not matter a great deal if one is primarily interested in political action, or the ways in which economic activity is intertwined with political institutions. If, however, one moves away from the realms of the political and the economic, scalar analysis offers an analytical language that is often of little use in trying to understand the patterns of interaction that define the contemporary world. Think, for example, of the kinds of ethno-, ideo- and techno-scapes described by the anthropologist Arjun Appadurai (1996). These describe relationships that certainly *have* a certain scale (they have a size and a spatial extension that could be measured and defined). Yet it makes little sense to try to locate them as being at (or even transcending) a specific geographical scale – 'the local', 'the urban', 'the national' or whatever.

2 *Scalar analysis does not adequately describe scalar transformation.* A key element of the scalar analysis developed by Brenner, Swyngedouw, Smith and others is the differential ability of social actors to move between geographical scales. The central insight is that this differential ability to manipulate the scale at which a social actor's legitimate scope of action is defined is the product of geo-historical power relations. As with much of scale analysis, this is fair as far as it goes. It does, however, rather beg the question of what happens when an entity moves from being defined by one scale to being defined by another. The point about moving from one

scale to another is that the thing being studied moves from one state of activity to another – that is to say, it behaves in different ways depending on the scale which it is at. (If it simply functioned in the way that it did at the lower scale, then there would be nothing to be gained by describing such scalar shifts.) In the analysis offered by writers like Brenner, Swyngedouw and Smith there is an implicit assumption that these shifts occur in ways that are essentially isomorphic with some already-existing (or about to be made existent) geographical scale. Actors, for example, 'jump' from being regional actors to being global ones (Glassman 2002; Smith 1992). Or they 'bend' scale (Smith 2004) to speak past the regional and national to the global. This kind of account describes something important about the way many (but by no means all) economic and political entities use geographical scale as a strategic resource, but it does not begin to exhaust the way scalar shifts can transform the way a collective is organized. If we trace out social relationships more closely we find a more complex and in fact more interesting topological ordering of scales. Things can, for example, start out being small-scale and global (or perhaps more accurately transnational) and then become large and local. Competitive running in the 1940s and 1950s provides a good example of this. While long-distance running was a minority – and generally obscure – practice, a small number of internationally oriented coaches such as the German Ernst van Aaken, the New Zealander Arthur Lydiard, the Australian Percy Cerruty, the UK-based Austrian Franz Stampfl and the American Bill Bowerman, to name the most prominent examples, followed the training innovations that each was developing, forming – with those they trained – a loose community of running expertise. It was only later, when through what we might describe as a 'thickening' of the relations established by these international actors, that jogging emerged as a mass-participation activity and became the pervasive, everyday activity that we know today.

3   *Scalar analysis views the global as a master term that defines or orients other scales beneath it.* As we have seen, scalar analysis evolved out of a concern to understand the relationship between processes of globalization and the local. Much of the impetus behind this writing is to stress the continued – and indeed often heightened – importance of the local within a globalizing world. Neil Brenner's (1998, 1999a, 1999b, 2003) work, for example, highlights the increased importance of the local and regional state under what he calls a 'worldwide regime of . . . neoliberalism' (Brenner and Theodore 2002: vi). Eric Swyngedouw (1992, 1997a) seeks to do something similar with the term 'glocalization'. Yet in both cases it is the global that defines these analyses. Without reference to the global the narrative and analytical dynamic that drives these analyses disappears. There are few, if any, scalar accounts that explore the ways that, say, the relationships between neighbourhood and city, or neighbourhood and nation state, are being transformed without recourse to

the global. And this is despite the fact that actually a great deal of what configures urban life involves a shuttling between these spatial scales, with little or no recourse to the 'global'. Think about the uniqueness of the American and British liquor-licensing laws (Valverde 1998) or the distinctive ways that American law defines the relationship between the automobile and the pedestrian (Jain 2004).

What, then, might a flat or non-scalar approach to the mass-participation urban marathon look like? Or, to put things another way, what happens if we refuse to assume in advance that we know what shape the social world has? What happens if we try to follow Bruno Latour's (2005: 173) instructions that 'we have to lay continuous connections leading from one local interaction to the other places, times, and agencies through which a local site is *made to do* something'.

We want to suggest two answers to this question. The first answer is that it would force us to stop making the analytical short cuts that make scale analysis so attractive and force us to ask, what exactly is being assembled through the marathon? And through what materials does the assembling take place? Now certainly one of the collectives that is being assembled through the marathon is an organization that knows how to put together a functioning marathon, and which creates all sorts of relationships with other collectives within the city. But if we were to adhere to Latour's injunction 'to lay continuous connections leading from one local interaction to the other places, times, and agencies through which a local site is *made to do* something' we are going to need to pay a lot more attention to the work of maintenance and repair this collective puts into the making of a marathon: the issue of devising systems that allow for a safe and orderly start to the marathon, the technologies that allow for the accurate measurement of the hundreds of runners per minute who cross the finish line at the marathon's end, and the way in which the marathon organizes traffic closures and crowd control, to name just three examples. And, in fact, what is clear when considering these issues of maintenance is simply the degree to which the sheer physical scale of the marathon – the number of runners and spectators – presents an organizational limit. Thus, in an important sense the scale of marathons like New York, London or Berlin is not defined by how global they are, but the sheer inability to get more than 40,000 runners through their streets.

The second answer to the question of what happens when we keep the social flat is that other forms of association and assemblage come into view (Amin and Thrift 2002; DeLanda 2006; Latham 2002; Latham and McCormack 2004). Perhaps the most striking example of this is the question of the kind of collective that the marathon involves. In a mass-participation marathon the size and shape of the organizers' overview of the event map only approximately and untidily upon the shape of the event itself. What is clear from most published accounts of mass-participation urban marathons, and indeed from participating in them, is the extent to which they are enormous

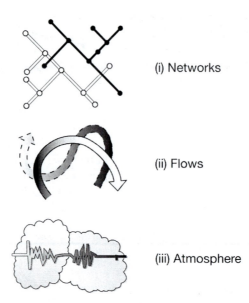

(i) Networks

(ii) Flows

(iii) Atmosphere

Figure 2.2 Alternative morphologies of spatial organization: networks, flows and atmosphere (source: the authors, after Thrift and Olds 1996: 321)

machines for the generation of affect. This affectual atmosphere is generated through all sorts of kinaesthetic involvements – the rhythm of the individual runner, the collective sound of thousands of feet striking asphalt, the in-out, in-out, in-out of the runners' breath, the shouts of the crowds, the rhythm of the bands that line the streets. So this is an affective atmosphere co-produced by both spectators and participants. This is not just a matter of the runners and their audience/supporters transforming ordinary streets into a kind of mobile carnival. It is also about the establishment of relational affective economy in which there is a mutual and resonant feedback loop between the affects and emotions of the participants and those of the spectators. The spectators respond to the visible signs of affective intensity – sweat, grimaces, smiles – with their own gestures – waves, cheers, applause.

CONCLUSION

There is, of course, much more that could be said about the methodological usefulness of working to hold the world flat. However, we want to end this chapter by returning to the productiveness of the concept of scale. Like any concept, scale is an abstraction. And the problem with abstractions is that they do not necessarily *reflect* the reality of the world: hence the often acidic nature of critiques of contemporary work on geographical scale from a range of thinkers, including Richard G. Smith (this volume), Sallie Marston *et al.* (Marston *et al.* 2005; Jones *et al.* 2007) and indeed Bruno Latour (1993,

2005). But, while concepts are abstractions, abstractions do not just aim towards representing the world. Abstractions also allow us to draw out aspects of the world that matter: abstractions are generative and pragmatic. And so, while the notion of scale may be problematic in a whole host of ways, it is important to recognize that there is something insistent about scale as an abstraction that allows us to grasp and think through the qualities of spaces. It is because of this insistent quality that we think that scale needs to remain a part of the lively conceptual vocabularies that social scientists use to think about and through spaces. It is difficult – indeed one might go so far as to say implausible – to imagine a convincing social science that had no notion of scale. Indeed, drawing from the argument developed in the preceding sections of this chapter, there are at least three reasons to think that potentially ANT could have a great deal to gain from a sustained encounter with the concept of scale.

First, while ANT may well suggest that scale is the effect of a multiplicity of factors, this does not mean that it is any less important. Effects make differences that matter. Following the networks and connections through which effects emerge does not make the effect any less real. We can trace these networks without reducing the power or reality of the effects they create. If we take Latour (1993: 120) seriously and argue that scale is nothing more than the effect of relations of 'length or connection' we are still obliged to understand what happens when the thickening, dilution, extension or shortening of relations alters the dynamics of how an entity functions.

Second, there is something about the 'this-ness' of scale that exceeds the language of networks, connections and lines that sometimes emerges from readings of ANT. This excess can be understood as the affective quality of the world. As the example of marathons suggests, there is something tangible about collective events that is both intensive and extensive at the same time. We might call this tangible quality an atmosphere. Such atmospheres have a certain size and duration even if they cannot be mapped on to a neat hierarchy of scales. That is, they have a sensed scalar quality that can be felt in the movement of large numbers of bodies.

Finally, our sense of scale as a generative abstraction should involve an understanding that the quality of spaces it captures has as much to do with intensity or density as it does with extension. That is, a sense of scale is not simply about reach: it is also about how resonant affects move and circulate between closely packed bodies moving together and differently. And the intensity of scale is also a matter of duration: not just a matter of how long an event lasts, but of how the temporality of an event registers differentially in moving bodies.

NOTES

1   It is worth stressing that, while this account suggests the profound originality of Lebow and his co-organizers, little of what they did had not

already been done elsewhere. Street races had been run in France, Spain, Brazil, America and elsewhere from as early as the start of the twentieth century. Indeed, a number of road-based marathons were held in New York in the 1900s, 1910s and through into the early 1940s. Later, in the 1950s and 1960s, clubs such as the New York Pioneer Club and the New York Road Runners Club (established in 1958) maintained this tradition of marathon running. What was most original about Lebow was his emphasis on the marathon as a mass-participation event, and his realization that the putting together of a mass-participation event demanded careful attention to publicity and staging. The primary sources for the history of the New York, London and Berlin marathons are Lebow (1984), Cooper (1998), Rubin (2004) and Bryant (2005).

2   Throughout the chapter we will refer to both running and jogging. Although in many ways the two terms are homologous, there are a number of important differences between the two. The three most important differences for the purposes of this chapter are: 1) jogging is meant to be entirely non-competitive whilst running is defined in some sense by an explicitly competitive element, whether that be racing others or attempting to beat one's own personal best; following from this 2) jogging is defined as being primarily about the health benefits generated from jogging, while running is much more explicitly oriented towards the process of physical movement in and of itself; 3) jogging tends to emphasize moving at a relatively low intensity, while running emphasizes, and places a great deal of value on, high-intensity physical effort usually in combination with sessions spent exercising at a similar intensity to jogging. Following from points 1 and 2, a fourth difference between the two sets of practices is that it is possible to locate a point in the late 1950s and early 1960s where jogging was invented and defined as a practice (see for example Bowerman and Harris 1967; Gilmour 1965; Melleby and Burrus 1969). Running has a much more variegated historical and geographical lineage which arguably reaches back centuries.

REFERENCES

Amin, A. (2007) 'Re-thinking the Urban Social', *City*, 11: 100–14.

Amin, A. (2008) 'Collective Culture and Urban Public Space', *City*, 12: 5–24.

Amin, A. and Thrift, N. (2002) *Cities: Reimagining the Urban*, Cambridge: Polity.

Appadurai, A. (1996) *Modernity at Large: Cultural Dimensions of Globalization*, Minneapolis: University of Minnesota.

Benjamin, W. (1977) *Illuminations*, London: Fontana.

Borden, I. (2001) *Skateboarding, Space and the City: Architecture and the Body*, Oxford: Berg.

Bowerman, B. and Harris, W. (1967) *Jogging: A Medically Approved Physical Fitness Program for all Ages*, New York: Grosset & Dunlap.

Brenner, N. (1997a) 'Global, Fragmented, Hierarchical: Henri Lefebvre's Geography of Globalization', *Public Culture*, 10(1): 135–67.

Brenner, N. (1997b) 'State Territorial Restructuring and the Production of Spatial Scale: Urban and Regional Planning in the Federal Republic of Germany, 1960–90', *Political Geography*, 16: 273–306.

Brenner, N. (1998) 'Between Fixity and Motion: Accumulation, Territorial Organization and the Historical Geography of Spatial Scales', *Environment and Planning D*, 16: 459–81.

Brenner, N. (1999a) 'Globalization as Reterritorialisation: The Re-scaling of Urban Governance in the European Union', *Urban Studies*, 36(3): 431–51.

Brenner, N. (1999b) 'Beyond State-Centrism? Space, Territoriality and Geographical Scale in Globalization Studies', *Theory and Society*, 28(1): 39–78.

Brenner, N. (2000) 'The Urban Question as a Scale Question: Reflections on Henri Lefebvre, Urban Theory and the Politics of Scale', *International Journal of Urban and Regional Research*, 24(2): 361–78.

Brenner, N. (2001) 'The Limits to Scale? Methodological Reflections on Scalar Restructuring', *Progress in Human Geography*, 25: 591–614.

Brenner, N. (2003) 'Metropolitan Institutional Reform and the Rescaling of State Space in Contemporary Western Europe', *European Urban and Regional Studies*, 10: 297–325.

Brenner, N. and Theodore, N. (2002) 'From the "New Localism" to the Spaces of Neoliberalism', in N. Brenner and N. Theodore (eds), *Spaces of Neoliberalism: Urban Restructuring in North America and Western Europe*, Malden, MA: Blackwell, pp. v–xi.

Bryant, J. (2005) *The London Marathon: The History of the Greatest Race on Earth*, London: Arrow Books.

Cooper, P. (1998) *The American Marathon*, Syracuse, NY: Syracuse University Press.

Crawford, M. (1991) 'The Fifth Ecology: Fantasy, the Automobile, and Los Angeles', in M. Wachs and M. Crawford (eds), *The Car and the City*, Ann Arbor: University of Michigan Press, pp. 222–33.

Davis, M. (1990) *City of Quartz: Excavating the Future in Los Angeles*, London: Vintage.

DeLanda, M. (2006) *A New Philosophy of Society*, London: Continuum.

Delaney, D. and Leitner, H. (1997) 'The Political Construction of Scale', *Political Geography*, 16: 93–97.

Dickens, P. (2004) 'Geographers and "Globalization": (Yet) Another Missed Boat?', *Transactions of the Institute of British Geographers*, 29: 5–26.

Ehrenreich, B. (2007) *Dancing in the Streets: A History of Collective Joy*, London: Granta Books.

Gilmour, G. with Arthur Lydiard (1965) *Run for Your Life: Jogging with Arthur Lydiard*, Auckland: Minerva.

Glassman, J. (2002) 'From Seattle (and Ubon) to Bangkok: The Scales of Resistance to Corporate Globalization', *Environment and Planning D: Society and Space*, 20(5): 513–33.

Herod, A. (1997) Labor's Spatial Praxis and the Geography of Contract Bargaining of the US East Coast Longshore Industry, 1953–89', *Political Geography*, 16: 145–69.

Herod, A. and Wright, M. (eds) (2002) *Geographies of Power: Making Scale*, Oxford: Blackwell.

Jain, S. (2004) ' "Dangerous Instrumentality": The Bystander as Subject in Automobility', *Cultural Anthropology*, 19(1): 61–94.

Jones, J.P., Woodward, K. and Marston, S. (2007) 'Situating Flatness', *Transactions of the Institute of British Geographers*, 34(2): 264–76.

Latham, A. (2002) 'Re-theorizing the Scale of Globalization: Topologies, Actor-Networks, and Cosmopolitanism', in A. Herod and M. Wright (eds), *Geographies of Power: Making Scale*, Oxford: Blackwell, pp. 115–44.

Latham, A. and McCormack, D.P. (2004) 'Moving Cities: Rethinking the Materialities of Urban Geographies', *Progress in Human Geography*, 28: 701–24.

Latham, A. and McCormack, D.P. (2008) 'Speed and Slowness', in T. Hall, P. Hubbard and J. Rennie-Short (eds), *The Sage Companion to the City*, London: Sage, Ch. 17.

Latour, B. (1993) *We Have Never Been Modern*, Hemel Hempstead: Harvester Wheatsheaf.

Latour, B. (2005) *Reassembling the Social: An Introduction to Actor Network Theory*, Oxford: Oxford University Press.

Lebow, F. (1984) *Inside the World of Big-Time Marathoning*, New York: Rawson Associates.

McCormack, D.P. and Latham, A. (forthcoming) 'In Active Cities: Spaces of Exercise, Sport, and Fitness in Contemporary Urban Life', Available from authors.

McDowell, L. (2001) 'Linking Scales: Or How Research about Gender and Organizations Raises New Issues for Economic Geography', *Journal of Economic Geography*, 1(2): 227–50.

McMaster, R. and Sheppard, E. (2004) 'Introduction: Scale and Geographic Inquiry', in R. McMaster and E. Sheppard (eds), *Scale and Geographic Inquiry: Nature, Society, and Method*, Oxford: Blackwell.

Marston, S. (2000) 'The Social Construction of Scale', *Progress in Human Geography*, 24: 219–42.

Marston, S. (2004) 'A Long Way from Home: Domesticating the Social Construction of Scale', in R. McMaster and E. Sheppard (eds), *Scale and Geographic Inquiry: Nature, Society, and Method*, Oxford: Blackwell.

Marston, S., Jones, J.P. and Woodward, K. (2005) 'Human Geography without Scale', *Transactions of the Institute of British Geographers*, 30: 416–32.

Marston, S., Woodward, K., and Jones, J.P. (2007) 'Flattening Ontologies of Globalization: The Nollywood Case', *Globalizations*, 4(1): 45–63.

Melleby, A. and Burrus, B. (1969) *Jogging Away . . . from Heart Disease and toward a New, Better Life*, New York: Volitant Books.

Mitchell, D. (2003) *The Right to the City: Social Justice and the Fight for Public Space*, New York: Guilford.

Mumford, L. (1938) *The Culture of Cities*, London: Secker & Warburg.

Rubin, R. (2004) *Anything for a T-Shirt: Fred Lebow and the New York City Marathon, the World's Greatest Footrace*, Syracuse, NY: Syracuse University Press.

Ryan, Z. (ed.) (2006) *The Good Life: New Public Spaces for Recreation*, New York: Van Alen Institute.

Schivelbusch, W. (1979) *The Railway Journey: The Industrialization of Time and Space in the Nineteenth Century*, Berkeley: University of California Press.

Schwartz, H. (1992) 'Torque: The New Kinaesthetic of the Twentieth Century', in J. Crary and S. Kwinter (eds), *Incorporations*, New York: Zone, pp. 71–126.

Schwartz, V. (1999) *Spectacular Realities: Early Mass Culture in Fin-De-Siècle Paris*, Berkeley: University of California.

Sennett, R. (1977) *The Fall of Public Man*, London: Faber and Faber.

Sennett, R. (1990) *The Conscience of the Eye: The Design and Social Life of Cities*, London: Faber and Faber.

Sennett, R. (1995) *Flesh and Stone: The City in Western Civilisation*, London: Faber and Faber.

Simmel, G. (1950) *The Sociology of Georg Simmel*, New York: Free Press.

Smith, N. (1984) *Uneven Development: Nature, Capital and the Production of Space*, Oxford: Basil Blackwell.

Smith, N. (1992) 'Geography, Difference and the Politics of Scale', in J. Doherty, E. Graham and M. Mallek (eds), *Postmodernism and the Social Sciences*, London: Macmillan, pp. 57–79.

Smith, N. (1993) 'Homeless/Global: Scaling Places', in J. Bird, B. Curtis, T. Putnam, G. Robertson amd L. Tickner (eds), *Mapping the Futures: Local Cultures, Global Change*, London: Routledge, pp. 87–119.

Smith, N. (1996) *The New Urban Frontier: Gentrification and the Revanchist City*, London: Routledge.

Smith, N. (1997) 'The Satanic Geographies of Globalization', *Public Culture*, 10(1): 169–89.

Smith, N. (2002) 'New Globalism, New Urbanism: Gentrification as Global Urban Strategy', *Antipode*, 34(3): 427–50.

Smith, N. (2004) 'Scale Bending and the Fate of the National', in R. McMaster and E. Sheppard (eds), *Scale and Geographic Inquiry: Nature, Society, and Method*, Oxford: Blackwell.

Smith, N. and Dennis, W. (1987) 'The Restructuring of Geographical Scale: Coalescence and Fragmentation in the Northern Core Region', *Economic Geography*, 63(2): 160–82.

Sorkin, M. (ed.) (1992) *Variations on a Theme Park: The New American City and the End of Public Space*, New York: Noonday.

Swyngedouw, E. (1992) 'The Mammon Quest: "Glocalization", Interspatial Competition and the Monetary Order: The Construction of New Scales', in M. Dunford and G. Kafalas (eds), *Cities and Regions in the New Europe*, London: Belhaven, pp. 39–67.

Swyngedouw, E. (1996) 'Restructuring Citizenship, the Re-scaling of the State and the New Authoritarianism: Closing the Belgian Mines', *Urban Studies*, 33(8): 1499–1521.

Swyngedouw, E. (1997a) 'Neither Global nor Local: "Glocalisation", Interspatial Competition and the Politics of Scale', in K. Cox (ed.), *Spaces of Globalization*, New York: Guilford.

Swyngedouw, E. (1997b) 'Excluding the Other: The Production of Scale and Scaled Politics', in R. Lee and J. Wills (eds), *Geographies of Economies*, London: Edward Arnold.

Swyngedouw, E. (2000a) 'Authoritarian Governance, Power, and the Politics of Rescaling', *Environment and Planning D*, 18(1): 63–76.

Swyngedouw, E. (2000b) 'Elite Power, Global Forces, and the Political Economy of "Glocal" Development', in G. Clark, M. Feldman and M. Gertler (eds), *The Oxford Handbook of Economic Geography*, Oxford: Oxford University Press.

Swyngedouw, E. (2004a) 'Scaled Geographies: Nature, Place, and the Politics of Scale', in R. McMaster and E. Sheppard (eds), *Scale and Geographic Inquiry: Nature, Society, and Method*, Oxford: Blackwell.

Swyngedouw, E. (2004b) 'Globalisation or "Glocalisation"? Networks, Territories and Rescaling', *Cambridge Review of International Affairs*, 17(1): 25–48.

Taylor, P. (1981) 'Geographical Scale within the World Systems Approach', *Review*, 1: 3–11.

Taylor, P. (1982) 'A Materialist Framework for Political Geography', *Transactions of the Institute of British Geographers*, 7(1): 15–34.

Taylor, P. (1984) 'Introduction: Geographical Scale and Political Geography', in P. Taylor and J. House (eds), *Political Geography: Recent Advances and Future Directions*, London: Croom Helm.

Taylor, P. (2004) 'Is There a Europe of Cities: World Cities and the Limitations of Geographical Scale Analyses', in R. McMaster and E. Sheppard (eds), *Scale and Geographic Inquiry: Nature, Society, and Method*, Oxford: Blackwell.

Thrift, N. (2008) *Non-Representational Theory: Space/Politics/Affect*, London: Routledge.

Thrift, N. and Olds, K. (1996) 'Refiguring the Economic in Economic Geography', *Progress in Human Geography*, 20(3): 311–37.

Valverde, M. (1998) *Diseases of the Will: Alcohol and the Dilemmas of Freedom*, Cambridge: Cambridge University Press.

Zukin, S. (1995) *The Culture of Cities*, Oxford: Blackwell.

3 Urban studies without 'scale'

Localizing the global through Singapore

Richard G. Smith

There is no general recipe. We are finished with all globalizing concepts.

Deleuze and Guattari (1983: 108)

I observe the world as it unfurls, I thought; proceeding empirically, in good faith, I observe it; I can do no more than observe.

Houellebecq (2003: 286)

INTRODUCTION

Actor-Network Theory (ANT), 'a sociology of associations', has in recent years become an important source of inspiration – along with poststructuralism and non-representational theory – for the development of a new approach to urban studies (e.g. see Amin and Thrift 2002; Smith 2003a, 2003b, 2006, 2007; Smith and Doel 2010). This chapter advances this new approach by demonstrating how the concept of 'scale' is problematical, rather than axiomatical: a redundant invention for seeing the spatiality of economic relations, and unhelpful for describing how Singapore has sought to manage the lengthening of its legal services to aid its position as a financial and services centre.

Through a fieldwork-led empirical discussion of the millennial (1999–2002) managed restructuring it is demonstrated how any approach that would try to explain the transformation of Singapore's legal services as a process where 'scalar processes' had intersected or interacted would miss the most important aspect of the whole process, namely the intermediary arrangements (the networks) that, no matter what their length, are always 'local'. Furthermore, the chapter demonstrates how, when no extrinsic explanation (an a priori framework, a blueprint or a context, such as 'scale') is imposed on actors to try to explain the networks in which they are involved, different findings emerge to those that the 'scalar social scientist' has foreseen. Indeed, through being attentive to how lawyers (actors) describe the formation of networks (or work-nets) in their terms, using their own dimensions and touchstones (rather than those of the trained social scientist), it is clear that 'scalar processes', or 'frontier zones', or 'border zones', or 'regulatory fractures', or 'analytic borderlands' (see Smith 2003a, 2003b), are all inadequate for accurately describing how Singapore has lengthened its legal services since the turn of the millennium.

QUESTIONING 'SCALE' IN URBAN STUDIES

> There exists no place that can be said to be 'non-local'. If something is to be 'delocalized', it means that it is being sent from one place to some *other* place, not from one place to *no* place.
>
> Latour (2005: 179)

Whilst it is true that the organization of 'global finance' means that hubs such as Manhattan (New York), Ginza district (Tokyo) or the City/Canary Wharf (London) are to some degree denationalized, disconnected from their 'national contexts' because they are so highly connected together rather than with their respective 'national urban hierarchies', it is also blindingly obvious that states do play an important role in the life of those financial centres, e.g. by providing legal frameworks, taxation regimes, basic services and infrastructure, amongst many other things. Indeed, states are especially important in shaping major primate 'global cities'.

With globalization, relations between cities and states have been challenged, and an interesting paradox is now evident. On the one hand, states' are very important for the restructuring of so-called 'global cities'. For example, cities like London, Paris and Singapore enhance and maintain their connectivity and 'global position' through the receipt of substantial public investment from their respective national governments. On the other hand, a state's minor cities are now less 'national' because they lack public investment and so *need* to be neo-liberal and entrepreneurial to compete and survive in the 'global economy'. A consequence of this paradox is that many contemporary urban researchers are keen to think of cities in terms of their degree of freedom (i.e. more or less autonomous) from states: 'we need to conceptualize the city at the multiple and interacting scales of global, national and local' (Short 2006: 219).

Short (2006) argues that we are witnessing a profound rescaling as cities are impacted by, on the one hand, 'global forces' and, on the other, 'national systems of regulation'. Indeed, Short is certain that 'scalar processes' exist:

> Global, national and urban processes are affecting individual cities around the world, while globalizing cities are the site and platform for shifts in national and global articulations. A new urban theory will be sensitive to cities as sites of intersecting scalar processes. As a (probably unworkable) reminder, we should henceforth place the term 'city' in the middle of *global–city–national*.
>
> (Short 2006: 220)

However, the notion that 'scales' are, or have, processes is anathema to actor-network theorists. ANT does not need a 'national scale' to recognize that states are important for urban and economic restructuring ('scalar thinkers' do not have a monopoly on recognizing the importance of states), precisely because ANT is always already attuned to that which has hitherto been subsumed under the neat category of 'scale': namely, the tireless work of the actor-network where actors and networks become one and the same in the construction of joint actions. States, cities and city-states are not sites where 'scales' intersect, interact, interface or overlap; they are all actor-networks, continuums (see Smith 2003b), scattered lines of humans and non-humans, that are not by nature local, national or global, but are more or less long and more or less connected.

In this chapter it is demonstrated how thinking of Singapore as a 'scale', somehow between, below or intersecting the 'global' and the 'national', is unhelpful for appreciating how the city-state is strengthened as a financial centre through the lengthening of its legal services. Indeed, it is not just 'the trap of scale' which is avoided in this empirical account of the lengthening of Singapore's legal networks; there is also no appeal to any other a priori construct, no shift to explanation by imposing a Context, a Structure or a Framework (see Latour 2005). Singapore is not placed into 'the scale/context/structure/framework of the global', precisely because one is always forced to jump from the 'local' to the 'global', the particular to the general, to pretend to explain what has happened. In other words, when the term 'scale' is used by social scientists it is done so to indicate an established state of affairs, a foundation, a given, an unquestionable framework from which one can subsequently begin to account for what is happening in a given situation. Hence, 'scale' is what a poststructuralist such as Lyotard terms an *exteriority*, a concept that is imposed on events before any empirical investigation has even started. That approach is avoided by this empirical account as it seeks to do no more than describe the forging of some longer networks at the turn of the millennium according to those lawyers who were involved.

The approach that is followed in the next section, in the empirical case study, is one that is attentive to the assemblage: the step-by-step restructuring, the sociology of associations, the formation of an actor-network. Overall, the empirical finding is that the restructuring of legal services through Singapore is solely the product of the formation of actor-networks that are always local whatever their length. To put it bluntly, this chapter does not follow a 'scalar view' of globalization, but instead confirms Peter Charlton's (a London managing partner of the law firm Clifford Chance) observation that '[Globalization] is not about scale, it's about having the right network' (2003: 1).

THE MILLENNIAL RESTRUCTURING OF SINGAPORE'S LEGAL SERVICES

At the Millennium Law Conference annual dinner the senior minister, Lee Kuan Yew,[1] noted that 'A major international financial centre needs to have high-quality and competitive legal services.' He noted Singapore's need to ease its restrictions in the face of globalization to attract more foreign lawyers and build its capacity as a financial centre. 'To build a critical mass of good lawyers to service the financial and business community, he [Lee Kuan Yew] said, it is necessary to draw in more top-quality offshore law firms and get more in-house lawyers of multi-national corporations and global financial institutions to base themselves here' (Boon 2000: 1). However, it was also evident that simply building capacity in legal services was not enough to enhance Singapore's position as a leading financial centre. The restructuring of existing capacity through the formation of collaborative networks between foreign and domestic law firms was also important. Only through collaboration will Singaporean law firms gain knowledge in the practical use of English and US (New York) laws (which are the preferred governing laws for documenting and executing financial transactions funded in US dollars) and consequently play a significant part in off-shore financial transactions. Let me start this story about the lengthening of Singapore's legal networks back in the early 1990s.

Immediately prior to the millennial liberalization and restructuring of legal provision through Singapore there were two distinct providers of legal services.[2] First, there were off-shore law firms. There were *circa* 60 international firms from over 15 different legal jurisdictions (from the UK, the USA, Australia, Canada, Indonesia, Japan, Hong Kong, China, Germany, Holland, Austria, Italy, France, Sweden and Norway – Sek Keong 1999: 12, 162–69) operating in Singapore, dominating the market for top-end international financial work (in the 1980s there were more international law firms with a presence in the city-state than in the 1999 to 2002 period). The strongest presence was that of London law firms (see Table 3.1), partly because the City firms began to establish offices there as far back as 1980[3] (see Figure 3.1). Second, there were on-shore law firms. There were *circa* 800 local firms, but only a handful (fewer than ten) had more than 50 fee-earners (see Table 3.2).[4]

Table 3.1 Ten largest off-shore (UK and US) firms prior to JLVs

Firm name	Fee-earners	Partners
1 Baker & McKenzie	49	8
2 Linklaters & Paines	28	6
3 Clifford Chance	28	7
4 Freshfields	21	4
5 Norton Rose	19	5
6 White & Case	18	5
7 Allen & Overy	16	4
8 Sinclair Roche & Temperley	12	4
9 Milbank Tweed Hadley & McCloy	12	2
10 Herbert Smith	8	2

Data source: World Legal Forum, 9 March 1999.

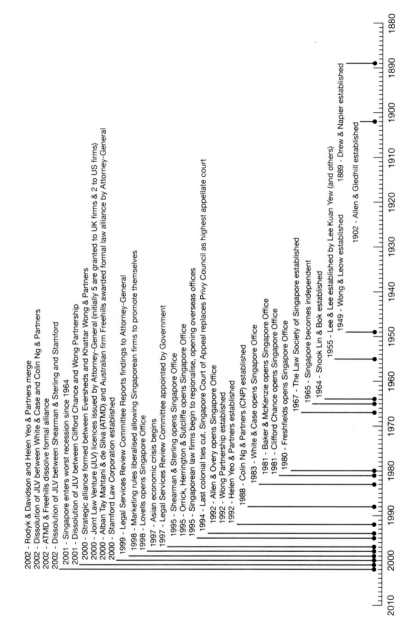

Figure 3.1 Time-line of key dates, 1889–2002 (source: the author)

2002 - Rodyk & Davidson and Helen Yeo & Partners merge
2002 - Dissolution of JLV between White & Case and Colin Ng & Partners
2002 - ATMD & Freehills dissolve formal alliance
2002 - Dissolution of JLV between Shearman & Sterling and Stamford
2001 - Singapore enters worst recession since 1964
2001 - Dissolution of JLV between Clifford Chance and Wong Partnership
2000 - Strategic alliance formed between Eversheds and Khattar Wong & Partners
2000 - Joint Law Venture (JLV) licences issued by Attorney-General (initially 5 are granted to UK firms & 2 to US firms)
2000 - Alban Tay Mahtani & de Silva (ATMD) and Australian firm Freehills awarded formal law alliance by Attorney-General
2000 - Stamford Law Corporation established
1999 - Legal Services Review Committee Reports findings to Attorney-General
1998 - Marketing rules liberalised allowing Singaporean firms to promote themselves
1998 - Lovells opens Singapore Office
1997 - Asian economic crisis begins
1997 - Legal Services Review Committee appointed by Government
1995 - Shearman & Sterling opens Singapore Office
1995 - Orrick, Herrington & Sutcliffe opens Singapore Office
1995 - Singaporean law firms begin to regionalise, opening overseas offices
1994 - Last colonial ties cut. Singapore Court of Appeal replaces Privy Council as highest appellate court
1992 - Allen & Overy opens Singapore Office
1992 - Wong Partnership established
1992 - Helen Yeo & Partners established
1988 - Colin Ng & Partners (CNP) established
1983 - White & Case opens Singapore Office
1981 - Baker & McKenzie opens Singapore Office
1981 - Clifford Chance opens Singapore Office
1980 - Freshfields opens Singapore Office
1967 - The Law Society of Singapore established
1965 - Singapore becomes independent
1964 - Shook Lin & Bok established
1955 - Lee & Lee established by Lee Kuan Yew (and others)
1949 - Wong & Leow established
1902 - Allen & Gledhill established
1889 - Drew & Napier established

Table 3.2 Ten largest on-shore (Singaporean) firms prior to JLVs

Firm name	Fee-earners	Partners
1 Drew & Napier	170	47
2 Allen & Gledhill	149	48
3 Khattar Wong & Partners	128	50
4 Shook Lin & Bok	60	18
5 Wong Partnership	60	16
6 Rodyk & Davidson	52	16
7 Colin Ng & Partners	50	12
8 Helen Yeo & Partners	40	19
9 Harry Elias & Partners	36	16
10 Haridass Ho & Partners	32	14

Data source: World Legal Forum, 9 March 1999.

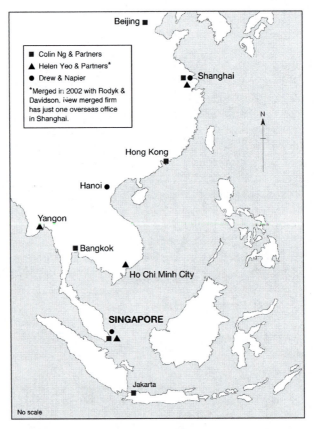

Figure 3.2 Locations of Singaporean law firms' overseas offices prior to their involvement in JLVs (source: the author)

The on-shore firms lacked the resources, specialized skills and client base of the international off-shore firms, whilst the off-shore firms were not permitted to practise Singapore law. Prior to liberalization there were a handful of informal alliances between off-shore (foreign) and on-shore (domestic) firms in order to provide legal services for cross-border financial transactions.[5] However, the small number of such collaborations highlights the inefficiency and extra costs that could be involved in conducting cross-border transactions from Singapore. Examples include: Deacons Graham & James and Yeo Wee Kiong & Partners; Sinclair Roche & Temperley and Colin Ng and Partners; and Norton Rose and Lee & Lee.[6]

Since the early 1990s the Singaporean government has actively encouraged companies to expand into South-East Asia and beyond. However, the so-called 'regionalization' of Singaporean law firms that followed that call was very limited. The three most significant attempts to practise beyond the city-state (see Figure 3.2) were by Colin Ng & Partners (established in 1988), Helen Yeo & Partners (established in 1992) and Drew & Napier (established in 1889). First, Colin Ng & Partners (CNP) grew quickly to 23 partners, over 80 lawyers, and total staff strength of more than 150 throughout Asia. CNP followed an ambitious strategy of 'regionalization with a view to globalization of our services' (CNP 2000: unpaginated). CNP opened offices in Singapore (main office, 1988), Singapore (Science Park office, 2000), Bangkok (1998), Hong Kong (1997) and Beijing (2000). Furthermore, the firm followed several other avenues to increase its geographical reach: 1) opened correspondent offices in Shanghai (1997) and Jakarta (in co-operation with Remy & Darus, 1997); 2) created practice groups to cover Myanmar (Burma), Vietnam and Cambodia; and 3) formed an informal alliance with the UK firm Sinclair Roche & Temperley (which specializes in international trade and transportation) to increase its capacity to service clients on multi-jurisdictional and international matters. Second, from just nine lawyers Helen Yeo & Partners (HYP) grew to more than 50 lawyers. HYP opened offices in Singapore (main office), Shanghai, China (1997), Ho Chi Minh City, Vietnam (1996) and Yangon, Myanmar (1996). Furthermore, the firm established practice desks to cover Cambodia and Laos. Third, the long-established Singaporean firm of Drew & Napier (DN), with more than 140 lawyers, opened offices in both Hanoi (Vietnam) and Shanghai (China). DN also established a trademarks and patents office in Malaysia and country-specific desks for India, Thailand, Greater China and Indonesia.[7]

Despite these few exceptions most Singaporean law firms remained firmly rooted in the city-state. There were perhaps several reasons for the limited 'regionalization' of on-shore (Singaporean) law firms. First, there was a reluctance amongst lawyers to be posted overseas to manage a branch office (which could be remote and involve a lower standard of living). Second, Singaporean firms had a poverty in expertise and resources in comparison to the international law firms that operated in these locations and with which they would have had to compete. Indeed, the few Singaporean practices established overseas came about primarily to serve the direct investments of Singaporean clients. Third, there was the negative and unprofitable experience of those domestic firms that had tried to set up offices in places such as Hong Kong, Vietnam, Malaysia and so on. Fourth, the 'regionalization' of Singaporean law firms had been restricted because of the limited use of Singapore law even in the Association of Southeast Asian Nations (ASEAN) region.[8] In short, it is the lack of success of Singaporean firms to 'regionalize' and the failure of Singapore law to establish itself as a governing law for ASEAN which are central to this story about the networking of Singapore's legal services, because a consequence of staying at home was that Singaporean firms lacked expertise in the two relevant governing laws of globalization, namely English and US (New York) law. And that lack was a restriction to Singapore's development as a financial centre, and to the success of the liberalization of its financial sector, because Singapore's domestic law firms were unable to provide convenient and efficient legal services for both off-shore financial transactions and both aspects of cross-border financial transactions. To quote the Attorney-General, Chan Sek Keong, 'Liberalizing the financial market will not make Singapore competitive unless the supporting services are also competitive.' He noted that 'There was a need to examine the legal services sector to see how it can better serve the growth of the financial services sector' (Meyer 1999: 18). Consequently, the Legal Services Review Committee was appointed by the government in September 1997 'to review Singapore's strategic legal needs in the financial sector, and the conditions under which foreign law firms and foreign lawyers are allowed to operate in Singapore, in the context of ensuring Singapore's competitiveness in financial services' (Sek Keong 1999: 1).

The Committee was faced with three options for the liberalization of Singapore's legal services. The first option was a restricted liberalization (closed door) with tight controls on the operations of law firms in Singapore. For example, one might have a rule obligating firms to base their top regional partner in Singapore if they are to be permitted to participate in a joint venture with an on-shore firm. At this time protectionist countries such as China place strict restrictions on the activities of foreign law firms (Tromans 2000). The second option was a semi-restricted liberalization (halfway house) that would allow some collaboration between on- and off-shore firms (Meyer 1999). The third option was unrestricted liberalization (open door); the option of full market liberalization (which is found in Hong Kong's and Thailand's markets) is what the UK legal profession was lobbying for (see Sek Keong 1999) at this stage in the process.

The Legal Services Review Committee reported in June 1999 and recommended the second option of a halfway house. The introduction of two new types of collaboration for law firms operating in Singapore was recommended. First, there was the joint law venture (JLV) where on-shore and off-shore firms may apply to join together through joint venture. Off-shore lawyers in a JLV firm may provide legal advisory services in Singapore law (including the drafting of legal documents) for on-shore, cross-border and off-shore financial transactions as long as they are a part of and remain in the JLV firm (Sek Keong 1999). In other words, JLVs practise English law, US (New York) law and Singapore law. Second, there was the formal alliance (FA), which is a less integrated form of collaboration than the JLV. Off-shore lawyers may draft legal documents regulated by Singapore law for cross-border financial transactions only. In an FA, advisory services and opinions relating to Singapore law have to be given by lawyers qualified to practise Singapore law (Sek Keong 1999).[9] Both JLVs and FAs were seen by the Committee as 'suitable vehicles to promote the regionalization of Singapore's law practices and may enable Singapore law firms to play a bigger role in offshore financial transactions' (Sek Keong 1999: iv). The Committee recommended a maximum of just five JLVs and an unrestricted number of FAs. Both require at least five lawyers resident in Singapore, including two equity partners who, for the off-shore firm, have five years' experience in off-shore legal work and, for the on-shore firm, have five years' experience in banking, finance and corporate work. Furthermore, in a JLV the number of local partners must always exceed the foreign partners and must be jointly managed (Sek Keong 1999). The government acted on the Committee's recommendations, which meant that JLV firms could be created and would be entities that could practise English and/or US (New York) law, and Singaporean law.[10]

As shown in Table 3.3, nine JLVs[11] were formed between large UK/US law firms and smaller Singaporean firms in the period from 1999 to 2002. The initial award of seven licences involved five UK firms and two US firms;[12] subsequently two more licences were awarded.[13]

Table 3.3 Joint law ventures, 1999–2002

	Global firms	*Singaporean firms*
Live	Allen & Overy	Shook Lin & Bok
	Baker & McKenzie	Wong & Leow
	Freshfields	Drew & Napier
	Linklaters	Allen & Gledhill
	Lovells	Lee & Lee
Dead	Orrick Herrington & Sutcliffe	Helen Yeo & Partners*
	Clifford Chance	Wong Partnership (Dissolved in 2001)
	Shearman & Sterling	Stamford (Dissolved in 2002)
	White & Case	Colin Ng & Partners (Dissolved in 2002)

Source: the author.

* In November 2002 Helen Yeo & Partners merged with Rodyk & Davidson and ended JLV with Orricks.

FORGET SCALE, FOLLOW NETWORKS: THE LENGTHENING OF SINGAPORE'S LEGAL SERVICES

Networks . . . are systems that create themselves.

Riles (2001: 173)

In the middle, where nothing is supposed to be happening, there is almost everything.

Latour (1993: 122–3)

I conducted interviews in Singapore in 2000 with the senior domestic and foreign lawyers of those law firms involved in the new JLVs.[14] The interviews led me to five significant conclusions.

First, it was clear that the lawyers described the JLV process in terms of *network formation and the lengthening of their networks* to serve clients and increase referrals. Even what motivated the law firms, both domestic and foreign, to enter into joint ventures in the first place was the opportunity that they saw the JLVs affording them to improve and lengthen their relationships, associations and networks to their collective advantage through Singapore. For domestic (on-shore) firms the key advantages to participating in a JLV with a foreign law firm were: 1) exposure to the expertise (i.e. in international capital markets, asset securitization, mergers and acquisitions, project finance work and other structured financing) and experience of foreign-firm lawyers who are well travelled and expert in negotiating long networks; 2) knowledge and use of the foreign firm's established non-human immutable mobiles (e.g. databases, documentation systems, finance and support systems, practice manuals), and the opportunity of direct exposure to the latest cutting-edge technologies and knowledges (so-called 'tier one legal software') being created in the legal profession; and 3) the possibility of increased business (more clients) through referrals from their partner foreign firm's extensive overseas office network.[15] For foreign (off-shore) firms the key advantages to JLV participation were: 1) to maintain good relations and connections with the Singaporean government; 2) to increase their lawyer numbers in Singapore without inducing significant costs (also see Page 1994) – by training domestic lawyers rather than having to fly in lawyers from the UK or US, firms can reduce costs because there is no need to pay foreign lawyers an overseas allowance (one partner explained how this is a substantial cost because of the numbers of staff who can be relocated); 3) the possibility of an increased Singaporean clientele by benefiting from having an 'Asian face' (an interviewee said that clients like to see a mixture of foreign and domestic lawyers) and the contacts of the domestic firm, which has greater and different connections into the Singaporean legal community; and 4) to tap into those practices of domestic lawyers which 'are not written down' (interviewee) but are a product of the participation of on-shore firms in different networks to those ordinarily accessible to off-shore firms. Thus, both domestic and foreign firms entered into JLVs for their own benefit, anticipating that through collaboration and integration they would be better networked and so a stronger actor-network in the marketplace for servicing cross-border transactions through Singapore.

Second, it was evident from the interviews that the lawyers (foreign and domestic) were concerned with having the *right network*. And a key part of having the right network was to be in not just any JLV but the right JLV. Firms were concerned with building networks through Singapore that enhanced their capacity both to attract or access clients and to service those clients seamlessly. A partner in an off-shore law firm said that:

> The companies here [Singapore] are globalizing, so one of the things we think the joint venture will help us with is, given that there is always a mix of Singapore and international law components [in any deal], that we think the domestic firm will give us more direct access to these [Singaporean] clients; and number two, enable us to provide one-stop shopping for these clients and for the banks, because in a lot of transactions, when you get down to the end, it is a question of what law is going to govern. Is it going to be Singapore law? Is it going to be New York law? Or is it going to be English law? And yes, there are some differences but it depends on what the deal is ... we don't really use Zs or Ss in globalization [laughter].
>
> (Partner, off-shore firm, 2000)

All the lawyers were concerned with creating the right networks, precisely because, for them, the world is not 'scalar'.

Third, it was notable that firms do not organize their operations according to a 'scalar logic'. The lawyers did not have a 'scalar view of globalization'. Below a partner at an off-shore firm describes how his law firm's spatial organization is one that prioritizes relations and networks; 'geographical scales' don't get a look-in:

> We don't carve the world up into sort of this is my country, this is your country, kind of thinking, but we do have areas of concentration ... It falls out fairly naturally. I mean most businesses speak of Japan and non-Japan Asia. We have a large office in Tokyo ... We [Singapore] work with the Tokyo office quite a bit ... Hong Kong, most of the Korean work is handled out of Hong Kong. That's natural. The head of our Korean practice group is in Hong Kong ... so there is interaction back and forth. Most of China is handled out of Hong Kong, and the head of the China practice group is out of Hong Kong, but there is a fair amount of China work that's done out of here [Singapore], large IPOs and debt deals for China-based companies ... We will do more China work here [Singapore], particularly with the joint venture. We don't have a regional office. I guess that is the bottom line. The Asia head is based in Hong Kong. The

regional business development marketing person is in Hong Kong. The regional IT person is in Singapore. Our capital markets group is based here [Singapore]. Project finance is split. M&A is split. We have no formal regional office.

(Partner, off-shore firm, 2000)

In their words, we hear how Singapore is, for them, not a regional hub, not an operational command centre in the 'global city' sense. Instead, we are told that their law firm is organized in the form of a network; the threads of that network are listed by the interviewee, who also indicates how his firm's new JLV will serve to lengthen and strengthen their China network.

Fourth, we learnt from the lawyers that they do not have a 'scalar view of globalization' and consequently we should not be surprised to learn that their firms' networks are also not organized as a nested 'hierarchy of scales' – with 'global', 'national', 'regional' or 'local' offices (hubs) – where power and knowledge flow from top to bottom. Rather the firm's networks operate at the 'ground level' (flat, horizontal), incorporating any number of localities (work spaces *and* rooms *and* offices *and* buildings *and* cities *and* countries, etc.) which are purely relational (i.e. dependent upon any firm's whole actor-network), not absolute. The law firms' actor-networks are always connecting, assembling, distributing and centring because they are held together only by these relational forces. Thus, it is evident that the lawyers have a *relational* understanding of power. Within their networks the lawyers talk about 'areas of concentration', what ANT scholars refer to as 'centres of calculation', in their networks. And this is precisely how they think about the JLV process in Singapore. The creation of the JLV entities facilitates longer service provision through Singapore, one effect of which is the enhancement of Singapore's position as both network *and* point, a financial and service centre (a 'global city') with an increased power to 'act at a distance'.

Finally, the lawyers described, in different words, the JLV restructuring as a work of 'heterogeneous engineering' rather than a consequence of intersecting 'scalar processes'; there is no 'production of geographical scale' here. In other words, they describe and acknowledge how the formation and lengthening of their firm's networks – through JLV participation – are the product of a number of actors (human, non-human, material, discursive) coming together in joint action. The Legal Services Review Committee report and the subsequent JLV entities are hybrid collectives, products of the coming together of a variety of actors: banks and financial institutions (the lawyers noted the importance of their demand for one-stop legal shopping through Singapore as a driver for the JLV experiment); foreign law firms (the lawyers talked of the need for them to meet demand from financial services for longer legal provision through the Lion City) and domestic law firms (the lawyers commented on their desire to participate in the longer fee-earning networks needed to service the cross-border demands of clients); the government of Singapore and its desire to make Singapore into a 'global city' based on financial services like London and New York (the lawyers described how the government came to be convinced that the managed restructuring of legal services through the city-state was an essential step if Singapore was to be furthered as a hub for financial services); and the Singaporean legal system and the UK/US legal systems (the lawyers spoke of how the separate legal systems needed the JLV vehicle to enable them to be brought together into a single legal provision), to name just some of the myriad of elements that were assembled through the JLV process to enable the lengthening of legal services through Singapore.

CONCLUSION

> The world, our world, is depleted, impoverished enough. Away with all duplicates of it, until we again experience more immediately what we have.
>
> Sontag (1966: 7)

ANT is a form of analysis that advances an intransitive understanding of the spatial, the temporal and the network. ANT's intransitive conception takes nothing as a priori, as pre-given, stable, fixed or absolute. ANT is a constructivist approach, so all spatial-temporal formations and relations are conceptualized as constructed within networks: length, distance, location, power, dimension, size and scale are accomplishments that are never fully guaranteed. Thus, following a network means describing how its topology is made, unmade and remade. It is not an exercise in restating a phenomenon by inventing an equivalent for it: in replacing the network with a duplicate, such as 'scale', to pretend to be able to provide a geographical overview by stepping 'outside' of the network. Indeed, a key advance of ANT over other theories is its commitment to fieldwork and case studies which expose the actor-networks that have hitherto been hidden behind categories and terms such as 'structure', 'system' and 'scale'.

Susan Sontag (1966) once noted how one only needs to invent a 'shadow world' of invisible forces and structures – i.e. levels, layers, tiers, territories, spheres, categories, structures, systems, scales, etc. – if one's intention is to interpret the world. ANT's aim is to describe, not prescribe, and so consequently all the devices of interpretation (duplicates, equivalences) which have served to short-circuit description, to impoverish it, to make it manageable and comfortable, are now abandoned. In other words, ANT does not reduce the world to, or deduce the world from, anything (visible or invisible), but instead insists that 'everything may be allied to everything else' (Latour 1988: 163). Abandoning all a priori dualisms and categories, including 'scale', ANT changes urban studies by demonstrating – through a description of networks of heterogeneous associations – how cities are relationally constituted entities: temporary assemblages made through associations with other 'actants'.

With neo-Marxist theory, neo-structuralist Marxism, the networks of firms were neglected in favour of a focus on 'the big picture' and something called 'total', 'world' or 'global' capitalism. However, Latour has pointed out that such an interpretation of Marx's work is highly suspect:

> The capitalism of Karl Marx or Fernand Braudel is not the total capitalism of the Marxists. It is a skein of somewhat longer networks that rather inadequately embrace a world on the basis of points that become centres of profit and calculation. In following it step by step, one never crosses the mysterious *lines* that should divide the local from the global. The organization . . . is a braid of networks materialized in order slips and flow charts, local procedures and special arrangements, which permit it to spread to an entire continent so long as it does not cover the continent. One can follow the growth of an organization in its entirety without ever changing [scale] levels.
>
> (Latour 1993: 121–2)

In other words, Braudel is anti-Marxist, and Marx had no 'world view' until after his death! They knew that the 'global' is always 'local', and consequently that any organization's growth and reach can be followed empirically as a network with no shift in register or 'scale' from 'micro' to 'macro' or 'local' to 'global'. Similarly, ANT is grounded in primary evidence and so knows that any firm's network is continuously 'local': unlike neo-Marxists with their universal abstractions ANT never takes the risk of losing touch with the city.

In conclusion, following the lengthening of legal services through Singapore requires no change in 'scale' to understand how so few lawyers seem to cover the world; we have seen how Singapore's lawyers build their worlds in ways that defy scalar categorization. In fact, invoking 'scale', rather than following networks, could lead one to some rather dubious conclusions about the restructuring process. 'Scale' refers to the size or dimensions of an object; thus if one was to think of foreign UK/US law firms as large ('global') and so 'powerful', and the domestic Singaporean law firms as small ('local') and so 'weak', then one might read the restructuring as an overbearing process of the 'local' surrendering to the 'global' (e.g. an article in *World Legal Forum* (Anon 1997: 1) begins with the sentence, 'The City's invasion of Singapore began in 1981'). What is more, one might get really carried away, imagining that the Lion City is powerless in the face of 'global' firms and 'global' capital, imagining the city-state as a 'victim', a mere 'pawn' in the great game of 'globalization'. Indeed, in this chapter's description of the events that unfurled, no lines were crossed. There was no switching between 'geographical scales', no 'scalar jumping' forwards and backwards from 'local' to 'national' to 'global', no 'hierarchies of scales'. There was only the creation, the furthering, the lengthening of relationships and connections to form actor-networks through Singapore. What is more, there was not any 'foreign invasion' and 'domestic surrender'; domestic Singaporean law firms did not see the arrival of foreign law firms as threatening, but rather as an opportunity to forge collaborations and participate in longer, higher-fee-earning actor-networks: 'Capitalists of the World Unite' (advert for Forbes Global, *Financial Times*, 20 October 1997, p. 25).

ACKNOWLEDGEMENTS

Thank you to the British Academy Committee for South East Asian Studies for funding this research. Thank you to the Centre for Advanced Studies at the National University of Singapore for hosting me as a visiting scholar. Thank you to all the lawyers who generously gave up their time to be interviewed. I presented several of the ideas in this chapter to Leicester University Management School, a colloquium at the University of Ghent, the Geography Departments at Swansea University and the University of Sydney; I would like to thank all those audiences for their interest and questions.

NOTES

1   The law firm Lee & Lee was founded by Lee Kuan Yew, Lee Kim Yew and Kwa Geok Choo in 1955.
2   Another provision is that by the in-house counsel of banking and financial institutions. In 1999 there were *ca.*140 in-house counsel in companies and statutory boards in Singapore (Sek Keong 1999). Other providers are: law firms based outside Singapore, some global accountancy firms, and QCs providing advice or opinions on particular aspects of transactions (Sek Keong 1999).
3   The Asian Dollar Market was established in Singapore in 1968 and created a demand for off-shore legal services because US dollars became the primary source of finance for investments in Southeast Asia.
4   According to Sek Keong (1999) less than 10 per cent of Singapore lawyers were providing services in banking- and finance-related work in 1999.
5   The phrase 'cross-border financial transactions' refers to those financial transactions that are regulated or affected by two laws (i.e. by Singapore law and at least one other national law). This definition is used in Sek Keong (1999).
6   In the 1980s Freshfields was granted a licence for seven years to practise Singapore law. Also in the early 1990s Wong Meng Meng and a small US firm (Shaw Fairweather & Geraldson) were permitted to form a joint venture to handle on-shore and off-shore work (Page 1994). With hindsight these permissions can be interpreted as a signal from the Singaporean government as to its vision for the city-state's future legal landscape.

7   Another firm that lengthened its network in the mid-1990s was Khattar Wong & Partners (KWP). Established in 1974, KWP had some 110 lawyers and 40 partners by 2000. At that time KWP had offices in Singapore (1974) and Hong Kong (1995), and an associated practice with another firm in Kuala Lumpur and Johor Bahru (Malaysia). The firm also had country-specific desks for Indonesia, China and India. KWP also had a formal alliance with Eversheds.

8   Another reason why Singaporean law firms have not significantly 'regionalized' may be down to risk. Financial risk would be partly reduced if firms were private companies (rather than partnerships), as this would allow limited liability.

9   The distinction between an FA and a JLV is that within the former entity a firm can advise on cross-border deals but not on Singapore's legal issues directly, whilst with the latter foreign law firms can advise clients directly on Singapore law and hire local lawyers. Combining both foreign and domestic practice a JLV can undertake on-shore, cross-border and off-shore work.

10   JLVs are necessary for Singaporean firms to attract more complex legal work because 'SingLaw' is an insignificant governing law compared to English and US (New York) law for cross-border corporate and financial transactions. Note that JLVs are restricted to just a few areas of practice (i.e. financial, banking, Internet and corporate legal services). Foreign lawyers in the joint ventures are allowed to practise Singapore law only in these areas. They cannot represent clients in Singapore's courts, handle property transactions or provide specialist services such as litigation and conveyancing. The monopoly enjoyed by Singaporean firms remains in many areas.

11   In the 1999 to 2002 period four of the nine JLVs had already ended, indicating the difficulty of maintaining networks (see Ferguson 2002). The Clifford Chance and Wong Partnership JLV was relaunched at the end of 2002 after dissolving in 2000 (Evans 2002).

12   The selection of firms from just the UK and the USA was controversial. For example, the inclusion of no Australian firms annoyed the president of the Law Council of Australia (see Cronin 2000). However, the most important issue was the lack of interest from Wall Street firms in the JLV process. The liberalization and opening up of the Singapore legal market is a selective one concerned not so much with opening up to competition but with enhancing Singapore's role as a financial centre. And that is why the participation of just one Wall Street law firm in the JLV restructuring process somewhat embarrasses Singapore's ambition. Indeed, the participation of Shearman & Sterling was an exception rather than the rule, as Wall Street firms such as Sullivan & Cromwell, Milbank Tweed, and Davis Polk & Wardwell have not been involved in the process. However, this is perhaps not surprising, because US law firms, and Wall Street firms especially, tend to shun the obligation of associations and joint

ventures abroad. To some extent the need to build a geographical presence to win business, as with London firms, is circumvented by having leading US investment banks as clients (such as Goldman Sachs, Chase Manhattan, CitiGroup or Morgan Stanley). Some Wall Street law firms do have offices in Tokyo and Hong Kong (unlike Singapore these financial centres have captive and substantial economic hinterlands) because their investment bank clients require them to be located in those cities.

13 The number of JLV licences awarded increased from five to seven to nine because the Attorney-General did not want to exclude any top UK/US firms from the process.

14 Interviews were conducted in Singapore with Freshfields Bruckhaus Deringer, Linklaters & Alliance, Allen & Overy, Lee & Lee, Colin Ng & Partners, Lovells, White & Case, Shooklin & Bok, Helen Yeo & Partners, and Orricks. An interview was also conducted with Khattar Wong & Partners.

15 Some domestic lawyers viewed the JLVs as the end of the restructuring process, whilst others saw their JLV as an important step toward a merger with their foreign partner. Whilst one should not confuse sequence with consequence, a common path to full integration seems to be emerging: strategic alliance to FA to JLV to full merger (if permitted by the regulatory framework).

REFERENCES

Amin, A. and Thrift, N. (2002) *Cities: Reimagining the Urban*, Cambridge: Polity Press.

Anon. (1997) 'The Role of Offshore Law Firms,' Online, available at: http://www.worldlegalforum.co.uk (accessed 9 August 2000).

Boon, T.O. (2000) 'Build a Vibrant Legal Sector: SM', Unpublished manuscript, pp. 1–3.

Charlton, P. (2003) 'The Globalisation of Law', Online, available at: http://www.lboro.ac.uk/gawc/rb/al5.html (accessed 9 August 2004).

CNP (2000) *Firm Profile*, Singapore: Colin Ng & Partners.

Cronin, C. (2000) 'Australian Legal Chief Condemns Awards', *Legal Week*, Online, available at: http://www.lwk.co.uk/archive.php3?page=archive&item=5587&printable=true (accessed 11 October 2000).

Deleuze, G. and Guattari, F. (1983) *On the Line*, New York: Semiotext(e).

Evans, M. (2002) 'Clifford Chance Relaunches Singapore Joint Venture', Online, available at: http://www.legalmediagroup.com/news/print.asp?SID=11579&CH= (accessed 3 December 2002).

Ferguson, N. (2002) 'The Failure of Singapore's Joint Venture Compromise', *International Financial Law Review*, 21(9): 77–80.

Houellebecq, M. (2003) *Platform*, London: Vintage.

Latour, B. (1988) *The Pasteurization of France*, London: Harvard University Press.

Latour, B. (1993) *We Have Never Been Modern*, New York: Harvester Wheatsheaf.

Latour, B. (2005) *Reassembling the Social*, Oxford: Oxford University Press.

Meyer, R. (1999) 'Singapore's Reforms: Too Little, Too Late?', *International Financial Law Review*, September: 17–19.

Page, N. (1994) 'Competition Intensifies in Singapore's Legal Marketplace', *International Financial Law Review*, May: 20–22.

Riles, A. (2001) *The Network Inside Out*, Ann Arbor: University of Michigan Press.

Sek Keong, C. (ed.) (1999, June) *Legal Services Review Committee Report*, Singapore: Attorney-General's Chambers.

Short, J. (2006) *Urban Theory: A Critical Assessment*, New York: Palgrave.

Smith, R.G. (2003a) 'World City Actor-Networks', *Progress in Human Geography*, 27(1): 25–44.

Smith, R.G. (2003b) 'World City Topologies', *Progress in Human Geography*, 27(5): 561–82.

Smith, R.G. (2006) 'Place as Network', in I. Douglas, R. Huggett and C. Perkins (eds), *Companion Encyclopedia of Geography*, 2nd edition, London: Routledge, pp. 57–69.

Smith, R.G. (2007) 'Poststructuralism, Power and the Global City', in P. Taylor, B. Derudder, P. Saey and F. Witlox (eds), *Cities in Globalization: Practices, Policies and Theories*, London: Routledge, pp. 258–70.

Smith, R.G. and Doel, M. (2010) 'Questioning the Theoretical Basis of Current Global-City Research: Structures, Networks, and Actor-Networks', *International Journal of Urban and Regional Research*, 34.4, December.

Sontag, S. (1966) *Against Interpretation*, New York: Farrar, Straus and Giroux.

Tromans, R. (2000) 'Firms Must Lobby for Liberalisation', *Legal Week*, 28 September, p. 1. Online, available at: http://www.lwk.co.uk/archive.php3? page=archive&item= 5648&printable=true (accessed 11 October 2000).

4 Assembling Asturias

Scaling devices and cultural leverage

Don Slater and Tomas Ariztia

INTRODUCTION

This chapter reports a very concrete case of assembling social scales, of stabilizing connections at different levels of proximity and distance. A projected international cultural centre in Aviles, Asturias (northern Spain) – the Centre Cultural Oscar Niemeyer (CCON) – aimed to link global cultural flows to regional development through a now conventional strategy of place marketing and culture as regional regeneration policy. Participants clearly understood the process as one of linking a pre-given 'global' to a pre-given 'local', and in fact employed the authors as members of another global institution – the London School of Economics – that was considered expert in understanding and engineering local–global connections. The authors, by contrast, were fascinated by such a clear case of the performance of locals and globals, one in which people were reimagining and reconstructing a city and a region through the invocation of scalar concepts of culture and cultural flow. What we believed ourselves to be observing was processes of mapping, and acting upon maps, that appealed to entities (e.g. Asturian culture, global culture) which were brought into existence by that very appeal: pure performativity.

At the same time, there is a certain futility – certainly an arrogance – involved when social theorists declare the death of concepts that are quite alive in the world. 'Local' and 'global' cannot be legislated out of existence because they are analytically dubious, particularly when they are paying for the research. What ensued therefore was an attempt to move from realist critique to a negotiated relativism: less theory and more standard ethnography; we would not allow terms like 'local', 'global' and 'culture' to have any meanings other than those that emerged *in situ*, from their mouths or ours. As intellectual and political intervention, the research aimed simply to bring the fact of scaling activity to light, topicalizing it or putting it on the table, as it were, so that different scales could be treated as constructions rather than as objective frameworks within which practices (including research and the building of cultural centres) were to be carried out.

The chapter therefore foregrounds the research encounter as much as it aims to present any research 'findings', and we hope this emerges as warranted rather than self-indulgent. It is simply that we are reporting a situation in which relationships between research institutions, governmental agencies, local actors and knowledges were even more obviously central than normal to an urban assemblage: the relationships assembled in and through the research – more than its truth value or even its instrumental use – were central and highly charged aspects both of the form the research took and of its capacity as an actant. We frame the chapter in this way not to critique – let alone undermine – our clients, ourselves or any other participants (indeed, we were all fairly happy with the various outcomes) but simply because this is a story about scaling devices, and that is not only what our research studied but also precisely what our research itself *was*.

The chapter is divided into three narrative sections. The first part presents something like the official version of the CCON initiative and its place in Asturian regional development policy. It is in many respects a conventional story of contemporary place marketing and of regional development through cultural policy. In our terms, CCON was conceived as a 'scaling device' that could both reconfigure and position Aviles in relation to global cultural flows that it could divert into local development, capturing a river of economic, cultural, social and political capital. The official version produced a range of technical issues (e.g. how to project plausible employment gains) which preoccupied most participants; however, it also produced an analytical concern that involved the more messy business of understanding social and cultural processes: how could captured global cultural flows be embedded, anchored or rooted in the locality? In fact, this concern was potentially worrying not because it was an intractable problem but because it pointed to the instability of the official version: if one scratched the surface there was actually very little agreement as to the meaning of culture, of locality, of globalization, and there were very different imaginations as to how these differently defined terms might be connected up. As a policy of regional development, there was little shared understanding as to what was being developed or how.

Secondly, this 'fudged' quality points to the essentially 'fudged' nature of the initiative and the alliances which it mobilized. The prime instigators of the project pursued a policy of 'cultural leveraging': using available networked cultural capital to raise very much larger sums, and borrowing against smallish but valuable connections to buy into the very biggest. This was a process of enrolment which required skilled exploitation of social capital: connections with global cultural 'stars' could be used to leverage local government funds; Spanish charitable funds could bring in tourist networks, and so on. The ability to grow and converge heterogeneous networks both appealed to the 'facts' of global and local cultures and also sought to assemble and construct these cultures so as to position CCON within them. The authors, the research and the LSE were brought into the leveraging process at a strategic point: CCON assembled a meeting of the C8 (like the

G8, but for culture), at Aviles, to stabilize their leveraged cultural capital into an enduring network; the presentation of research about local–global cultural connection, by a global academic institution, would fix the entire picture of CCON as one of many elite institutions contending with the issue of how to embed global culture in local contexts.

The research itself, thirdly, was defined as a study of local youth culture, with a specific focus on new media and technology use. The official and fudged version of cultural development policy, and the leveraging strategy of CCON, pointed to an analytical structure in which global flows had to be connected to local culture, to be measured in terms of variables such as youth participation. This had the immediate effect of framing CCON as global and youth as local. We report research findings that we used to challenge this framework and to present Aviles youth as analytically entirely equivalent to both CCON and to governmental agencies: firstly, Aviles youth demonstrated use of multiple scaling devices to construct diverse maps and scales; secondly, they were 'multilingual' in their scaling, reasonably fluent in both youth and official cultural maps; thirdly, in contrast to the official discourses of CCON (but much like CCON in its private practice), youth treated global cultural connections as mundane, practical, routine – what we call 'cool globalization' – rather than as spectacular, extraordinary or transformative. This led us to propose reimagining CCON as a different kind of scaling device: rather than using it to construct the global, or anchor the global in the local, it could be conceived as a space in which the very process of scaling itself – as central to the making and mapping of 'culture' – could be topicalized, discussed and negotiated.

CULTURAL POLICY, REGIONAL REGENERATION AND THE ART OF 'FUDGING'

The Niemeyer Centre (CCON) is now, as it was when we conducted our research (November–December 2007), virtual. It exists as an architectural model and computer graphic, as an organization with a couple of employees and numerous trustees, as an element in a regional regeneration strategy represented in policy documents and pronouncements and, powerfully, as a collective vision and ambition. The object in formation is a cultural 'box', an iconic building designed – literally and figuratively – to house performances, exhibitions, archives and events. There are many such boxes in the world, from older and established ones (the Barbican, the Tate, the Pompidou Centre) to newer and trendier ones (Tate Modern, the Bilbao Guggenheim), and CCON is projecting itself into a crowded market. CCON is being positioned at what might be called the 'funkier' end of the market in world-class culture, slightly closer to, or in dialogue with, popular cultural forms: notably, it will house a film archive, and one of its notable allies is Woody Allen.

However, cultural boxes are only one manifestation of a broader movement that underwrites and legitimizes CCON: it is one form of regional regeneration through creative and cultural industries, which might otherwise, or also, be accomplished by strategically situating a region within a global division of cultural production (e.g. the film and CGI industry in New Zealand – Neitzert 2008) or by the transformation of urban space into entertainment, leisure and retail facilities (e.g. the 'museumification' of many formerly industrial cities, or production of 24-hour cities). What is common to these regeneration strategies is a belief in a globalization process which is intimately connected to a shift from industrial to cultural or informational production. This narrative is purveyed through canonical texts (e.g. Florida 2005), through management education and conferences, through networks of administrators and, of course, through academic knowledge production and dissemination, both directly (specific studies of place marketing) and indirectly: these development narratives rely heavily on well-established meta-narratives concerning post-Fordism, information and network society, globalization and mobilities. More specifically, all those associated with CCON could point to a series of comparable 'success stories' (above all the Bilbao Guggenheim), which they could also interpret through a shared development narrative. Hence, Asturian actors could unproblematically adhere to the now-conventional conviction that their regional (mis)fortunes could only be reversed by successfully tapping into global cultural flows, and that these – once diverted to the region – will naturally transfer into economic and other stimulation.

On the one hand, the level of commitment to this narrative and its localization is fairly awesome. Historically a coal and steel region, with significant working-class and trade union traditions that earned it a dramatic place in both the Spanish Civil War and the post-war period, Asturias – from the 1980s onwards – could be described as 'rustbelt', and the local narratives are indeed of inexorable economic decline, mirrored in major out-migration. CCON is to be built in the town of Aviles (population *circa* 200,000), which is part of a conurbation also comprising the cities of Gijon and Oviedo, which together represent industrial Asturias. The siting of CCON involves unavoidable symbolism: it will be built on industrial wasteland which – like the polluted river between the site and the town – requires extensive detoxification to remove its trade and manufacturing past. The site stands next to an old steel plant, part of an enterprise that once employed 28,000 Asturians and now employs about 800 and is owned by the Indian steel conglomerate Tata.

Not unusually, Aviles spent the entire modern era turning its back on the river and the manufacturing spaces that lay on its other side: these are not visible, street and building layout is designed to cut off views to the river, and a highway and railway run along the river so that it has been difficult and unusual for visitors or bourgeois residents to cross it by foot. The most recent previous regeneration strategy accentuated this stance: the old medieval town centre was pedestrianized in the hope of attracting tourists to a quaint premodern locale untainted by the industrial. By contrast, the CCON strategy involves not only building a gigantic cultural icon but also turning the city around 180 degrees to face the river and the icon on its far shore: rail and roads will be moved, bridges built and streets straightened to open up all the lines of sight and movement that were historically blocked. This is in addition to the incalculable and unimaginable reconstructions that many people expect and hope for as new hotels, restaurants, businesses and services arise to meet the demand of visitors to CCON.

This massive reassemblage of the town has secured considerable and largely fulsome support: the city governments of all three cities are solidly behind it, and the president of Asturias has tied his political career and considerable funding to the vision. There is either a profound belief in the narrative of regional regeneration through cultural industries or – given the level of public commitment and economic deprivation – a considerable need to believe. However, what was most striking about this apparently unanimous commitment was the fuzziness about just how the strategy works and the unacknowledged lack of any agreement over the meaning of its basic terms. We will call this feature 'fudging': silences or ambiguities over crucial issues that did not preclude – and were probably essential to – securing wide enrolment of actors in a regional initiative.

'Fudging' was most notable in the great vagueness and unacknowledged differences firstly in how actors defined and identified 'culture' and secondly in the closely related matter of specifying the mechanisms by which 'cultural' flows might translate into economic gains. In one typology we could distinguish three different but overlapping and allied versions. Firstly, typified by the CCON organizers, culture could be defined in terms of translocal values, standards and achievements embodied in the people and objects that circulate in elite networks. This sense of culture is profoundly modernist in that it follows its own internal logic of aesthetic development, shared by practitioners, critics and audiences who participate in it. This logic is relatively autonomous of particular locations and located traditions; CCON is there to recognize authentic exemplars and entice them to Aviles. This places Aviles on this cultural map, as another participant, and as a node in the circulation of these artists, critics and audiences. For example, great expectations were placed on getting tour operators to offer tickets that included flights and admission to both the Bilbao Guggenheim and the Aviles CCON as part of one great cultural excursion. The upshot of this version of culture should be tourism and cultural prestige, with local employment arising essentially from leisure services rather than cultural participation.

A second, alternative, notion of culture was eloquently articulated by the president of Asturias: culture means local heritage, history and tradition, roughly translated as 'Asturian culture'. The president's vision, shared by many political actors, was of tourists initially attracted to the region by CCON then discovering the delights of Asturias – its (truly) fantastic cuisine, its beautiful landscape and numerous national parks and forests, and its quaint (and Celtic) customs. This was also fundamentally a tourism-driven image of regeneration, and connected to similar developments such as increased British and German purchases of holiday homes in the region, but one far more focused on promoting rather than subsuming 'local' culture within the 'global'.

Finally, a range of actors promoted an 'institutional' or 'behaviourist' view of culture: here, culture takes the form of myriad institutions (concert halls, academies, organized exchanges, local and foreign groups, educational programmes) that add up to a kind of cultural infrastructure, and of observable and hence measurable behaviours (ticket sales and attendance, group memberships, number of cultural objects travelling to and from the region, and so on). This version of regeneration strategy had the narrowest vision of development, with its eyes largely focused on the cultural segment itself. It also had the most traditional definition of culture, closely identified with the high arts as defined within nineteenth-century bourgeois culture, and attended by a conventionally passive audience. Our sense was that both CCON and the president found this version of culture both boring and strategically fruitless: indeed it could be considered 'off-message'; nonetheless it was the unshiftable language of culture bureaucrats both in government and in voluntary sector organizations.

The mechanisms that might relay these different notions of culture into economic revival were obviously different, and the claim of unanimity therefore required considerable fudging. There was audible, albeit good-humoured, muttering behind the scenes. Given the high-stakes commitment to a strategy without clear mechanisms, the role of research – ultimately including ours – was interesting. While negotiating a role for our own research, CCON responded to almost every suggestion by claiming that that research had already been done. In fact, there were only a couple of economics-based projections that calculated employment and income growth out of commonsense assumptions about increased tourism. It seemed important that research should not threaten a fundamental agreement over the importance of local–global connections in a globalizing world.

Two points can be highlighted already. First, concepts of culture already have a spatial dimension – they each indicate different scales and dynamics of cultural exchange (of objects, people, tastes, etc.) – which ties culture to different development models. Secondly, however, this did not preclude differing participants from agreeing on a fundamental formulation: Aviles, and Asturias, *must* 'scale up' in order to survive, and it must do so because its survival depends on global flows which are increasingly cultural in character. In response to these points, very early in the research we began to think of CCON as a cultural 'scaling device': CCON – as both virtual and ultimately real object – was a device through which diverse actors could imagine, negotiate and perform different versions of the global, and of the local in the global. The fact that they were imagining and performing different versions through the same device just seemed to make it a more interesting and possibly effective machine – or else the stage set for a looming disaster.

CULTURAL 'LEVERAGING'

Fudging seemed to be a condition for enrolling motley actors, but the dynamic behind the enrolment process was provided by what we came to call 'cultural leveraging'. This is best conveyed through narrative. The idea of CCON started from a couple of young administrators of the Prince of Asturias Trust in Madrid. The Trust, which is indeed under royal patronage, offers a prize – often referred to as the Spanish Nobel prize – to major figures in global culture, household names such as Stephen Hawking, Woody Allen, Daniel Barenboim, Norman Mailer and Bob Dylan. The young administrators argued that, having secured through past awards a network of figures who indisputably make up a version of global culture, the Trust could now be more proactive, moving from simply giving prizes to playing a role in the production and circulation of this global culture. The young men already had significant expertise in networking based on their elite educational background and their service in diplomatic and other high-status government jobs. The idea of 'leverage' was there from the start: the Trust possessed a good measure of valuable capital (networks, reputation, positioning, historical cultural allegiances) that could be set in motion to generate a lot more capital, at higher levels of global participation. In essence, they could borrow huge stakes against a small base of cultural collateral, and a rolling process of accumulation could be set in motion. They scored their first success by enlisting Oscar Niemeyer – a Prince of Asturias Trust prize winner – to donate architectural plans for a cultural centre. Niemeyer, the architect of Brasilia, not only designs iconic buildings but – particularly as he approached his 100th birthday – was himself an icon of iconic architecture. His enrolment would allow the scaling up of the entire emerging project, and CCON was in fact energetic in leveraging this connection.

Further expansion followed from leveraging the princely connection. Just as the British Prince of Wales has a feudal connection to Wales, the Prince of Asturias is historically connected to Asturias. The cultural connection clearly joined up with regional development needs, which thereby not only enrolled local political actors, as above, but also attracted massive EU capital. Indeed it is at this point that the project shifts from dream to virtual reality.

Our own research enters the narrative as part of the same leveraging process. CCON planned a symposium of what they called the 'C8' – the cultural equivalent of the G8 – comprising representatives from bodies such as the Pompidou, Lincoln and Barbican centres. This gathering had multiple functions: bringing CCON to the notice of the networks of which it wanted to be a part; signalling its global status to Spanish and European actors; and overcoming its virtual status by bringing social connections into being. Overall, the gathering was to make manifest precisely that notion of 'global culture' that legitimated the CCON concept and gave the regional development strategy its realist status ('this global culture on which our future depends really does exist, and you can see it meeting in Aviles'). This initial gathering was to be stabilized into an ongoing network, so that the story about global cultural centres might take on an objective organizational form.

The London School of Economics was invited to be one of the C8, and it was clear that both the institution and the research were regarded from the standpoint of the leveraging process. CCON had no interest in the substantive content of any research; their concern was solely that the LSE, as a global cultural brand, should present a paper to the other assembled global brands. The LSE is not only a prestige brand in international flows of cultural capital but its previous director – Anthony Giddens – and several of its professors were major names in the production of the entire concept and public agenda of 'globalization'. At the same time, as mentioned above, we were repeatedly told that they had already conducted any research that we suggested. They simply wanted an LSE academic to address a global cultural gathering on the subject of local regeneration through cultural initiatives. However, this was incompatible with the LSE's own cultural leveraging strategy, as well as that of an individual academic career: protection of personal and institutional cultural capital, as well as simple academic self-respect, dictated that the LSE couldn't just appear at the C8 to sound off pompously – some real research had to be done.

It took three months to negotiate a brief, and the ensuing contract was itself a fairly classic case of fudging. CCON wanted a study that addressed the connection of the global to the local, the embedding or anchoring of the global in the local, and the participation of the local in the global. This could be seen as specifying what kind of scaling device CCON might be in relation to both the cultural demands of local constituencies and the desired economic gains for the region. More specifically, however, the focus was to be on local youth, an entirely sensible choice. Young people have a strategic position in such discussions, signifying significant and leading-edge cultural actors and consumers, the future of culture and society, resources who need to be retained in the region after school, the future of employment structures and regional development, and much more (and of course they are equally, negatively, signifiers of social disintegration, public order issues and political ill-discipline – the subjects of moral panics which are just as institution-defining as the hopes pinned on them). In entirely practical terms, many countries actually tie public funding of cultural centres directly to metrics of youth participation. It therefore made great sense for CCON to focus on young people's cultural lives and understandings as a precondition for an informed approach to connecting global culture to local life through CCON as a scaling device.

However, this framing of the research, however sensible, constituted a travesty of the entire history by which we had arrived at the research, as described so far in this chapter. Firstly, where we have been describing a performance of scaling practices through fudging and leveraging within a conventional narrative of regional development, the research brief simply treated local and global as pre-given and objective entities that needed to be connected up, analytically and practically. The process of its emergence was to be obscured, such that the 'local–global' divide appeared as an unchallengeable research framework, rather than as the social accomplishment of volatile coteries of actors.

Second, and more specific to our research, this framing of the research relegated the researchers and the researched to polarized positions dictated by the analytical categories: simply, CCON (and LSE) represents the global, whereas 'youth' belong to the sphere of the local. While our story is of diverse actors scaling at diverse distances, the brief applied specific 'scopes' to specific constituencies. This consigned youth to a particular role in the story from the start, and could only narrow our understanding both of their spatial and cultural practices and of their potential spatial and cultural practices.

And this is the crux of our account: the story of CCON – as told by participants – is a realist narrative of adaptation to global dynamics, whereas the story as told by the researchers is a narrative of performativity, of the construction and materialization of 'the global' and 'the local', which includes the research itself (just another scaling device). To be very clear: the argument is not about dishonesty, manipulation or bad faith; it is about the framing of the forces one invokes, alternatively as objective structures ('we are responding to globalization') or as outcomes of one's own practices ('we are performing "the global" as part of the process of building associations, given that all associations necessarily must be organized at different scales').

The situation seemed almost too straightforwardly 'Latourian'. As he puts it, the problem with conventional social scientists (and – we need to add – policy makers) is that they tend to 'use scale as one of the many variables they need to set up before doing the study, whereas scale is what actors achieve by scaling, spacing, and contextualizing each other' (2005: 183–4). Adopting the notions of 'global' and 'local' as analytical framework or established 'variables', rather than as emic categories that participate in assembling situations and interventions, would be firstly to make the research complicit in universalizing one participant in the situation, and subsuming all others within that participant's viewpoint – our research would, effectively, become one cog in a machine that is globalizing the concept of globalization. Secondly, this would take a very specific policy and political form which would in fact undermine some of the more creative and progressive intentions of CCON: to adopt the framework of 'globalization' was to position CCON as representing 'global culture' or at best mediating this thing to 'the local' population. That is to say, the various peoples of Aviles and Asturias are figured as 'local' actors with the limited options of being allowed to access or participate in global culture through CCON, of ignoring it, or of benefiting economically rather than culturally.

The alternative analytical strategy offered by ANT seemed simply to acknowledge (and work from) the obvious: CCON did not represent an objective truth of cultural globalization; entirely to the contrary, they – and our research – constituted a clear narrative of constructing global and local culture as part of fudging and leveraging a virtual project into a real, and really big, building. By the same token, we could reasonably expect different 'local' constituencies to be constructing all manner of different scales as part of their various social projects; and they might or might not categorize some or all of them through the notions of local and global. What seemed obvious was that nothing was to be gained by formulating them – at the level of research framework, as opposed to observable behaviours – as 'the local' in relation to CCON as 'the global'.

The research strategy therefore sought to move as far as we could from studying how 'local' Avilans might participate in 'global' CCON. We would rather explore the different cultural scaling practices and devices of two constituencies – young people in the town and CCON – and think about how their conjuncture might produce some new and creative openings. The research brief allowed for this fudge; we'd deal with the outcomes later.

CULTURAL AND SCALING PRACTICES OF YOUTH IN AVILES

A bit of methodology before discussing the research itself: given constraints of time and money (this research could be considered part of an emerging tradition of 'quick and dirty ethnography'), the task was not exhaustively to map young people's cultural scales but to provide sufficiently rich material for us to be able to argue for the complexity and perhaps unpredictability of their cultural maps (they are not simply 'local'). The notion of 'scaling practices' worked well for this purpose. In terms of fieldwork, it sensitized us in an open-ended way to the diversity of means through which scale is produced. We needed to attend to:

- How people make and stabilize connections (including their aesthetic form – networks? groups? alliances?).
- The ways in which connections at different distances are represented – what do the various 'maps' look like? And how do these maps re-enter the making of connections at different distances?
- The use of a range of scaling 'devices' – objects which might objectify different scales (as embodiment, as material culture) and might be instrumentally useful in performing different scales. CCON was clearly a scaling device. In addition, our research was specifically asked to look at Internet and mobile phones as part of the cultural globalization of local youth; our reformulation of this question was: how are these devices used to scale up, down, sideways, whatever? How are these scales categorized by young people when experienced in some measure through these devices?
- The ways in which different scales and scaling practices connect to diverse projects and strategies, including the project of being a particular kind of young person in Aviles, or living in a town which is likely to be imminently reconstructed in terms of a specific scaling practice and device (CCON)?

- The different ways in which people construct narratives involving scale – e.g., the history of Aviles or Asturias as part of a story of globalization.

More pragmatically, we used our two and a bit weeks to focus on a tightly delimited group – late secondary school students (16–18 years old) and young university students (18–21) who had lived most of their lives in Aviles, with roughly equal men and women. The selection of schools near the town centre allowed for a reasonable mix of social classes. We obviously did *not* treat 'youth' as a homogeneous category (though it is hard to reflect their underlying diversity in this brief chapter; we will talk rather too much about 'youth' in general), and aside from demographic differences the range of youth subcultures was vastly more extensive and consequential than we – perhaps naively – expected for a smallish town. The young people were initially accessed through focus groups organized in and through their schools. A few individuals from each of these groups were then taken on 'walkabouts', which provided the real core of the material: they went with us on extensive walks around Aviles, armed with cameras, in which they narrated their spatial connections and practices and could extend this into commentaries on Aviles's place in other scalar organizations. These engagements allowed for some more extensive ethnographic encounters, informally observing and chatting in some youth culture venues ('hanging out'). Connected to this we investigated websites and social networking systems that young people pointed us to. Finally, we conducted a range of 'expert' interviews with figures who had a stake in cultural and development policy.

In presenting our 'findings', we will stick closely to those which address the narrative we have been presenting rather attempting a full exploration of cultural scaling practices and devices: how far was the research able to demote privileged scalar terms (local and global) from analytic resource to empirical topic; and to what extent did this impact on the ways in which the relationship between CCON and youth in Aviles might be performed?

Young people produce multiple scales

Young people articulated, and mapped out for us, numerous scales which had different relevances for their lives. Each was related to different senses of culture, and none was easily mapped on to the notions of local or global (and those terms were only selectively used). At least four kinds of 'locality' emerged from most of the research encounters.

Firstly, 'the local' is considerably smaller than Aviles for most young people: on walkabouts, participants took us to specific streets, parks and clubs, and generally in a sequence that added up to the evenings spent out with their friends. These local pathways would obviously differ, but overlap, depending on subculture; all shared the sense that 'locality' is a scatter of culturally relevant oases in a desert of other people's cultural maps, connected by a sociable trek through their evening.

Secondly, however, 'the local' is also rather larger than Aviles because by common consent there's not enough happening in the town. Only by combining Aviles with its neighbouring cities of Gijon and Oviedo can one sustain a reasonable youth lifestyle. Hence discussion of lived culture generally maps a wide distribution of friendship networks and parties, gigs and dances, and involves extensive knowledge of the bus schedules and ticket prices through which culture must be coordinated in order to reproduce this scale.

Thirdly, there is a seasonal pattern of spending summers and holidays in one's family village, often with grandparents or other close relatives. The 'local' in this case traces a history of mobilities and a temporal rhythm, while also identifying locality with an Asturian identity rather than with the town.

Finally, young people talk easily and readily of 'Aviles' as a locality that is framed in official terms, just as CCON or policy makers would: on the one hand, 'Aviles' is a series of sites that one would show to a visitor – cathedrals, parks, quaint town squares – as opposed to the one that a young person claims as their own; on the other hand, young people could fluently narrate the same regional history of economic decline and renewal that was being recounted by official adults, and by researchers like us. In both these cases, the name 'Aviles' clearly denoted a space that corresponded to the official locality.

Young people articulated equally various 'globals', but none of them related very closely to the sense of 'global culture' that underpinned CCON and its many backers. The global geographies of young people were not generally organized around their cultural consumption: they consumed their fair share of American film and Anglo-American music, but they had little detailed awareness of where specific cultural goods came from; they did not map distant connections in terms of the origin of their cultural goods and rather thought of them as emanating from an undifferentiated cultural space ('Hollywood') which they did not map as they did other large-scale connections. Above all, they had no sense of a global cultural sphere marked by universal standards of excellence, prestige and recognition.

Rather, large scales were generated out of three ways of thinking. The first logic was very pragmatic: there is a global map of desirable travel locations made up of places your family might go on summer holiday, places you might visit because of a school exchange or penpal friend, or places to which relatives have emigrated. The global is simply a network of potential travel destinations.

Secondly, there are 'cool cities'. Cities are 'cool' because they are multi-cultural and cosmopolitan, places of stimulating difference. This is often summed up simply and concretely in terms of music: a cool city has a lot of different musical genres, events and active fans and performers. Supposedly global cities like London and New York did have a special status as cool, but Madrid and Barcelona were mentioned just as much and given very nearly the same status. The point seemed to be that, whatever the origin of different musics, Madrid and Barcelona are now just as cosmopolitan as anywhere else and accomplish the same convergence of global cultural flows into a dynamic multicultural space. For example, the intense interest in hiphop culture did not involve an identification of (and with) New York or LA as hiphop Meccas, points of rap origin and authenticity, however much young Avilans enjoyed US hiphop. To the contrary, the most extensive Avilan hiphop network online was most densely linked to a Madrid-based website which linked onwards not to the US or the UK but to Parisian, Peruvian and Colombian hiphop scenes.

Thirdly – and just as in the case of 'the local' – youth endorsed and articulated a kind of official version of large scale, usually glossed in terms of 'what's on the map'. They were very aware of – and fluent in – what they recognized as a generic story as to what places are 'important' in the world, and on a map which was not theirs but which they understood to be consequential for their lives. In this sense they were completely comfortable in arguing that CCON will place Aviles 'on the map' and bring in foreign visitors, and they did so with cool critical distance: those tourists lived on a map they did not share but could nonetheless theorize as a development strategy.

Young people are spatially multilingual

We stress the diversity of scales articulated by young people not simply to destabilize naturalized versions of local and global, and the identification of young people with the local, but also to indicate the range of different logics, resources and practical contexts that are involved in assembling these scalar structures. Quite simply people are thinking through many different kinds of connection simultaneously, sometimes entirely independently from each other, sometimes interrelated. We can usefully think of them as *multilingual*.

One research vignette tells the whole story. Pedro, a 16-year-old, was on walkabout with one of the authors, showing us the park where he meets his friends and the two pubs on the same street which (surprisingly) mapped his sense of local culture – one was a heavy metal club and the other was Asturian nationalist, providing bagpipes and cider. During the walkabout, we ran into Pedro's father; after it was explained what was going on, he said – forcefully – that he hoped Pedro was showing us the Aviles a visitor should see. The most striking thing was that Pedro had indeed already done so: alongside the subcultural mapping, he had shown off the cathedral, medieval palaces and touristically important squares. Pedro could move adeptly between personal and official cultural maps, with his own understandings of the logic of each. This involved a highly reflexive understanding of different pragmatic contexts and types of conversation, different audiences, and different involvements.

Indeed this is a social skill that would be just as important in communicating with someone from a different youth subculture as it is in dealing with foreign visitors or cultural policy officials, if only because all cultures involve different places and spatial organizations. However, the ability to engage with an 'official' version – even the very ability to identify one scaling practice as 'official' – points to an additional dimension of Pedro's story: Pedro, like all the young people we engaged with, was and had to be a development theorist or development economist. While none of them personally bought into the idea of 'global culture' that informed CCON and the very idea of regional regeneration, they all understood it at an intellectual level and were all concerned to speculate on how Aviles might firstly tap into it and secondly convert this access into economic growth, jobs and opportunity. Quite simply, they were no different from CCON or the president of Asturias, and probably no more or less clear about these connections, but also no less enthusiastic about the prospects and urgent about the necessity of this strategy. The irony of course is that CCON and its backers were rather more monolingual than the young people, committed to a single narrative and unable to imaginatively take on the youth perspective in the same way that the young people were able to take on theirs. The strength of this claim is underlined by our final point: the divergent notions of culture deployed by CCON and the young people.

What is 'culture'? Everyday life and 'cool globalization'

Across our research, youth claimed that culture that is 'real' and valued has at its core *participation* – intense involvement, a sense of 'being there'. Clearly music culture is exemplary, and the gig is paradigmatic: it is about active involvement, interactivity, events; even if you are not producing the music you're always more than a passive audience because as a fan you may be dancing all night, travelling hundreds of miles for a concert, emotionally investing in extreme ways, and making it your own, with others.

This view of culture generated the most common response to CCON amongst young people, as it should: asked if they would participate in CCON, young people generally felt it would depend on the form of participation, and not particularly on what was being offered or where it came from. They would not be likely to go to CCON if they were being asked to be a passive audience. The issue was not global culture or a high/low culture division (many indicated a serious, and informed, interest in some pretty demanding cultural products) – the issue was the mode of cultural participation itself. On this basis, many young people strongly supported CCON as development theorists who felt it would be good for their region's economy, while at the same time clearly indicating that CCON would have nothing to do with their cultural lives.

This attitude shows a sophisticated and reflexive understanding of multiple cultural scaling, but we need to go a bit further. Young people's participatory version of culture – pretty much shared across the subcultures we encountered – could easily be misconstrued as indicating a preference for localization, for the primacy of the face-to-face, embodied event, for something that is by definition 'local'. Yet that would be to enforce an opposition – local/global

– that didn't particularly feature in their life-world. The most vivid feature of the data was not locality but rather something more like practicality, or everyday practices. This was most evident in studying their new media uses, precisely because as scaling devices the direction in which they were scaling was not very predictable. Internet and mobile phone use was ubiquitous if not universal amongst young people (95.8 per cent of 16- to 24-year-olds use the Internet according to 2006 figures). In some respects that use could be fairly unproblematically described as 'local', particularly amongst the school children: the most discussed use of the Internet was MSN Messenger as a means of maintaining local social networks and coordinating everyday social life. Messenger lists could be very extensive: many claimed 500-plus contacts. These could be international (though tending to reflect the idea of global places as destinations rather than cultural connections – e.g., friends met on summer holidays). But the active contacts were the people they met every day anyway; the facility was used to chat about everyday events (homework, gossip) and to coordinate meeting up. This is a common pattern amongst youth (Livingstone 2002; Miller and Slater 2000; Slater and Kwami 2005). So too was the use of mobile phones, either on their own or in association with MSN. Phones are expensive for young people to keep in operation; they rather relied on a system of 'miscalls' or 'lost calls': by calling a friend, letting it ring once and then hanging up, you have given the friend a signal to log on to Messenger for an urgent chat. At this point you can arrange meetings or discuss homework for free.

Equally in keeping with more general material on youth and new media, international contacts (via Fotolog or Messenger) had a very local feel: chats with those in Latin countries, or the US, were likely to be about what they did today, what they are planning to do tonight, problems with girlfriends or boyfriends or family, how they're feeling – everyday stuff. As in many places, the social and technical connections afforded by the Internet are not necessarily regarded solely as means to obtain information or global cultural capital; certainly in the case of chat and Fotolog, the opposite is the case: a little knowledge of international music and film culture is rather a means to make and sustain social connections online.

While Fotolog and MSN are clearly the main uses of the Internet, young people nonetheless consider the Internet as the natural source for any cultural content such as movies, software, music or games. Most of the respondents use the Internet for finding movies, music and software. This practice was not individualist, and was often linked to their Fotolog and MSN use: it was mediated through other social relations. For example, most youth linked Internet content with active networks of friends and relatives who share the movies and music they downloaded. Getting content from the Internet is, thus, not an individual practice but a highly social activity.

This pattern of use does not correspond to the kind of global scaling up that many – including CCON – expected of young people's use of new media. However, it also does not square with simple localization: if so much of their new media contact was with people they also met at school, this was because use followed their participatory culture rather than because a higher intrinsic value was placed on the local as such. By the same token, they generally had no worry about extending their contacts over greater distances, but these activities replicated the mode of participation they valued rather than indicating a leap into a 'global culture'. For example, the most popular social networking facility was Fotolog, a Spanish and Portuguese language version of MySpace; the densest connections entered into by our young people were to other kids in Aviles, then Asturias, then working outwards the rest of Spain, and then Latin America. Connections to the Anglo-American space of global culture was, as we have seen, largely for sourcing cultural goods, which were then heavily mediated through their own participatory cultural arrangements.

Over the course of the research we came to refer to this stance as 'cool globalization': it is not so much that global culture is localized; it is rather that the possibility of more distant connection is valued in terms of ongoing cultural projects. Internet access to everywhere, all the time, is simply a mundane resource for everyday life rather than offering spectacular access to a prestige sphere of global culture. Another anecdote exemplifies this. A very active Avilan hiphop network had the chance to invite and perform with a major, indeed legendary, figure in 'global' music – a founding member of Wutang Clan. Rather than leap at the chance to catapult themselves into global cultural space, they decided on reflection to cancel the planned event. As the coordinator of the network explained – and indeed 'coolly':

> We almost brought Fura here. He has a representative in Spain and she contacted us through MySpace – this is the good thing about MySpace, man. She said she wanted to take Fura here. The only issue was that her plans only allowed two weeks for organizing everything. We explained to her the way we work, our resources and she even tried to help with the expenses. But my problem is that we only had three weeks, so it was difficult to find sponsors, to find a venue and also to organize something to allow local bands to play with him. We could have brought Fura without local bands and make a good concert but this is not our aim; we wanted local bands to play with him.
>
> (Pedro, hiphop producer, November 2007)

'Cool globalization' points to an important symmetry between CCON and youth in Aviles. The official version tended to present CCON as embodying a global culture that is spectacular, extraordinary and inaccessible from the standpoint of the local, and which CCON promises to mediate to them. There is an almost priestly role being performed. In the private life of CCON organizers, by contrast, 'the global' is a mundane feature of life: they routinely email and otherwise maintain contacts all over the world, unconcerned as to whether people are next door or in another continent; they source and consume cultural goods from anywhere, contextualizing disparate goods within their own cultural frame. Distant connection is a mundane feature, not spectacular, and their private attitude is as cool as the young people's: the fact that 'the world at large' is technically and culturally accessible to them is nothing to inspire awe; it is something to be routinely integrated within one's portfolio of scaling practices.

The young people, as in the Wutang case, are quite the same. On our last day of fieldwork, we came across a bunch of teenage boys skateboarding in the oldest town square, videoed by their friends; they repeatedly re-staged the same stunts in order to get a good take. The videos are destined for YouTube, of course. The lads are well aware that the clips could be viewed – theoretically – by anyone anywhere, but their intended audience is other kids in Aviles; another day it might be different. That's cool.

'CONFIGURING THE CLIENT' [1]

In breaking down the local/global opposition into a multiplicity of scaling practices and devices, the research aimed to reposition both CCON and the researched young people as cultural actors. Intrinsic to this was an analytical symmetry: we could study CCON and young people in exactly the same terms, as actors who assemble diverse scales, connections at various and variously understood distances. Rather than interpreting one within an analytical frame imposed by the other, the research pointed to a dialogue between multiple scaling actors. We put this in terms of reconceptualizing CCON as a scaling device. Instead of seeing it scaling up Aviles, or mediating global cultures down to the local, could we rather see it as a machine for making scaling practices visible and bringing them into dialogue? As a device, it could perhaps be seen as a 'laboratory', a place in which implicit cultural maps are made explicit and the results of their interactions, conflicts and discussions could be explored creatively. Dialogue itself rests on reflexivity and transparency: CCON itself would have to present itself as one scaling device amongst many, making clear and public its own scaling narratives. In all these senses, CCON – like our research – could shift from resource to topic, from framework to contested space.

This move involves shifting other terms in the encounter between cultural centres and 'local' constituencies. Above all, 'local' involvement, 'outreach' or 'participation' needs to be based on acknowledging that people in Aviles are not simply local, and that we are dealing with a meeting between various actors who are scaling at many different levels and distances. As part of this we also have to acknowledge 'local' constituencies not just as potential cultural consumers but also as development theorists, and recognize that these two roles are connected.

The form of these conclusions – dialogue, transparency, reflexivity – points to our analytical conclusion. As we have signalled at various points, a performativity position must focus less on a critique of terms like local and global – and culture – than on a shift of their social status. Given the role that they played in assembling a regional regeneration initiative – with all its powerful fudging and leveraging – the researcher responsibility was to establish a theoretical position that did not simply endorse these terms by replicating them, but which also did not legislate them out of empirical existence because of our own analytical distaste. Quite simply, they are part of the story – for client, researched and researchers – of assembling urban strategies, and the difficulty is in analytically and practically acknowledging and acting on this appropriately.

NOTE

1   This phrase was used, we believe, by Steve Woolgar at a seminar in Cambridge: his famous account of technology 'configuring the user' is paralleled by a sense in which research can or should configure the actors who commissioned it.

REFERENCES

Florida, R. (2005) *Cities and the Creative Class*, London: Routledge.
Latour, B. (2005) *Reassembling the Social: An Introduction to Actor-Network Theory*, Oxford: Oxford University Press.
Livingstone, S.M. (2002) *Young People and New Media: Childhood and the Changing Media Environment*, London: Sage.
Miller, D. and Slater, D. (2000) *The Internet: An Ethnographic Approach*, London: Berg.
Neitzert, E. (2008) 'Making Power, Doing Politics: The Film Industry and Economic Development in Aotearoa/New Zealand', Unpublished Ph.D. dissertation, London School of Economics and Political Science.
Slater, D. and Kwami, J. (2005) 'Embeddedness and Escape: Internet and Mobile Use as Poverty Reduction Strategies in Ghana', ISRG Working Paper 4, University of Adelaide, Adelaide.

Interview with Nigel Thrift

Ignacio Farías

Ignacio Farías: Let me start with a question about this ubiquitous and elusive object, the city. I have often read you saying that we are dealing with distinctive spatial formations and imaginaries, that, despite urban sprawl, urban divides, distantiated communities, and the multiplicity of sociotechnical networks proliferating in urban spaces, cities are a legitimate object of study and analysis. Why do you think it is important to stay with the city as an object and not to just say: 'forget the cities; follow the networks'?

Nigel Thrift: That's a good point. I would say that there were three reasons. One of them is an empirical point: you know that you are in particular cities rather than in others. I know I am in New York. It's great; it's wonderful. It is distinctive in comparison with other cities. Not all cities are wildly distinctive, but sufficient of them are that we can say that there is an object that I think most of us are able to experience or feel one way or the other. So that's one reason. Another reason: if you do follow the networks, they often end up in very particular cities. And they end up in those places because that's where a conglomeration of different kinds of actor-networks actually do come together. You can think of a lot of the work done recently on industries like finance or the creative industries. So follow the network and you will still end up in the city. And the third thing is that, to me at least, cities are illustrations of the fact that big results don't need big forces to achieve them. If we reference Gabriel Tarde, he believed that small changes could build up and have an impact unseen by the average social scientist. These small changes produced a propensity to move in a particular direction and that is still more likely to happen in concentrated populations, even though relationships are more distantiated than ever before. Though it is true that we now have the means of, if you like, simulating some of the effects of cities at a distance it is still true that cities can be powerful actors in terms of producing small changes that can move on to become big changes.

IF: We will come back to the issue of scale later on. So let me continue saying that there are only a handful of scholars that have done what you and Ash Amin have done in *Cities: Reimagining the Urban*, namely, providing a fully new vocabulary and also a systematic description of the urban along the lines of non-representational theory, especially Actor-Network Theory and also post-structuralism. I remember an article you wrote at the beginning of the 1990s called 'An Urban Impasse?', where you were saying: well, the descriptions we get of cities are too conventional and, while Sassen and Zukin show some ways out of this impasse, we are still stuck. Was the aim of your book *Cities* to propose a sort of definite way out of this impasse?

NT: It was an attempt. I make no apologies for that. Ash and I consciously tried to be experimental. And, like all these things, when you try hard to be different some things don't really work, but then, if you don't give it a go, you don't get anywhere. But there are forebears to our work in urban theory. Just yesterday I was reading a wonderful article by Everett Hughes, the famous American sociologist, in the *American Journal of Sociology* from 1961, and he was looking at why Tarde's work was ignored. And what he had done was to go back to Robert Park's original library. Park had Tarde's *Economic Psychology* in his library and had read a bit of it. Hughes could tell that he hadn't read all of it because he wrote in the margins of the books and he had stopped about half way, presumably because he felt he had got as much as he needed. But the fact of the matter is that Park too was interested in imitative flow, which was one of the reasons he was drawn to Tarde to some degree, as a lot of people around that time were. And what happened subsequently is that this kind of tradition was lost until it was rediscovered, partly by Bruno Latour, partly by modern biology. But it is fascinating to think about who the potential forebears in urban theory might be. And Park might well be one. It is worth remembering that Park himself had a very interesting background. He was not just an academic but had also been a journalist. So there are all kinds of ways in which the practical spatial arts of the diffusion of thought might have infiltrated into urban theory at that point, which were then forgotten later.

Going on from that to talk about the second part of your question: yes, we do need to push the boundary. And the reason is because we do not have a good vocabulary at the moment for describing cities. I don't think that anyone would really deny that. A good example: every now and then I read around the literature on building cycles. We generally do not have a good vocabulary for describing these kinds of spaces at the moment. Another example: we don't have a very good vocabulary to describe a lot of modern consumer spaces. Or what we have has become repetitive. So we think we know what's there, which of course is the worst possible thing that can happen. Perhaps we shouldn't be too bothered by that. After all, Latour and other sociologists of science argue that most science is localist. It finds some patterns which are roughly the case for a period of time. But for other patterns we do not really know what is happening. So science is a sort of map of intensities, in which some understandings are connected and others are not. And I think that's perhaps the way we need to look at urban theory at the moment. We need to actually map out which are the understandings that are connecting and what are the things that still need really quite serious attention.

IF: To what extent can and should these perspectives be connected with more conventional critical urban geography or everyday urbanism?

NT: Well, I think they can be connected. Let me say straight away that I don't think anyone thinks that you can just think anew. It doesn't happen. It's ridiculous. Everyone starts somewhere and there are all kinds of connections backwards. But I'm still convinced that one of the big things we probably do need to do is think about the methodologies we use to actually make these kinds of maps of understanding. And one of the problems we face is that the methodologies we use have been too pedestrian and it is possible to be much more interesting in choosing methodologies we are working with. And I think a lot of different kinds of people are coming to this conclusion at the moment. Roughly speaking, what they are working towards is the idea of an experimentalist methodology. Experiments have never been a big thing in the social sciences – with some exceptions. But they could be. And that could be very interesting. If you think about the arts and humanities, they effectively experiment. And then you go to scientists with their experimental procedures. So a different kind of methodological take becomes possible.

IF: There is this discussion nowadays in economic sociology about real experiments, Chile being for example a sort of huge laboratory for economics. Are you also thinking in those kinds of experiments?

NT: I think of all kinds of experiments that might be thought about, and they exist on several levels. We could take economics first, which has always been the 'other' of social sciences. But actually of course nowadays it is a very different beast from what it was 20 years ago. Not enough people understand that. But the fact of the matter is, if you look there, you can see all kinds of things going on. So there are experimental procedures being used in behavioural economics, in neuroeconomics, and these kinds of areas. Similarly, there are experiments in economics in the wild, where you actually try to build markets, and there are examples of that actually happening and their proponents often don't know exactly what the outcome will be. That tells you much more a lot of the time than anything else you could think of.

Now, in the arts and humanities of course these kinds of things exist in a different format with less control. People set out to do things only roughly knowing what the outcome will actually be, but actually being able to get an effect that allows them to think about what it is they actually did. In a sense, they post-construct the methodology, but in doing so they can achieve some very interesting things. I would love to see Master's courses which gave people the ability to do these kinds of things. For example at the moment we do not often teach Master students to construct things, to make things. We don't teach them to draw; we don't teach them to program; we don't teach them all sorts of things which might be interesting if you believe that what we need is an experimental social science.

IF: Do you see urban studies today as a field where theoretical or analytical innovation is going on? My sense is that while that might have been the case in the 1970s or 1980s it kind of faded away . . .

NT: I guess the best way of putting it is by referring to the social studies of science. I think the 1970s and the 1980s look good because they had a strong programme and that's always a reassuring moment for all. When most people have a rough idea of what they think they are doing and say: 'Yes! We are going to do this' . . . I mean, good luck, I am not complaining about that. All I am saying is that it seems to me that once the excitement of having a strong programme starts to fade, it starts to feel to most of the participants as if they were missing something. A certain kind of nostalgia grows up. But some of the things I am seeing at the moment, for example, in the kind of areas I am working, are just fascinating. I am very interested in locative technologies, for example. And I think people there are experimenting away like crazy doing all kinds of wonderful and interesting things: building things, seeing if they work, flying around like demented birds trying to find pretty things to put in their nests. But the fact is that this often works rather well and can produce moments of surpassing interest and also things which might well take us on in various ways. So I am really quite keen that we keep these kinds of things going. And let's face it: no one really knows when an area is really doing well until quite a long time afterwards . . . I do believe to some degree in the judgement of history.

IF: I would like to pose some questions about this connection between ANT and the city . . . I think one could fairly argue that, while most ANT studies have tended to focus on quite sophisticated sociotechnical settings and processes, be they laboratories, trading rooms or urban oligopticons, there has been a certain reluctance in dealing with the kind of issues you deal with in conventional urban studies, like urban poverty, urban development and resurgence, touristification, new forms of governance, entrepreneurialism, etc. Is dealing with unspectacular but quite complex and also highly political issues the challenge urban studies pose to ANT? Is there a new way to go that maybe ANT hasn't gone yet?

NT: I think ANT works best in strongly defined situations and I think that's difficult to deny, truth to tell. You talked about the laboratory, where in a sense ANT started out. It's moved into the trading room; it's moved into other milieux where you can be very sure of what you are getting, if I can put it that way. There are set apparatuses, set procedures, and you can use ANT to see how these things are built up. I think it is more difficult for it to work when you are looking at, if you like, everyday life as a whole, or even when you are looking at political movements, which often take on a kind of life of their own as they go on, and which by definition don't have bounded spaces, because they don't know in what spaces they are operating as they move on. The one exception I would probably note is the work of Michel Callon, who has been trying to do some very interesting work on all kinds of patients' groups. But, to repeat myself, I think it is fair to say that ANT does work best in well-defined situations. I don't think Bruno Latour would agree with me, but I think, if you like, the empirical evidence supports such a contention.

We mentioned right at the beginning of the interview Gabriel Tarde, and I was always quite interested in why Bruno Latour was interested in Gabriel Tarde. At least one of the reasons I think is because it allows Bruno to move to an area he wanted to move to in which ANT is not that easy to work with and this is the whole domain that we might call the psychological, because classically Tarde was interested in what he called interspiritual communication, mind-to-mind communication. He was able to look at vast imitative flows moving backwards and forwards through society. And in a sense Tarde is both the forerunner and the opposite of ANT. And then both ANT and Tarde do come up against the fact of how to deal with things like opposition, violence, and these kinds of things. And I can think of ways you can do that with ANT and with Tarde, but I don't think that they cover the whole field. There are other things you have to add in to be able to make their programmes work properly. One of the problems with process theories is that things do precipitate out every now and then and can be worked on almost en bloc and that is something they are not so good at doing. It's almost the corollary of seeing the world as process.

IF: Probably *Paris Invisible* was Latour's attempt to deal with the city in a more Tardean way . . .

NT: Absolutely, but one of the interesting things is how little attention has been given to that book. I never quite understood this, because after all it is easily available, isn't it? But actually I can think of hardly anyone who has really worked with it. Now that's interesting in itself. There may be two reasons for that. One is that only Bruno could have done it and that's a real possibility. The second is, well, it works rather well in Paris! But actually Paris is a very specific case. Paris in a sense almost is as a laboratory: it's small, it is quite controlled, and it works reasonably well as a city. You can't say that for a lot of other cities. I am not convinced it would have worked as well for Mumbai, for example.

IF: But still in this book you do have this problem of identifying what is the object of urban studies, of identifying the city as an object. Basically what Latour is suggesting there is that the city is invisible – the title of the book is indeed quite important – and what we get of the city, the way we experience a city, is through its enactments in different settings. But in a way, I think, the book sort of undermines the whole project of urban studies. Be it a sociology, an anthropology or a geography, the aim was always to make a study *of* the city, and what Latour is doing is a study *in* the city, looking at how the city, which is a circulating reference, is actualized in different places, but the city as a whole, as an object, as a spatial formation, is not being studied . . .

NT: He is studying a circulation, in effect. I don't think there is anything wrong with that. I think we should see Bruno as being a kind of pioneer at looking at cities like that. I particularly like the way he has become caught up with new technologies of various kinds. The new mobile technologies, for example, allow us to start seeing in a way we never could before (a lot of enactments). We can actually follow what's going on. As things increase in terms of density and in terms of the ways these technologies work, I think, we will start to have a very different perception of the city. It will be one which is much more mobile, where it is possible to follow all kinds of currents, which is something that was always difficult for urban studies people. Not to say they haven't tried. You only have to think of the early Chicago School; you only have to think about some ethnography, which is effectively about following people. The difference now is you can actually start doing it en masse. And these kinds of technologies are very much like the invention of the photograph, which allowed us in the late nineteenth century to start having a very different view of cities. But now we find that we have a new technology which allows us to see the city in another way altogether.

IF: I would like to come back to this book *Cities: Reimagining the Urban*, which is quite important, I guess, also for this volume. One of the quite new accents you set in this book is on the hybrid ecology and ethology of cities. Biological flows, natural formations, animal life and so on, they all appear as an integral part of city life. But I think one should also note that these hybrid ecologies and ethologies are quite characteristic of non-urban spaces. So, are there any sort of special features that could distinguish the urban nature-cultures, put together, from non-urban ones? Where lies the specificity of the urban hybrid ecologies?

NT: One of the main answers would have to be interactions with human beings in one way or the other. If you look at, say, animal communities, they adapt to what's going on in cities and they start to respond and change. We can see this in all sorts of ways. Urban foxes, seagulls, rats, squirrels, a lot of different kinds of animals actually start to change their own habits as a result of the urban environment. I was reading over the weekend that seagulls have started to resist migrating anywhere any more: they hang around in cities . . . because they are just such good places to be! So why go anywhere else? There is at least some evidence that the form of the jaw of urban foxes is starting to change so that we are going to have to have a separate line of urban foxes over time. That's not a particular surprise in some ways. It applies to humans too. If you look at the Neolithic, you can start to see the changes that happen there, in particular the discovery of settled agriculture and cities, have already been inscribed in our bodies, neurally and genetically. And the point can be made in other ways too. For example, humans have to react to animals, and what is extraordinary is the degree of effort that has to be put into keeping animals from destroying large parts of the cities. Animals themselves quite clearly have what we might call agency in this.

IF: Another important accent posed in *Cities* is on the notion of cities as virtualities, as involving sets of potentialities and tendencies that can be actualized in sudden ways. And you suggest that this is connected with a politics of hope. But then at one moment you talk about and speak in favour of a form of participatory and deliberative democracy for cities in which the urban planner, the figure of the urban planner, would play a leading role. She would, and I rephrase what you were saying, mobilize the voices from the borderlands, arbitrate between stakeholders, never lose sight of social justice as the binding goal, and so on. I was wondering whether and to what extent this figure could rather involve a constraint for the actualization of the unexpected. Or, differently put, how much planning does a politics of capacities resist?

NT: I think the answer to that is that no one can really know. But it doesn't strike me as terrible to have planners and organizers. It is not by itself a bad thing. Some of the greatest art has been produced by people who brought together large numbers of people in quite planned and organized ways. And they have still been able to do wonderful things. There is no absolute veto on having those kinds of things happening. Then it comes to whether you see these kinds of people as being dictators or enablers. And the fact of the matter is that they probably have to be both at different times. And there are reasons for that. One is that sooner or later you have to make a decision to do something. I think it is unlikely that you could do that in a serious sense through absolute deliberative democracy each and every time. So, you have to have these people that would act as conductors. But I don't know how to proceed further than that. One of the things I have also been trying to do is to make sure that people can pull in all sorts of things into their lives and be open to all sorts of things. And you can see that's happening in some sorts of artistic practices; some people are doing that. We need a kind of urban correlate, if I can put it that way. There are very brave exercises in planning which have tried to do precisely this. Sometimes it has worked; sometimes it hasn't.

IF: Your last major books do not focus especially in cities. This is particularly noticeable for me at least in *Knowing Capitalism*, perhaps because the whole tradition of critical geography put late capitalism in direct relation to cities, saying that these are not just sites of production and accumulation, but also sites of control and command. So if you want to talk about capitalism you have to talk especially or specifically about cities. But you don't in this book, at least not particularly. Was this an intended omission in a way?

NT: No. I was talking about something different. What I was talking about was already too large, and carrying it on by ending in cities would not have been a good thing to do. *Knowing Capitalism* was really about things going on in capitalism at the moment. But I have another book coming out with Paul Glennie about the history of time which has quite a bit on cities within it actually. So I am coming back to cities to some degree. And I am also doing another book with Ash Amin. It's about politics. It will have a lot to do with cities too.

IF: I asked you this because one of the ways nowadays capitalist economic activity is connected to cities is in terms of the thesis of embeddedness. Probably the most common assumption is that of the urban embeddedness of economic production. So, this has been a major issue for the new economic sociology, Granovetter and other economic sociologists, but also for economic geography. But in a way it seems to me that you are trying to go beyond . . .

NT: I think the problem has become how studies of embeddedness are also able to acknowledge flow between places. Ash Amin and I have thought long and hard about this precisely because so many so-called 'closed economic areas' actually tend to have high levels of contact with the outside world, in fact usually more contact with the outside world than within their own area. So, the problem becomes how you actually think about embeddedness when actually it is a kind of heterarchical embeddedness most of the time, in which a lot of contacts will be outside of the particular area and a lot of what is actually being thought about will be outside of the particular area. There is a lack of empirical material. But more and more people are doing work that tries to show these kinds of things. In the old days it would have been the line of difference between industrial clusters and large firms organized over many sites and how you amalgamate those. But now we have more technologies for amalgamation than used to be the case in the past. It's actually genuinely a lack of empirical work which makes it difficult to operate in this area. Some of the best work I know has been done through quite small ethnographies.

IF: Do you think the obduracy of scale and scalar distinctions might also be connected to a lack of empirical work?

NT: Well, it might be; you are quite right. I am hardly the person to talk about this, because I never really understood scale and I still don't. One of the problems you do get into if you decide that there are scales is that you start allocating things to one scale or another, to one territory or another. Once you start doing that you almost predetermine the conclusions in ways which are really quite problematic. They are problematic in terms of the distinctions you use: big or small, flow or static, all these kinds of distinctions. Once you start using scale you start to foreground conclusions. I can agree with the literature on scalar shift to the extent that sometimes I think there are ways in which an operating system goes from being of one size to being of another size. I can see that; I can see that the boundaries are important. But that's not quite the same thing as scale. For me, it is a term we can do without.

IF: One of problems is also the way this whole debate on scale has been framed as there would be a simple alternative: scale or flatness, and flatness being connected with a quite neoliberal . . .

NT: Well, that can often be the case. There are some people who want to see something bad going on. But these two strike me as non-viable choices. You can't say it's all about scale and you can't say it's all simply flat. It doesn't make sense. There are all sort of ways in which boundaries could actually be crucial, but this is not the same thing as scale.

IF: But would it make sense to you to speak of scaling practices?

NT: There are ways in which actors assign themselves to particular scales, and there are scaling practices in that sense. I don't have any particular problem about that. Why you would call it scale, I don't know, but I am not averse to calling it that. One of the interesting things is what are the kinds of worlds that people think they are actually moving in.

IF: Well, behind the whole discussion about scale is a discussion about power, and the whole idea that some actors could scale the world is discussed because it is argued that these scaling practices might have structural consequences for others: 'your scaling practices, my scalar structure', and structure in a strong sense as underlying practices. So, if it is not through scale, as you think, how do we connect spatial distinctions and the issue of power?

NT: The first thing to say is that there is no one way of doing it. There are different ways in which power is exerted. You can move from an extreme example, such as an army which is invading a territory, to something as simple as the fact that one person has control over another person's work, or something like that. These are not the same kinds of ways of exerting power, and they involve very different kinds of ways of producing space. To argue that you could have one link between space and power strikes me as unnecessary. So that's the first thing to say. Going on from that, that's not to say that I don't think that power is important. What I have been trying to do is to look at particular kinds of power and the ways they are exerted. So I became very interested in issues such as imitative flows and what they mean for how we look at things like politics and economics. I was trying to pull out from the general issue a very specific mechanism, if I can put it that way, about the way people in modern life are trained to react, in the semiconscious domain. New mechanisms allow them to cope with things like imitative flows but also direct them in particular ways. Now, that must involve trying to manipulate space in various ways. It involves trying to do that in ways which are quite different from what happened before, in that you are moving with the person, you are behind their shoulders, acting as a kind of subconscious, trying to get them to buy something, trying to make them make particular political decisions, etc., the point being that there are now new technologies that are becoming increasingly able to do these kinds of things. So, for me at least, most of what I am doing is precisely about these new kinds of power and how they can be used. And it does make a big difference. We are talking here about how people vote, about what goods people buy and if they buy them at all, and so on.

IF: I am interested in this idea of boundaries, because in a way the whole discussion on scale was a way of saying, well, it is not just about spatial boundaries but about scale. So it started in a way. But you are saying that we should rather stay with boundaries. Why do you think they are crucial?

NT: I think the reasons for that probably come from the fact that many modern spaces are spaces in which boundaries are drawn quite strictly and are drawn strictly not just to keep people out, but also to keep people happy within. That's why I think Peter Sloterdijk's work on 'worlding' does make some kind of sense. If you look at modern homes, for example, one of their functions is to act to bound the world, to build inside a particular atmosphere, as he would put it, which can be controlled. And that strikes me as correct. If you look at consumer spaces, they are bounded in order to produce some particular kind of experience. I think it is very difficult to deny the existence of things like that, nor would I want to. But you can argue about how bounded these places are and about how controlled they are. That's another thing altogether, I think.

IF: And this idea of boundaries could be also useful for defining what cities are?

NT: I think it is useful for defining what's going on within them in the sense that it's quite clear that it is part of the enactments you were talking about. People draw boundaries in many ways, whether it is the case of a gang that has a very specific turf in the city that is trying to defend or trying to expand, whether it is a middle-class home where people would be outraged if anyone thought about walking into their property, or whether it is a corporation that is trying to control particular spaces in one way or the other. And no one could say that boundaries are not important in that sense. I suppose what's interesting though is the way in which there are more opportunities for subverting boundary-making than in the past. My latest project is to convert these boundaries into doors.

University of Warwick, 2009

Part 2

A non-human urban ecology

5 How do we co-produce urban transport systems and the city?

The case of Transmilenio and Bogotá

Andrés Valderrama Pineda

INTRODUCTION

This chapter addresses the question of how we co-produce urban transport systems and the city. The first part of the question states a reflective aspect that is central to the analysis: the agent of co-production is plural and it includes a number of planners (politicians, engineers, economists, lawyers, communication experts, journalists, consultants, sociologists, historians), citizens, operators, investors *and* non-human actors *and* the author of this text among many other analysts (some of them listed in the references) (Callon 1986). The question emphasizes that the object of study is the action of co-producing: neither the city nor the transportation system pre-exists the other. In other words, it is important to avoid any analytical choice that assigns ontological precedence to either entity, which is a consideration also known as the principle of symmetry (Latour 1987). The city and the transport systems are both produced simultaneously, and my task is to propose a valid description of that process. I prefer 'co-production' to 'co-evolution' because this last term conveys the idea that there are some distant natural causes behind the process. Additionally, the produced entities, the transport system and the city, are not bounded and single objects. They are heterogeneous, multiple and contested. And they change. And it is that change that I propose to account for with this analysis.

I will develop the analysis through a case study. I will analyse how Transmilenio and Bogotá have been co-produced since 1998. These two entities are interesting because they have become popular worldwide, especially among architects, urban and transport planners and city politicians during the first years of this century. This celebrity is due in great measure to the activism of the former mayor of Bogotá, Enrique Peñalosa, who tours the world, financed by the Institute for Transportation and Development Policy and other organizations, presenting the highlights of a transformation in which he himself played a central role (Ardila-Gómez 2004; Hidalgo and Hermann 2004; Shane 2006). This chapter is also a critical review of this promoted story of Transmilenio and Bogotá.

For the purposes mentioned, I will base my analysis in two main theories of the field of science and technology studies: Actor-Network Theory and Large Technological Systems. These theories have been elaborated by a number of scholars to account for the dynamics of co-production of scientific knowledge, technologies and society during the last three decades. In line with other chapters of this book, I attempt to use these analytical tools to contribute to a development of a more robust field of urban studies, one that pays attention to the dynamics of knowledge and technology development as part of city life (Graham and Marvin 2001).

TRANSMILENIO: A LARGE TECHNOLOGICAL SYSTEM

What is Transmilenio? The standard answer to this question is that Transmilenio is an urban transportation system that began operating in the city of Bogotá on 4 December 2000. The system is a world innovation because it is the first bus rapid system designed for mass transit. In other words, it is the first bus system that achieves capacities which it was formerly believed only heavy rail systems could provide (Ardila-Gómez 2004; Hensher 2007). The system, in technical terms, basically consists of high-capacity buses that run in the middle of the road in dedicated lanes (Figure 5.1). These buses stop at stations that are situated at 400- to 600-metre intervals along the city main corridors. The bus has four sets of doors on the left side, which match automatic glass sliding doors at the stations, allowing passengers to get on and off. Passengers pay to enter the system at the station entrance, as in some rail systems (for example the Paris Metro or the London Underground).

Figure 5.1 Transmilenio in Bogotá (source: the author)

The coverage of the system is enhanced at end stations with a system of feeder routes with conventional buses to concentrate demand, in a fashion common to some rail systems (for example in the Caracas Metro). From the organizational point of view Transmilenio is a mixed institution. It is governed and coordinated by a public agency, Transmilenio SA. This agency coordinates the action of private operators that own and operate the buses and the fare collection system (www.transmilenio.gov.co).

The standard normative technical and organizational descriptions of Transmilenio do not mention many aspects that are crucial to its very existence. Three important aspects are revealed when Transmilenio is conceived of as a Large Technical System (Coutard 1999; Hughes 1993; Summerton 1992, 1994). First, Transmilenio is also a political system in at least two important respects. The CEO of Transmilenio is appointed by the elected mayor of the city, and thus Transmilenio is part of the political government of the city; and the system is designed to function as a public–private partnership with a set of rules that guarantee a delicate balance of power between the private operators and the public coordinating agency. Second, Transmilenio is the result of a planning process aimed at confronting the increasing inefficiencies of the previous collective system. The collective system had reached a level of stagnation that can be characterized as a *reverse salient*, that is, a critical situation whose causes cannot be isolated and solved by conventional rational processes (Hughes 1993). The concentrated efforts of a number of people produced a solution that generated a whole new system. Paraphrasing Thomas Hughes (1993), I can say that confronting the reverse salient led to radical innovation. And, third, the development of the new Large Technological System was steered by engineers, urban planners and designers during the development phases. After the system began operation the initial innovators were replaced by another type of system builders, with competencies in economics, finance and politics, to steer the further stabilization and growth of the system. In other words the system builders and their competencies changed as the system moved from the innovation to the maintenance and growth stage (Hughes 1993, 1998).

The main emphasis of the Large Technological Systems theory is that all technical aspects of any system are political at the same time. They are intertwined in a seamless web (Hughes 1986). How do we account for that process, and what are its implications and consequences? To answer this question I will analyse two sides of the planning process: first, how innovators decided on the main technical aspects of the new system; and, second, how this process implied a reconfiguration of the city of Bogotá. To do so, I will use some concepts of Actor-Network Theory.

THE 'DE-SCRIPTION' OF LARGE TECHNOLOGICAL SYSTEMS

One of the problems that many analysts face when they attempt to account for processes of production of reality is the fixity of language. The majority of the concepts available are static. Additionally they refer to finished objects. There is an overemphasis in denoting what things are rather than accounting for the process that produced them: too many subjects, too few verbs. Actor-network theorists have attempted to overcome this limitation by developing new concepts and even entire new sets of concepts that consciously avoid the limitations of positivist, modern, objectivist terminology (Latour 1999). Madeleine Akrich (1992) made an early attempt at importing from the field of semiotics a number of concepts to enable this analytical task. The two fundamental Actor-Network Theory points of departure are: first, that the social and the technical are not distinguishable during the design process; and, second, that the inside and the outside of any object, or in other words the boundary of the object, are a consequence of the design process. Therefore, neither the social, nor the technical, nor the external, nor the internal aspects or features of any object are suitable to account for the process that produces that very object.

Akrich builds on the proposition that designers (and engineers and urban planners and politicians) producing new objects, in other words *innovators*, are applied sociologists (Callon 1987). Therefore, the task of designing an object necessarily implies a proposal for the setting in which that object will exist: a scenario. Akrich suggests that this process can be regarded as a process of *in-scription*. Put in her own words, 'A large part of the work of innovators is that of "*inscribing*" this vision of (or predictions about) the world in the technical content of the new object' (Akrich 1992: 208). Analysing an object, thus, is a process of isolating the different inscriptions in an object: it is a *de-scription*. However, this process is not simple, because scripts are not stable. They change in time and through the interaction of innovators among themselves, with the imagined users first, and after with the real users, with constraints of all kinds (legal, physical, normative, cultural, technical), through the process of design, operation and maintenance. Therefore, at different points in time during the design, construction and operation processes of any technical object there will necessarily be a clash 'between *the world inscribed in the object* and *the world described by its displacement*' (Akrich 1992: 209). Special actants in all these processes, the final users, will accept the object only to the extent that they consent to the specific kind of world the object is proposing. This process Akrich regards as 'to make a *pre-inscription*' (Akrich 1992: 215).

The language proposed by Akrich provides a tool to scrutinize the complex, contested and vexed process of design of any object. To illustrate the analytical capacity of this language she presents various examples of how objects designed in Europe by agencies of international cooperation failed when operated in their final setting, which according to her accounts are various places in the less developed parts of Africa. Her point is that the world proposed by the designers was at odds with the world in which users actually lived. This set of examples, however, conveys two suppositions that do not apply when the designed object is an urban transportation system: first, that objects are designed in one location and then transferred to the place of actual use; and, second, that the new technologies are presented to the final users at one specific point in time, that is, that they suddenly propose a radically new set of relations. Although a good deal of the case studies of actor-network theorists among other constructivist approaches aim at overcoming this analytical limitation, many of the metaphors end up emphasizing the novelty of the new actor-world or system and its internal story. The laboratory and the socio-technical ensemble are cases in point (Jørgensen and Sørensen 2002).

I will carry out a *de-scription* of some aspects of Transmilenio in Bogotá to suggest ways to overcome these limitations. I will concentrate on four main features of Transmilenio: the design of the buses; the fact that both the buses and the stations are designed with high platforms; the location of the stations in the median of the trunk roads; and the use of discriminated traffic. All these scripts are systemic: they are features of the system as a whole and they play a role in the relation of the components of the system. In contrast, other aspects of certain components can be changed: for instance, buses can be powered by diesel *or* natural-gas engines. In other words, the fuel and the engine do not play a role in the way the bus interacts with the stations, for instance.

Buses

The buses of Transmilenio are articulated, with a maximum capacity of 160 passengers: 48 seated and 112 standing (Figure 5.2). They have four sets of doors on the left side of the vehicle to allow rapid boarding and off-boarding

Figure 5.2 Transmilenio articulated bus (source: the author)

of passengers, similarly to rail systems. The layout and all dimensions are standardized.

There are three scripts in this vehicle design that are interesting for this analysis. First, there are few seats. Unlike in the collective system, in Transmilenio it is expected that passengers will not remain inside the vehicle for long periods of time, which depends on the velocity of the vehicle, which in turn depends on having as few vehicles as possible on the road (450 new vehicles replaced 1,140 conventional buses approximately). Second, there are four sets of doors on the left side of the vehicle. When the bus stops at a bus station the bus doors match sliding doors in the station. Passengers board or leave the vehicle in a very short time, which also improves the average velocity of the vehicle. This depends on the fact that the collection of the fare is separated from the vehicle, unlike in the collective system. And, third, all red buses have the same exterior and interior design, unlike the vehicles of the collective system, where it is hard to find two that are similar.

The fact that the buses of Transmilenio are all standardized and with the features mentioned reflects that the process of definition of its features was successful in this respect. However, it was a hard process. It began in 1998 when the small initial planning team succeeded in convincing around 20 bus manufacturers around the world to bring a prototype to Bogotá for the double purpose of providing publicity for the project and for testing. The manufacturers accepted the invitation with the hope that the best would be awarded an order of 450 buses for the new system. All buses were tested by engineers of the Universidad de los Andes for mechanical performance, a key issue for a city built at 2,640 metres above sea level, which implies very particular conditions for machine operation (the most notable is an approximate 30 per cent lower density of the air, and thus 30 per cent less oxygen for combustion) (Huertas *et al.* 1999). The innovators soon discarded the small vehicles and concentrated on the larger ones, with the idea of reducing the maximum number of vehicles required, which in turn was a response to the collective system, which exhibited an excess of vehicles of more than 100 per cent. The city of Bogotá had in 1998 more than 22,000 buses serving the city, where transport engineers calculated that between 8,000 and 10,000 were enough (Ardila-Gómez 2004).

High platform

The buses and the bus stations are designed with a high platform at about 70 centimetres above ground level (Figure 5.3). Of course, all the vehicles and stations comply with the standard, in order to make the system work. This particular script was the result of the political process of design of Transmilenio. During the process, the planning team invited the current owners of bus companies to take part in the process. There were two strategic reasons: because the planners had a lack of knowledge with regard to how the collective system actually worked in organizational, political and economic

Figure 5.3 Transmilenio bus station (source: the author)

terms; and because it was part of a strategy to break the resistance that the current owners of the business might pose. This led to their inclusion as sources of information and possible bidders for the contracts of the new system. Given that the bus company owners might eventually purchase the buses and operate them, the planning team and the city authorities allowed them to decide if they recommended a high or a low platform. They finally decided on a high platform. Once in place, the high platform also became a critical actant for the new system because it played a crucial role in making the new system physically incompatible with the collective system. In other words, in the event of political crisis in relations between the city authorities and the private operators (divided since 2000 into those that became included in Transmilenio and those that remained in the collective system), the buses of the collective system would not use the dedicated lanes and the stations of Transmilenio; nor could the red articulated buses of Transmilenio operate as conventional buses, picking up passengers at any point and using the door on the right side of the vehicle (Ardila-Gómez 2004; Valderrama and Jørgensen 2008).

Bus-only lanes

Since 1988 the transport authorities and experts in Bogotá had been trying to solve the problems associated with the seemingly disorganized and fragmented collective system in Bogotá. Many would not regard it as a system, since both the property and the responsibility were too fragmented. The first attempt to reorganize the whole system was watered down to a proposal of discriminating traffic in the Avenida Caracas, the backbone of the transport corridors of the city. Following experiences in Brazil and with the participation of Brazilian consultants, the city authorities redesigned the avenue, providing for a corridor in each direction of two bus-only lanes (Figure 5.4). This arrangement began operation in 1991. Transit was only permitted to large conventional buses with a capacity of 80 passengers seated and a small aisle in the middle of the vehicle. The urban design of the corridor was poor, which attracted furious criticism from citizens and urban experts alike. However, the technical performance of the corridor was remarkable. Measures of capacity showed that this arrangement could make the bus system as efficient as some rail systems, with a peak of 24,000 passengers per hour per direction. When the city authorities, the planning team and Enrique Peñalosa designed the bus rapid transit project between 1998 and 2000, this particular script played on their side, providing both a technical background for the new design and an existing example of the benefits of discriminating traffic (Ardila-Gómez 2004).

Bus stations

The location of stations in the centre of the median is one of the scripts in which Transmilenio really stands alone. All the other reference projects, including the famous Curitiba in Brazil and the former Avenida Caracas in Bogotá, had buses circulating in the central lane, but the bus stops were on the right side of the vehicle, in the pedestrian separator of the bus lanes and the mixed traffic lanes. This script plays a crucial role because it allows the system to have a particular boundary. Passengers pay when entering the

Figure 5.4 The Avenida Caracas (courtesy of Transmilenio S.A.)

station and then can remain inside for as long as they want to, as in the Paris Metro or the London Underground and unlike many other transport systems where there is a limit to the time one can remain inside the system. Once passengers have paid, they can board and off-board any bus or change direction at bus stations, and, as with many train stations, the system allows for free transfers. The boundary is also enacted by a sophisticated (some claim exaggerated) computerized system with electronic cards, turnstiles and paying booths at the entrance of stations. This particular script also breaks clearly with the tradition of the city of having the collection of the fare inside the vehicle and the driver as responsible for this task.

Scripts: conclusion

The design of Transmilenio reveals three key aspects of the process. First, the process of inscription of the new system was performed in the location where the new system would operate, and the whole process was influenced by readings of the current situation, especially analyses of the problems of the collective system and proposals of how to produce a new scenario where these defects would be left out by design. This is why I have emphasized in the description of the scripts of Transmilenio that they were different from the scripts of the collective system. The innovators of Transmilenio consciously addressed many of the features of the new system to avoid what they considered to be the distortions of the collective system (Ardila-Gómez 2004).

The second key aspect is that the process is contingent on the interaction among different actors, including planners, politicians and the current bus operators. I must clarify that the planners' group is quite heterogeneous, as it includes: local planners in charge of the design of Transmilenio; local consultant firms and academic experts; and international consultant firms like Steer Davies Gleave, McKinsey, and Logitrans. Additionally, it is also revealed that many existing non-human actors or previous scripts, like the 'old' solo-bus Avenida Caracas and the bus stops in the median, support many of the inscription processes of the new design. In this sense, all new scripts are networked entities that emerge from a process of juxtaposition and translation (Callon 1987) of many other actors located physically inside and outside the city of Bogotá.

And, third, all these scripts were aligned to enact a clear boundary between the new system and the existing collective system in order to make the innovation process irreversible (Callon 1991). This is the most notable contribution of Actor-Network Theory to Large Technical Systems Theory: systems are made to exist; they are produced. The process that configures the system necessarily has to define a boundary. And this boundary is the result of distributing agency, causes and responsibility to a number of actants in the new system, including operators, drivers, city authorities, high platforms, sliding doors, discriminated traffic lanes and so on (Akrich 1992; Law 2002; Valderrama and Jørgensen 2008). And this boundary is the result of the action of those features inside the system that were made to oppose those features that ended up outside the system during the design and planning process. In other words, the design of the new system was performed in a world where other existing technologies or actor networks or Large Technical Systems were already functioning. To characterize the interactions of technologies or actor-networks of transport and how they are constitutive of the city of Bogotá I now turn to the concept of arenas of development.

ARENAS OF DEVELOPMENT: THE CONCEPT

Actor-network theorists have attempted to analyse the process of innovation following the workings of different actors. Callon (1986, 1987) followed the efforts of a group of engineers within the French state prestigious giant company Electricité de France to design a new transport system for France based on electric vehicles to replace the private car system entirely at the beginning of the 1970s. Callon describes the different strategies pursued by the engineers to interest and *displace* other existing human and non-human actors such as battery accumulators, car manufacturers, city authorities in France, drivers and so on. These engineers attempted to become spokespersons for these entities and thus become an obligatory passage point and therefore make the networked entity of the electric vehicle function. The whole process Callon denominates as *translation*.

Callon (1986, 1987) explains that this particular electric vehicle project failed because a number of actors resisted incorporation into the network: the accumulators did not become economical; Renault continued to be a combustion engine car manufacturer and so on. Because he focuses on the workings of one actor, the analysis does not account for what other actors did to preserve the other technologies in place: the combustion engine cars, the trains, the buses, the rail systems, etc. In a similar fashion Latour (1996) explains the failure of another group of engineers in Paris to develop the urban transport system Aramis between 1972 and 1989, in the lack of effort (or love, as he puts it) of the central actors to make the whole network cohere and hold in place. In both cases, the focus on following a specific set of central actors obscures the workings of the other existing or even projected technologies.

To overcome these limitations Jørgensen and Sørensen (2002) propose the concept of arenas of development, which are cognitive spaces for research that aim at improving our analytical tools to account for the dissimilar processes that produce conditions for innovation. The concept provides a basis for analysts to shift the focus from the new entity – the actor-network or the new Large Technical System – to the set of interactions that makes possible the emergence of the new entity and its relation with the existing ones throughout its becoming. It differs from other concepts such as a socio-technical landscape, which conveys a notion of natural order, or a technological regime (Geels and Schot 2007), which assigns an internal natural logic to a context of development. An arena of development is populated by many actor-networks or Large Technical Systems that compete, interact, interfere or complement each other. The arena is composed of: a number or elements such as actors, artefacts and standards; a variety of locations for actions, knowledge and visions; and a set of translations that shape the stabilizations and destabilizations in the set of relations of the arena (Jørgensen and Sørensen 2002: 198).

In this section I will argue that Transmilenio was designed to produce a major change in the arena of development of urban transport in Bogotá. The project was intended not only to solve a technical problem of mobility, but to reconfigure a whole set of relations, including power relations, spatial and distance relations and identity relations within the city of Bogotá and at national and international levels. The innovators in this process not only struggled with the existing technical systems of collective transport and private cars, but also with the non-built, but much discussed, project for the metro of Bogotá. This engagement with local technologies in use was also an engagement with the international networks that support these technologies (Edgerton 2006). The struggle also incorporated a discussion of different chronologies and images of the old and the new (Bender 2006).

The arena of development of transport in Bogotá

Ardila-Gómez (2004) has elaborated a very detailed history of the planning and decision-making process that led to the design and construction of Transmilenio in Bogotá. Based on his account, in this section I want to review the elements that position the new system within the context of the existing technologies in the city of Bogotá. The objective is to trace the inter-dependencies among technologies, which otherwise are normally conceived of as independent modes of transport (Lay 2005). To do so I will concentrate on three movements, each one related to the other three major technologies in the arena of technology of transport in Bogotá: the collective system, the non-built metro and the private car.

First movement: collective transport becomes dominant

By the beginning of the 1980s the arena of development of transport in Bogotá was composed of a collective public transport system, a public car system (taxis) and a private car system. The dominant mode of public transport was the collective system. This was a particular system organized around transport companies, which had the responsibility of managing routes for public transport in the city. This responsibility was delegated by the city authorities via the Secretariat of Traffic and Transport. The transport companies affiliated buses: this means that they did not necessarily own the buses, but allowed bus owners to serve their routes in exchange for a lump sum for affiliation and a monthly fee. Particular individuals owned the buses.

They were normally small investors, and on average there was more than one owner per bus. The bus driver was occasionally a hired person, but more often than not one of the owners of the bus. In practical terms this meant that the driver performed all the work: cleaning and maintenance of the bus, driving, and collecting the fee from passengers. At the end of each day the driver would give the owner a portion of the income. The owner would in turn pay the monthly fee to the transport company. In legal terms all the responsibility of operation was assigned to the bus driver. This meant that transport bus companies received money without being responsible for any service or failure in the service. It also meant that drivers competed in the street to pick up passengers, which resulted in an aggressive driving behaviour, with very high rates of accidents and mortality. This particular phenomenon was denominated *la guerra del centavo*, the cent war, and came to be regarded as the core of the problem (Acevedo and Barrera 1978).

This particular organization of the urban transport sector began in the 1920s when the city authorities allowed private buses to serve the new settlements of the city which could not be served by the existing tram system. The system grew rapidly owing to the flexibility of the then new technology, the fact that private operators assumed the risks, and the inflexibility of the tram. The inflexibility of the tram was due to the technology used and the fact that it was owned and operated by the city and was thus subject to political struggles among competing parties. Between 1930 and 1952 the new bus system grew, while the tram stagnated. Political and economic support for the tram diminished, and finally the tram was dismantled (Castañeda 1995). During the second half of the twentieth century all efforts by the city to regulate and structure collective transport failed, including: the purchase and operation of city-owned buses (both internal combustion and electric-powered trolley buses); strengthening regulation; and various attempts at reorganization. By 1998, 68 bus companies affiliated around 22,000 buses owned by more than 25,000 persons. The fact that the property was distributed among so many people made the system quite stable. In other words, any effort to change the whole system could not be undertaken by one company alone (Ardila-Gómez 2004).

Second movement: the various projects of the metro of Bogotá

The metro of Bogotá became a key player in the transformation of the arena of the development of transport in Bogotá from 1982, albeit that it was never constructed. Based on studies carried out during the 1970s, the national government made the decision to commit the nation to support any metro development in any city in Colombia by way of serving as the guarantor for loans taken in the international financial market. Medellín, the second city of the country, took the offer and began designing its own elevated metro. A number of transport experts in Bogotá opposed the project for the city on the grounds that it was too expensive and other solutions with buses were possible. In 1988 the president of Colombia, Virgilio Barco, a civil engineer from MIT, proposed to develop a metro for Bogotá using the old train corridors. The proposal evolved to the point where the Italian consortium Intermetro won the rights to design and construct the system. A growing number of experts with increasing influence in the city administration managed to delay this project, specially criticizing the forecasts for passenger demand, because the old rail corridors basically did not have much demand, as they traversed the city mainly across non-residential areas.

In 1989 the construction of the metro of Medellín was halted because of escalating costs. As a reaction to the critical situation in Medellín, but also considering the costs of a possible urban train system for Bogotá, the National Congress approved the so-called Metro Law at the end of 1989. The new legislation allowed local governments to raise taxes by 20 per cent to cover the costs of the infrastructure development. This law also established that local governments should cover at least 80 per cent of the total investment costs, and that the nation would cover at most 20 per cent. The law demanded that the fare income of the new systems should cover operation costs and depreciation of the equipment. This law made it ever more difficult for the city of Bogotá to aspire to develop a metro system. Andrés Pastrana, then mayor of Bogotá, followed the advice of the local transport experts then focusing the efforts of the administration on the development of a solo-bus system instead of a metro. During 1995 the Japanese government gave as a present to the city of Bogotá a comprehensive study of the transport situation of the city and made a recommendation for future development. The study basically proposed the development of a three-level city with elevated highways for private transport, a surface bus rapid transit system for passengers and an underground metro. In 1997 Enrique Peñalosa ran for mayor and won the elections with the promise of constructing a metro for the city of Bogotá. During his tenure he established a working group to develop the project and signed an agreement with the national government to partially fund a package for the development of a new bus *and* a metro system for Bogotá. During his administration Peñalosa negotiated part of the resources to build Transmilenio and strategically withdrew attention from the metro, eventually dismantling the planning team in charge of the project.

Third movement: containing the public space for private cars

Like many other cities in the world Bogotá did not develop a significant car dependency, as the number of vehicles never reached the proportions of the US or Europe. However, the car became a key player in the discriminated character of the city. In the first place the majority of car owners lived to the north in the wealthier neighbourhoods, where the road infrastructure was better. But, furthermore, during the 1970s and 1980s the car began using more and more areas of public space that were designed for other purposes. Therefore it became common for drivers to park their vehicles on the

sidewalk, green areas and even parks. Congestion rapidly became a problem in a city that never spread out too much and became quite dense. In 1982 the city administration undertook the first major project to contain the public space used by cars. On Sundays, 120 kilometres of main roads in the city were closed to traffic, allowing pedestrians, cyclists, people on skates and so on to enjoy the city for sport and recreation purposes.

In 1989 the city administration decided not to build a metro, but the solo-bus system on the Avenida Caracas. Although the idea was to reorganize in its entirety the whole organizational and technical system along the corridor, the idea was watered down to an infrastructure that discriminated bus traffic from the rest. The space allowed for private cars was thus reduced from three to two lanes, albeit the interference with buses was removed. During the 1990s the congestion in the city reached unexpected levels, with the traffic collapsing entirely on at least two occasions. During his first year in office Mayor Peñalosa enforced a car restriction programme. On weekdays 40 per cent of private cars were not allowed to run during peak hours (7–9 a.m. and 5–7 p.m.). With time the timeframe has increased. Peñalosa enforced another tough measure to remove cars parking on the sidewalk and completely forbade this practice, at least in the main corridors of the city. The containment of congestion became an integral part of prioritizing public transport developments.

Arenas of development: conclusion

The movements described in this section are not mere descriptions of the other actor-networks that populate the arena. They are overall an attempt to trace the ways in which each system configures the arena and acts and reacts to the other systems. The first movement traces schematically the ways in which the collective system became the dominant mode of public transport in the city, thus framing the type of technical problems that the planners of Transmilenio addressed in the design of the new system. The second movement relates the workings of the particular actor-network of the metro of Bogotá, an actant of variable geometries that was nevertheless quite effective in providing a scenario of development that triggered energetic responses from transport experts in Bogotá. During the 1990s it also became instrumental in mobilizing resources from the nation to the city, especially for the construction of the infrastructure of Transmilenio. And finally the arena of transport in Bogotá has also been populated by an ever-increasing number of private vehicles, which create problems of congestion on roads and other public spaces. The regulations designed to contain the private car are also part of a discourse of improvement of the public transport system and thus are linked to the development of Transmilenio.

CONCLUSION

The analysis of the arena of development of transport in Bogotá sheds new light on the agency distributed to the different scripts of Transmilenio. They are not just technical features that respond to the previous organization of the arena. They also become actors in the new order. The lanes, the stations, the high platform and the vehicles are conceived to enact a boundary to separate the new from the existing. They also perform to the stability of the system, disciplining future mayors and city authorities (Valderrama and Jørgensen 2008). They play a role in the physical reorganization of the city and thus in the lived experience of citizens, thus redefining space and time in Bogotá. And finally Transmilenio also co-produces the identity of the city and its citizens.

Throughout the text I have referred to the collective system as the 'previous' or the 'old' system. I have used those words because many analysts and Peñalosa himself uses them. The choice is not accidental or naïve. The stories told through these analyses participate in a narrative to construct the notion of Transmilenio as the future of Bogotá and the collective system as its past. It is a conscious enactment of a trajectory of development aimed at strengthening the further growth and extension of the system. This chapter plays its role in further producing both the system and Bogotá as networked entities that also exist in academic texts in different locations around the world.

REFERENCES

Acevedo, J. and Barrera, J. (1978) 'El Transporte en Bogotá: Problemas y Soluciones', Unpublished document, Bogotá.

Akrich, M. (1992) 'The Description of Technical Objects', in W.E. Bijker and J. Law (eds), *Shaping Technology/Building Society: Studies in Sociotechnical Change*, Cambridge, MA: MIT Press.

Ardila-Gómez, A. (2004) 'Transit Planning in Curitiba and Bogotá: Roles in Interaction, Risk, and Change', Unpublished Ph.D. dissertation, Massachusetts Institute of Technology.

Bender, T. (2006) *History, Theory and the Metropolis*, CMS Working Paper Series No. 005-2006, Berlin: Center for Metropolitan Studies, Online, available at: http://www.metropolitanstudies.de/workingpaper/bender_005-2006.pdf (accessed 25 February 2008).

Callon, M. (1986) 'Some Elements in a Sociology of Translation: Domestication of the Scallops and Fishermen of St Brieuc Bay', in J. Law (ed.), *Power, Action and Belief: A New Sociology of Knowledge?*, London: Routledge & Kegan Paul.

Callon, M. (1987) 'Society in the Making: The Study of Technology as a Tool for Sociological Analysis', in W. Bijker, T. Pinch and T. Hughes (eds), *The Social Construction of Technical Systems: New Directions in the Sociology and History of Technology*, London: MIT Press.

Callon, M. (1991) 'Techno-economic Networks and Irreversibility', in J. Law (ed.), *A Sociology of Monsters: Essays on Power, Technology and Domination*, London: Routledge.

Castañeda, W. (1995), *Transporte Publico Regulacion y Estado en Bogota 1882–1980*, Bogotá: CEAM, Universidad Nacional de Colombia, IDCT.

Coutard, O. (ed.) (1999) *The Governance of Large Technological Systems*, London: Routledge.

Edgerton, D. (2006) *The Shock of the Old: Technology and Global History since 1900*, London: Profile Books.

Geels, F. and Schot, J. (2007) 'Typology of Sociotechnical Transition Pathways', *Research Policy*, 36: 399–417.

Graham, S. and Marvin, S. (2001) *Splintering Urbanism: Networked Infrastructures, Technological Mobilities, and the Urban Condition*, London: Routledge.

Hensher, D. (2007) 'Sustainable Public Transport Systems: Moving towards a Value for Money and Network-Based Approach and away from Blind Commitment', *Transport Policy*, 14: 98–102.

Hidalgo, D. and Hermann, G. (2004) 'The Bogotá Model for Sustainable Transportation: Inspiring Developing Cities throughout the World', Paper presented at TRI-ALOG 82, Urban Mobility, Darmstadt, Germany, March.

Huertas, J., Loboguerrero, J., Báez, F. and Moreno, F. (1999) *Pruebas al equipo rodante del proyecto de transporte TransMilenio*, Bogotá: Universidad de Los Andes, Departamento de Ingeniería Mecánica.

Hughes, T. (1986) 'The Seamless Web: Technology, Science, Etcetera, Etcetera', *Social Studies of Science*, 16: 281–92.

Hughes, T. (1993) *Networks of Power: Electrification in Western Society, 1880–1930*, Baltimore, MD and London: Johns Hopkins University Press.

Hughes, T. (1998) *Rescuing Prometheus*, New York: Vintage.

Jørgensen, U. and Sørensen, O. (2002) 'Arenas of Development: A Space Populated by Actor Worlds, Artefacts and Surprises', in K.H. Sørensen and R. Williams (eds), *Shaping Technology, Guiding Policy: Concepts, Spaces and Tools*, Cheltenham, UK: Edward Elgar.

Latour, B. (1987) *Science in Action: How to Follow Scientists and Engineers through Society*, Cambridge, MA: Harvard University Press.

Latour, B. (1996) *Aramis or the Love of Technology*, Cambridge, MA: Harvard University Press.

Latour, B. (1999) *Pandora's Hope: Essays on the Reality of Science Studies*, Cambridge, MA: Harvard University Press.

Law, J. (2002) *Aircraft Stories*, Durham, NC and London: Duke University Press.

Lay, M. (2005) 'The History of Transport Planning', in K. Button and D. Hensher (eds), *Handbook of Transport Strategy, Policy and Institutions*, Amsterdam: Elsevier.

Shane, D.G. (2006) 'Urbanism: A Bus System Transforms Bogotá's Prospects', *Architecture*, 95(7): 54.

Summerton, J. (1992) *District Heating Comes to Town: The Social Shaping of an Energy System*, Linköping: Linköping University, Department of Technology and Social Change.

Summerton, J. (ed.) (1994) *Changing Large Technological Systems*, Boulder, CO: Westview Press.

Valderrama, A. and Jørgensen, U. (2008) 'Urban Transportations Systems in Bogotá and Copenhagen: An Approach from STS', *Journal of Built Environment*, 34: 200–17.

6 Changing obdurate urban objects

The attempts to reconstruct the highway through Maastricht[1]

Anique Hommels

INTRODUCTION: DEALING WITH OBDURATE URBAN STRUCTURES

This chapter deals with the clash between a variety of new ideas about urban development and the multifarious viewpoints that are already embedded in a city's existing urban structures and outlook. It addresses the unexpected or unforeseen societal developments that gradually give rise to a questioning of existing urban configurations. It is about urban design and challenges to renew it, a process in which the stakes are commonly so high that years of planning, debate and controversy may result in no changes at all in some cases, while in others they may eventually result in concrete and lasting urban reconfiguration.

The wide range of planning initiatives and activities confirms the image of city building as a continuous, ongoing process: cities are in a process of being built and rebuilt all the time; they are never finished but always under construction, always in a process of being realized. Many plans to redesign urban space assume an almost infinite malleability of the existing urban configuration. Urban historian Josef Konvitz claimed for example that 'Nothing may look less likely to change in a radical way than the status quo in city building, but nothing else may be more likely' (Konvitz 1985: 188). It seems counter-intuitive to change cities, but nevertheless they change continuously. But despite the fact that cities are considered to be dynamic and flexible spaces, numerous examples illustrate that it is very difficult to radically alter a city's design: once in place, urban structures become fixed, obdurate, soon securely anchored in their own history as well as in the histories of the surrounding structures. Objects and facilities that define urban space have a tendency to coagulate, to all become part of one amorphous whole. As a consequence, urban artifacts that are remnants of earlier planning decisions, the logic of which is no longer applicable, may prove to be annoying obstacles for those who aspire to bring about urban innovation.

I conceptualize cities as sociotechnical artifacts: huge constellations of human beings, infrastructures, buildings, political structures, values and so on. While STS has so far not paid much attention to cities,[2] I claim that STS concepts are useful when analyzing processes of urban obduracy and change.[3] The analysis will be focused on the attempts (between the 1950s and the 1980s) to reconstruct the highway that cuts through the Dutch city of Maastricht. Until now (2009), the attempts to redesign the highway and build a tunnel have been unsuccessful.

The highway stretch that cuts through Maastricht was built in the late 1950s. At that time, when cars were still comparatively sparse in the Netherlands, there seemed to be many good socio-economic reasons for building highways near downtown districts. Moreover, noise regulations did not yet thwart the construction of apartment buildings adjacent to highways. Between the early 1960s and the late 1980s, however, several interconnected processes radically changed the issues involved in highway design: car traffic rose dramatically; traffic safety and the quality of urban life became increasingly important issues; environmental concerns started to play a more important role in traffic projects; stricter environmental norms, regulations and standards for the design of highways were developed; activists and lobby groups began to influence urban redesign projects; and local, regional, national and international governments changed their policies on traffic circulation, in part because of other financial priorities.

Since the construction of the highway in Maastricht, there has been an ongoing effort of engineers, politicians and citizens to change and adapt it. From the beginning it was clear that the highway, which just to the south of the city serves as an important link between the Dutch and the Belgian highway network, basically split the city in two. Over the years, provisional adaptations were made, such as the reconstruction of intersections and the placing of sound baffles, but more radical new designs, such as overpasses or a diversion east of the city, failed to be adopted. The idea of a tunnel, as the definitive solution to the traffic congestion, the reduced traffic safety, and the limited quality of life for those who live near the highway, has been considered and reconsidered for as long as 50 years. Although this idea has figured prominently in the various municipal policies, proposals and strategies, so far this solution is not even close to being implemented, nor is any other solution. Given the concerns of my study, this raises a number of questions. How could this urban highway stretch build up and maintain its obduracy? What were the strategies employed by the various actors engaged in the effort to improve the highway's design? Why is it that the highway that cuts through Maastricht has remained virtually unchanged for almost half a century, despite all efforts to change its design?

DESIGNING AND BUILDING HIGHWAY 75 (1956–60)

During the 1950s, the city of Maastricht, like many Dutch cities, had to address questions such as 'Where should the new highway be situated in relation to the city center?' and 'How should it be designed?' The city board, in collaboration with Rijkswaterstaat,[4] had already considered plans for a

north–south highway near Maastricht before World War II.[5] Although
Rijkswaterstaat initially projected the highway's trajectory to be further away
from the city center of Maastricht, the road was eventually planned closer to
Maastricht's downtown area.[6] The chamber of commerce of South Limburg
effectively lobbied for a road trajectory near the city's main industrial zones,
which happened to be adjacent to the downtown area.[7] The companies, of
course, would benefit from enhanced nearby infrastructure. Moreover, it was
argued that, in the case of a diverted highway trajectory to the east, the new
highway would barely contribute to the local traffic circulation.[8]

The city council of Maastricht was well aware of the new highway's
significance to the city:

> There is no doubt that Highway 75 is of utmost importance for
> Maastricht. This era's traffic development depends on roads and high-
> ways rather than – as in the old days – on rivers and railroads, and now
> that Highway 75 passes Maastricht it is important to effectively integrate
> it into the city's traffic system.[9]

The city council also emphasized the necessity of good roads for the future
development of Maastricht as an industrial center and underscored the
significance of having a highway that connects the city to other parts of
the country. It believed that for Maastricht, 'as a city at the intersection of
cultures, good and fast connections are necessary . . . When this highway
is completed, Maastricht will eventually have a good and fast connection
with the central and western part of the country.'[10]

The director of the city's public works department, J.J.J. van de Venne,
argued that Maastricht fulfilled a major socio-cultural function for the
region.[11] He felt that the new traffic plan should facilitate the fluid passage of
international traffic through the city and that it should enable regional traffic
to come as close as possible to the city's industrial areas and its public
services and institutions. The public works director perceived a direct link
between the city's socio-economic ambitions and its traffic problems, and
argued that the inner city, with its commercial district, should be accessible
for cars by means of a circular system of roads. The projected ring road
was seen as the backbone of Maastricht's traffic circulation system, and the
urban stretch of Highway 75 constituted a part of it.[12] Clearly, the economic
and socio-cultural arguments of the city leadership and the chamber of
commerce prevailed and fixed the trajectory of Highway 75 at its present site:
right through the city.

The actual planning and building of Highway 75 took place between 1956
and 1959. Initially the plan was to construct overpasses at two of the main
intersections with local roads (Berggren 1956).[13] The city supported this solution. In June 1958, some council members voiced their concerns about the level
of road junctions, especially with regard to the safety of school children who
had to cross the highway. Moreover, they argued that the new highway would

add a third north–south barrier to the city, which was already divided in two by the Maas river and the railroad tracks. This is why they favored a highway that was built below ground level. Rather than opting for overpasses to allow for safe intersections, they favored lowering the new highway to achieve that same effect.[14] The Ministry of Transportation subsequently postponed the construction of overpass junctions (Bruijnzeels 1960; Thewissen 1958). This meant that for the time being the intersections would remain at ground level.

Another option that was put forward in the late 1950s was the idea of building a tunnel for Highway 75. One of the proponents, Council Member Schreuder, argued: 'Maastricht will only benefit from the immediate construction of a tunnel.'[15] The author of the 1958 Maastricht yearbook also referred to this issue: 'it is the intention of the government to build a tunnel at this place in the future' (Thewissen 1958). In a recent brochure of the city of Maastricht a similar plan was suggested: pictures of the road immediately after its construction reveal that there was an intention to build a tunnel.[16] However, in retrospect, Rijkswaterstaat engineer Jamin disputed the assumption that a tunnel would be built at that time. He suggested that the confusion about the tunnel was due to the fact that many people failed to make a clear distinction between a tunnel and a lowered road, indicating that some had articulated ideas about building a lowered highway with overpass junctions, not a tunnel.[17]

Whereas a lot of space was reserved on the northern edge, at the southern edge of the urban section of Highway 75 much less space was set aside: the highway was in fact planned immediately adjacent to the first high-rise building of Maastricht, named the 'Municipal Flat', which was built between 1948 and 1950 (see Figure 6.1).[18] The building, designed by city architect F.C.J. Dingemans (1905–61) and consisting of 90 apartments, was built in response to the great shortage of homes in the post-World War II era. From the beginning its design included elements that were seen as very modern at that time, such as elevators, central heating and refuse chutes (Bisscheroux and Minis 1997: 80). In general, it seems, the city's leadership viewed this building and the other apartment buildings that were planned to line the urban stretch of the highway as quite 'representative'; they were regarded as showcases of the city's modernity and, consequently, a lowered highway with slopes was basically not an option. It would take up too much space and obscure the buildings from view.[19]

Rijkswaterstaat, by contrast, preferred a spacious design and a less conspicuous presence of buildings along the highway. It argued that having a more spacious layout of the highway 'will prove to be the right opinion in the future'.[20] Notwithstanding this view, the apartment buildings were eventually placed relatively close to the highway.

Those in charge of the decisions involving the highway's construction in the 1950s deliberately tried to keep various redesign options open, but, ironically perhaps, in later periods it turned out to be extremely difficult to radically deviate from the original design. The ultimate plan was based on

Figure 6.1 Municipal Flat (source: the author)

a lot of considerations: opinions about traffic engineering, views of the role of automobiles in cities, socio-economic considerations, town planning motives, experiences in other Dutch cities, and earlier extension plans all became integrated and embedded in the design of Highway 75. This underlines the importance of understanding the highway as part of a larger sociotechnical structure, consisting of the highway itself, the city's traffic circulation system, the apartment buildings surrounding it and the many ideas and implications involved in its design. As will become clear in the remainder of this chapter, many features that were built into the highway's design maintained their obduracy for quite some time. In the remainder of this chapter, I will analyze how, under the influence of new ideas and developments, the sociotechnical structures of the urban section of Highway 75 became contested. I will focus on the phase between 1978 and 1982, when a trajectory study was done in which two redesign options played a major role: a tunnel and a diversion east of the city. But before I continue my analysis of the case, I will introduce three models of obduracy that can help us to explain the roots of the obduracy of the urban highway in Maastricht.

THREE MODELS OF OBDURACY

The concept of obduracy has been cast in three different ways. I will propose three conceptual approaches or heuristics that can be used to explain why (parts of) cities persist or why they don't.

Dominant frames

The category of 'dominant frames' consists of conceptions of technology's obduracy that focus on the role and strategies of actors involved in the design of technological artifacts, while the constraints posed by the sociotechnical frameworks within which they operate will be addressed in particular. The concepts of this category apply to situations in which town planners, architects, engineers, technology users or other groups are constrained by fixed ways of thinking and interacting. As a result, it becomes difficult to bring about changes that fall outside the scope of this particular way of thinking. The concepts in this category are generally used to analyze the *design and use* of specific technological artifacts. As an *interactionist* conception of obduracy, this category highlights the struggle for dominance between groups of actors with diverging views and opinions. In relation to specific technological artifacts, examples of this conception of obduracy include Wiebe Bijker's 'technological frame' (Bijker 1995a), Michael Gorman and W. Bernard Carlson's 'mental models' (Gorman and Carlson 1990) and the notion of 'professional worldviews' as found in work by Cliff Ellis (Ellis 1996). The concepts of 'technological frame' and 'professional worldviews' have also been applied to town planning design. Specifically, the concepts in this category highlight the significance of users, or 'relevant social groups', and inventors when it comes to explaining technology's obduracy.

Embeddedness

In this approach, the obduracy of technology can be explained precisely because of technology's embeddedness in sociotechnical systems, actor-networks or sociotechnical ensembles.[21] In this respect, historian of technology Thomas Hughes argues that the building of a system is accompanied by fewer difficulties when it has not yet become linked up with politics, economics or other value systems (Hughes 1994). This category involves a *relational* conception of obduracy: because the elements of a network are closely interrelated, the changing of one element requires the adaptation of other elements. The extent to which an artifact has become embedded determines its resistance to efforts aimed at changing it. Such efforts may be prompted by usage, societal change, economic demands, zoning schemes, legal regulations, newly developed policies and so forth.

Actor-network theorists such as Michel Callon, Bruno Latour, Madeleine Akrich, John Law and Annemarie Mol describe technological development as a process in which more and more social and material elements become linked up with each other in a network.[22] They investigate the attempts by actors to stabilize that network. But the larger and more intricate a network becomes, the more difficult it will be to reverse its reality. In this way, a slowly evolving order becomes irreversible (Callon 1995).

Persistent traditions

The category of persistent traditions comprises conceptions of obduracy that address the idea that earlier choices and decisions keep influencing the development of a technology over a longer period of time. Because of this focus on the longer-term persistence of traditions in sociotechnical change, I call this category *enduring*. Specifically, the notions of trajectories, path dependence (Garud and Karnøe 2001), momentum (Hughes 1983) and archetypes (Kitt Chappell 1989) embody this conception of obduracy. The crucial difference with the concepts discussed earlier is that they are less focused on interactions in local contexts than the other two models of obduracy: long-term, structural developments that transcend local contexts and interactions get more attention in this approach than in the previous two. One of the potential disadvantages of the frames approach, for instance, is that it is always focused on groups, and always emphasizing the local level. This makes it difficult to point at wider 'contextual' or structural factors that play a role in the construction of obduracy.[23] Generally, the notions within the category of persistent traditions put more emphasis on the wider cultural context in the explanation of obduracy in cities.

We now return to the case of Maastricht to see how these three models of obduracy can help us to understand why it has been so hard to change the highway's design. The aim of this chapter is not to argue which of the three models is the best, but to analyze their advantages and disadvantages in confrontation with this empirical case.

ATTEMPTS TO RECONSTRUCT THE MAASTRICHT HIGHWAY

By the 1970s, policy makers' opinions about the location of highways in relation to cities had become more pronounced: highways should preferably not be led *through* cities (where they cause all kinds of environmental, safety and livability problems), but *around* them. This shift took place in the context of a more general change of opinion among town planners, policy makers and citizens about large-scale urban interventions in the 1970s.[24] Whereas, in the 1960s, increased automobile traffic often served as a legitimization of drastic solutions, including major interventions in infrastructure (filling in canals) and urban districts (demolishing old towns) that had been around for centuries, this was no longer accepted in the 1970s. Local action groups increasingly voiced loud protests against huge urban redesign projects.

In the late 1970s, three policy reports mentioned the possibility of a diversion of the urban section of Highway 75 (now named A2) around Maastricht. The province of Limburg proposed the diversion as an option, but stated that putting in overpass junctions at the present trajectory still had priority. The national government's 'Structural Plan for Traffic and Transportation' (Structuurschema Verkeer en Vervoer 1976), a general planning report that summarized the most important policies of the Ministry of Transportation, also referred to a diversion of the highway in Maastricht, to be completed in the 1990s. The report argued that highways of the main national network should bypass urban areas, rather than go through them, because the absence of large flows of traffic in cities would improve urban living environments and reduce risks such as those associated with the transportation of hazardous materials.

These policy reports had a concrete effect for Maastricht in that they caused the city leadership to decide to set aside a strip of land that in the future might be used for the diverted highway east of the city. As a consequence, this area could no longer be used for other extension plans of the city. The municipal structural plan mentioned the possibility of a diversion of the highway east of Maastricht, but it also anticipated urban expansion at the eastern edge of the city. Although there was not very much (international) through traffic on Maastricht's highway yet, the city board considered the prospect of heavy trucks passing through the city 24 hours a day as unacceptable.[25] Because traffic was expected to go up and international standards for the design of highways were ever tighter,[26] the city board felt pressured to investigate the possibilities for adaptation of the existing trajectory or its diversion.[27]

After World War II, Maastricht, like many other Dutch cities, suffered from lack of space and scarcity of decent housing. In 1962, the director of public works, van de Venne, had already proposed the annexation of neighboring villages, to establish a 'Greater Maastricht' (Venne 1962). The planning effort aimed at the city's expansion was affected, of course, by the uncertainty about the highway's future. Although the mere mention in policy documents of a diversion on the eastern edge of the city implied a claim on this area, other actors advanced competing proposals for this area. These claims involved urban expansion, nature reserves and water procurement areas, as well as other possible road connections east of Maastricht.[28]

Because of the projected traffic growth, changed political and urban spatial policies, and international regulatory adaptations, discussions about the highway's design resurfaced. It was decided to study the various trajectory and design alternatives so that the city board could make an informed decision on how to use the available land. The project team 'E9 and Maastricht', which was going to be in charge of the trajectory study, was formally established on June 16, 1978 by the mayor and city board of Maastricht and the chief engineer and director of Rijkswaterstaat.

This team studied two design variants and a number of subvariants. One variant started from the existing trajectory of the highway and explored the closed tunnel option, the lowered highway option, and the lowered highway with acoustic fencing option. A distinction was made between designs that either did or did not integrate the Scharnerweg intersection. The second variant explored possible diversion trajectories along the eastern edge of Maastricht.

The idea of a tunnel to resolve the problems associated with the urban stretch of the highway had already been mentioned in the late 1950s – in the annals of the city as well as in the minutes of a city council meeting. In the 1970s, when Rijkswaterstaat and the city were thinking about reconstructing the highway's intersections, the idea of a tunnel turned up again. Residents of the nearby apartment buildings established an action committee, called 'E-9 Underground' (E-9 ondergronds), in which members of environmental action groups also participated. In a letter to the city council, the committee argued that the plans for reconstructing the Scharnerweg intersection were unacceptable. In their view, the planned reconstruction would not contribute to a reduction of the noise produced by the highway. They claimed that a tunnel could be built without demolishing any houses, while at the same time keeping all the existing connections of the highway to the local traffic network in place.[29] Moreover, they argued that the proposed modifications would make future adaptations increasingly difficult.[30]

Although a tunnel would be more expensive than the diversion, the city's leadership also preferred a closed tunnel: 'The only right solution in our view is a tunnel at the location of the present trajectory, especially now that its technical feasibility has been established.'[31] Since then, the city's leadership has been fairly consistent in its preference for a closed tunnel.[32] According to P. Jansen, a city traffic engineer, the city opposed all lowered road variants because none of them could sufficiently diminish the various forms of environmental damage produced by the urban highway stretch.[33] Only the closed tunnel option would meet the environmental standards (those regarding noise pollution in particular), while it also significantly reduced the highway's barrier function, something the city strongly supported.[34]

The city board rejected a diversion east of the city. A diverted highway along the city's eastern edge would form a barrier as well and would completely do away with the city's smooth transition into the surrounding landscape.[35] The diversion would negatively affect agriculture, drinking water supplies, recreation and the living environment of the residents of the eastern districts. Moreover, a diversion would conflict with the then prevalent 'city border philosophy' of the city's leadership that aimed at a gradual transition from city to countryside.[36] This philosophy had become increasingly important and was closely linked up with the city's identity. Former city manager Ad Lutters commented that in the eyes of the city leadership 'the eastern border has been a sacrosanct fringe that should not be touched,' a view that almost became 'a kind of dogma' over the years. Situated in a river valley and surrounded by rolling green hills that contain the characteristic yellow marlstone, Maastricht has always cherished its unique geography.[37]

Apart from the traffic and cost arguments put forward in the discussions about the reconstruction of the intersections in the mid-1970s, environmental arguments (landscape, water supplies, vistas) and the importance of a gradual transition from city to countryside were now advanced to discredit the diversion option. In the opinion of both the city leadership and the residents the city's eastern border was rigid – it should not be tampered with.

The idea of a diversion had nevertheless played an important role in Rijkswaterstaat's way of thinking for a long time; it had already been considered – and abandoned – twice: before World War II and in the mid-1970s. Although the amount of traffic did not necessitate measures at that time, Rijkswaterstaat opted for a diversion around the city for reasons related to traffic growth, traffic safety and the living environment. In Rijkswaterstaat's view, safety was paramount and therefore it did not favor a closed tunnel: the transportation of hazardous goods through tunnels was widely considered to be too risky. Furthermore, its wish to transform the urban section of the highway into a 'real' highway could be more easily realized by opting for a diverted trajectory. In general, highway engineers find it easier to build highways in relatively open spaces than in built-up urban environments, where existing town planning structures seriously limit the available design options.[38] A Rijkswaterstaat memo of April 1978 suggests that at that time the decision process was primarily geared toward the diversion variant rather than the existing trajectory.[39] Even though this preference was denied in an earlier letter to city board member Dols of January 27, 1978,[40] in retrospect Jamin, a highway engineer involved in the design process, emphasized the importance of the diversion alternative for Rijkswaterstaat:

> At that time, we were not at all working on a road trajectory that went through the city. We were only studying alternatives outside the city. At the very last moment, we received a directive from The Hague: 'You should also make a plan for a highway that cuts through the city.' There were, however, only broad outlines for such a plan, while the diversion option, by contrast, was thoroughly investigated, including zoning schemes.[41]

In 1981, the residents of Maastricht were publicly consulted about the proposed diversion. A majority appeared to be against the diversion, whereas most considered a tunnel below the existing trajectory the best solution (Damoiseaux 1981). According to the residents of the eastern city districts, the diversion of the highway would locally devastate the environment. Yet the residents of the apartment buildings along the existing highway viewed their situation as 'unbearable'. The smell of exhaust fumes and traffic noise made their lives utterly miserable. Remarkably, perhaps, the majority of them also viewed a tunnel as the best solution. The old highway would then become superfluous, they argued, and should be transformed into a green area or park.[42]

The Council for Water Works (Raad van de Waterstaat) had to advise the minister of transportation about the trajectory study. They concluded that: 'Reconstruction of the present passage through Maastricht is, when taking into account nearly all aspects, more attractive than the construction of a diversion around the city.'[43]

On January 13, 1982, the minister of transportation, H.J. Zeevalking, decided in favor of the trajectory through the city. He did not say when the solution should be implemented, nor did he stipulate which of the three tunnel options should be built – a closed tunnel, an open tunnel, or a semi-closed tunnel with acoustic fencing.[44] Unambiguously, however, the minister stated that a diversion was not a good option.[45]

The hesitation to build a tunnel immediately can be explained by a number of interrelated cultural, historical, economic and political factors. In 1982 there basically was no congestion on the urban section of the highway. The oil crises of the 1970s had culminated in economic recession by the early 1980s. This, in turn, had diminished the growth of traffic, thereby decreasing the need for immediate action.[46] Moreover, in the 1970s, Dutch people had become more aware of environmental problems; automobile usage was increasingly criticized for its contribution to air pollution and the depletion of fossil fuels. Governmental transportation policies became more focused on reducing the number of cars. In addition, owing to the then current noise regulation standards and tight budgets, full realization of the tunnel within a short time span could not have been possible.[47] Furthermore, the problems of Maastricht's highway infrastructure were considered minor in comparison to those in the more densely populated western part of the Netherlands. Finally, at that time Maastricht and the province of Limburg simply failed to have a strong highway lobby in The Hague, the seat of the national government.[48]

After the 1982 decision of the minister, the diversion option stopped playing a role: future alternatives would be defined in line with the idea of a tunnel and were limited to a reconstruction of the existing trajectory. The decision has since become more fully integrated in various municipal policies; as a result, a tunnel basically became a must in the eyes of the city. Clearly, the minister's decision contributed to the tunnel's further embeddedness in the city's plans and policies.

After 1982, the idea of a tunnel became more and more part of the expectations and ideals of residents and local politicians alike. Rijkswaterstaat and the city began to work on several tunnel variants in the Working Group Tunnel Design that was established in 1989.[49] For a while, the city strongly opted for an expensive drilled tunnel, since this was seen as the only possibility to avoid large-scale demolition. In the early 1990s, local real estate agents sold houses in nearby neighborhoods claiming that the city highway would soon disappear in a tunnel with the 'old' highway transformed into a city park. Residents of Maastricht also unfolded initiatives aimed at developing tunnel plans. In the mid-1990s, local town planner René Daniëls, director of Buro 5, proposed a tunnel design, to be developed in collaboration with project developer HBG Vastgoed and an engineering business. Their provocative plan played an important psychological role in the overall planning process.[50] According to Daniëls, the plan was largely in accordance with the city's viewpoints in the 1990s; the soon-to-be former highway, for instance, was designed as a boulevard with trees. Moreover, demolition of apartment buildings would not be necessary in Daniëls's solution.[51] Although Armand Cremers, the city board member in charge of traffic issues at that time, liked the scale model of the plan very much, the plan was never seriously considered as a definite solution by the city. Looking back, Cremers argued that the plan turned up 'too early' in the process of thinking about solutions and that it was 'too concrete'. He suggested the solution did not fit into the existing town planning structure and failed to solve the financial problems.[52]

EXPLAINING THE OBDURACY OF MAASTRICHT'S HIGHWAY

This analysis has shown that the highway's obduracy can partly be explained by its embeddedness in the larger town planning structure of Maastricht: its embeddedness in the local traffic system, and the apartment buildings adjacent to the highway. By stressing the highway's various modes of embeddedness, multiple explanations for the difficulty of changing its design can be advanced and supported. This case shows nicely how the road persists because of the node it occupies in a network of people, laws and regulations, buildings, investments and so on. In terms of Latour and Callon, it had become an 'obligatory passage point' (Callon 1986) that could not easily be ignored. That Maastricht residents were aware of this embeddedness and the limitations it put on the reconstruction options became clear in their recommendation to build a tunnel 'without demolishing any houses, while at the same time keeping all the existing connections of the highway to the local traffic network in place'.[53] Furthermore, they argued that the proposed modifications would make future adaptations increasingly difficult.[54] Moreover, Rijkswaterstaat wanted to avoid dealing with the embedded urban highway structures and rather preferred a diverted highway in a more spacious, and thus more flexible, non-urban environment.

It is perhaps surprising that the tunnel option, despite its popularity with the residents and leadership of Maastricht, is still (2009) not implemented. The idea of a tunnel played a central role in the expectations and ideals of those immediately involved. In time, the tunnel idea acquired robustness: it became difficult to neglect in the unbuilding process. Obviously, its obduracy had no material component, for the tunnel did not exist as a reality yet. The tunnel grew obdurate, in the first place, because it became linked with the expectations and ideals of residents: as early as 1958, the tunnel idea had been mentioned. That same year, council members proposed a tunnel as the preferred future design of the urban highway stretch. In the 1970s, residents who were represented in the 'E-9 Underground' action committee made a strong case in favor of a closed tunnel. The public reactions to the 1980 trajectory study showed that residents preferred a closed tunnel with a park on top of it. Most recently, the tunnel turned up in a new citizens' initiative: Buro 5 designed a tunnel and visualized it in a scale model. Second, the idea of a tunnel acquired more robustness when it became integrated into municipal policies and activities: the city established a working group particularly aimed at designing tunnel variants. After the decision of the minister in 1982, the idea of a tunnel in Maastricht became further integrated into local plans and policies.[55]

I have now made a paradoxical move by using the concept of embeddedness to explain the obduracy of a technology that is not even implemented yet. I argue that particular solutions, as in this case the tunnel option, can become embedded in urban, regional and national policies, as well as in the ideals and activities of citizens and politicians, *before* they even become a physical reality. Thus embedded, the tunnel, while non-existent, had important effects on the unbuilding process.

However, one of the disadvantages of the embeddedness concept is that it fails to address broader social and economic structures of power and interests. Because of the focus on local actors ('actants') and their efforts to create and sustain durable assemblages of human and non-human actors, the importance of long-term traditions and structural aspects remains, underexposed.[56] Actor-Network Theory in general is criticized 'for ceding too much power and autonomy to individual actors, rather than to existing structures of power and interests' (Williams and Edge 1996: 890). In the case of the Maastricht highway, I pointed at contextual developments, like the oil crisis, Limburg's weakness in The Hague, and the changing political context in the Netherlands. These factors, which played an important role in the attempts at unbuilding the highway, were difficult to reconcile with embeddedness as an explanation of the obduracy of the Maastricht highway. Such longer-term contextual and structural developments are better explained by my third model of obduracy: persistent traditions.

Such an approach would highlight, for instance, the tradition of tunnel engineering and the cultural values and metaphors that are associated with that tradition. In this case we might refer to the nineteenth-century tradition of 'putting the less glamorous aspects of civilization underground' (Williams 1990: 206), a tradition that has also influenced the effort of twentieth-century architects to solve particular redesign problems. In the discussions about the highway's redesign, the city of Maastricht supported its preference for a tunnel by suggesting it would hide the environmentally damaging flow of cars from view, an option equally favored by residents who advocated the creation of a public park to replace the old highway once it had become obsolete.

Rosalind Williams (1990) gives an interesting cultural interpretation of persistent traditions in her book *Notes on the Underground*. By analyzing artificial underground worlds as an 'enduring archetype', Williams shows how literary traditions from all over the world have always expressed a concern with the underground, which suggests the persistence of the opposition between surface and depth in our thought. Contemporary developments in town planning and architecture, in particular the trend to build under-ground, to construct tunnels and subways and to hide less attractive urban functions, can be understood in relation to the work of the nineteenth-century novelist H.G. Wells. In *The Time Machine* (1895), Wells wrote about an underground world inhabited by the Morlocks, who operated machines and utilities, and a paradise of nature and leisure (inhabited by the Eloi) situated above the surface. Williams shows that this tradition of 'putting the less glamorous aspects of civilization underground' reverberates in the work of twentieth-century architects (Williams 1990: 206). In their urge to deal with overpopulation and space-consuming distribution networks, roads, central heating infrastructure and factories, they have turned their gaze to the underground world, so that the surface may still be available for the more pleasurable aspects of life (leisure, recreation, parks, housing, schools and so forth). Such traditions can be enduring in the sense that they are likely to keep influencing choices and decisions of large groups of people.

This 'persistent tradition' of trying to hide less attractive urban functions from view in part explains why specific types of design solutions, in this case the tunnel, are proposed and supported time and again. An important episode in the Maastricht case can thus be explained using the persistent traditions model.

The 'frames' model in the explanation of technology's obduracy is particu-larly helpful when two (or more) groups are opposing each other and have strong and multiple interactions. The model is sensitive to varieties and differences between social groups, and it distinguishes two types of obduracy (closed-in and closed-out obduracy), depending on the degrees of inclusion of actors in specific technological frames. The model is strong in that it links group interactions to an opening of the black box of sociotechnical design. It is thus a symmetric approach that does justice to the social and material aspects of obduracy.

In the debates on the tunnel option versus the diversion option that took place between 1978 and 1982, the emergence of two technological frames can be discerned. Clearly, the actions and reasoning of Rijkswaterstaat were predominantly derived from the 'diversion' frame. In this frame, safety con-siderations were important and served as an argument against the tunnel option. Moreover, technical arguments played a key role in this frame: the diversion was preferred, because this would offer more space than the city to design a 'real' highway, not hindered by any urban obstacles. The city's views and arguments mainly relied on the 'tunnel' frame. In this frame, environ-mental concerns played a key role. A tunnel would be better from an environmental point of view, because noise and pollution could be better controlled. For a couple of years, these two frames guided the interactions between Rijkswaterstaat and the city board and made it harder to reach agreement on a common solution. However, the diversion option disappeared from the agenda of both actors after the decision of the minister of trans-portation in 1982. While the 'tunnel' frame continued to be important for the city leadership, alternatives were not excluded from the discussions. This shows that the role of these technological frames in the constitution of the obduracy of Maastricht's highway was only temporary.

Using the 'frames' model, I could highlight only a limited number of aspects of obduracy. A disadvantage of the frames approach is that it is always focused on groups, and always emphasizing the local level. This makes it difficult to point at wider 'external' or structural factors that play a role in the construction of obduracy. This aspect of the SCOT approach has been criticized by Klein and Kleinman:

> There is no attention for the ways in which institutionalized social values shape components of technological frames ... Bijker never considers the ways in which deeply institutionalized social values shape components of a technological frame or actors' interactions or practices more generally.
>
> (Klein and Kleinman 2002: 40)

I agree with Klein and Kleinman that frames 'are likely to draw on cultural elements with historical resonances in the society at large', but so far this aspect has never been fully developed in the context of the SCOT model.

My analysis has aimed at clarifying the apparent obduracy of the highway through Maastricht. The three models provide different perspectives on the reasons behind this obduracy. Choosing a single explanation will in most cases not sufficiently explain the obduracy of urban objects. Therefore I proposed a strategy of using the three perspectives simultaneously – allowing them to fill each other's blind spots.

Despite all efforts, no solution to the problems around Maastricht's highway has been implemented so far (2009). After a number of temporary measures in the early 1990s, a new trajectory study was initiated. A whole range of redesign options was studied again (including a tunnel and a diversion). In 1999 a new traffic lights regime was installed at the highway section through Maastricht. Some people believed that the problems were solved by that simple measure. But Rijkswaterstaat and the city board of Maastricht emphasized that the ultimate solution could only be a tunnel. Much effort has been put into lobbying in The Hague, and in November 2008 Maastricht citizens were consulted about three preliminary tunnel designs proposed by building consortia. At the moment, the city board of Maastricht is quite optimistic about the chances that the building activities for a tunnel will actually start in the second half of 2010.

NOTES

1   This chapter is based on earlier work by the author published in *Unbuilding Cities: Obduracy in Urban Sociotechnical Change* (Cambridge: MIT Press, 2005). Reprinted with permission.

2   There are some examples though of STS(-related) work on cities: Aibar and Bijker (1997); Graham and Marvin (1996, 2001); Hård and Misa (2008); Moore (2001).
3   See Hommels (2005a) for an overview of conceptions of obduracy.
4   Rijkswaterstaat is the national governmental body for road construction and road management. It has several regional departments, including one in the province of Limburg.
5   Interview J. Jamin, Rijkswaterstaat Limburg, highway engineer, involved in the A2 (re-)design process since the late 1950s, Maastricht, February 2, 1999. See also the memo 'Notitie voor wethouder openbare werken en sport', December 30, 1974 (Archive Stadsontwikkeling en Grondzaken – SOG).
6   Interview Jamin. See also Minutes city council Maastricht, June 3, 1958, No. 10–17 (Archive Sociaal Historisch Centrum (SHC) Maastricht).
7   The views of the chamber of commerce on Maastricht's traffic problems at that time are illustrated in an advisory report: 'Advies uitgebracht door de Kamer van Koophandel en Fabrieken voor Maastricht en Omstreken betreffende noodzakelijke verkeersvoorzieningen te Maastricht, aan zijne excellentie de Minister van Verkeer en Waterstaat en aan het Gemeentebestuur van Maastricht', Kamer van Koophandel en Fabrieken voor Maastricht en Omstreken, Maastricht, 1955 (Archive SHC Maastricht).
8   Interview Jamin. Interview P. Jansen, city of Maastricht, traffic engineer, Maastricht, March 4, 1999. Interview T. Jenniskens, city of Maastricht, expert on Maastricht's history and culture, Maastricht, April 15, 1999.
9   My translation. See Minutes city council Maastricht, June 3, 1958, No. 10-19 (Archive SHC Maastricht).
10   My translation. See Minutes city council Maastricht, May 7, 1962, No. 6-3 (Archive SHC Maastricht).
11   J.J.J. van de Venne was director of public works in Maastricht between 1956 and 1977.
12   Minutes city council Maastricht, June 3, 1958, No. 10-25 (Archive SHC Maastricht). See also van de Venne (1964).
13   Interview Jamin.
14   Minutes city council Maastricht, June 3, 1958, No. 14 (Archive SHC).
15   Minutes city council Maastricht, June 3, 1958, No. 10-21 (Archive SHC Maastricht).
16   See the brochure by the city of Maastricht: 'A2-Traverse Maastricht: Geen eindpunt van Nederland, maar startpunt voor Europa zonder hindernissen', Gemeente Maastricht, Maastricht, 1998 (Personal archive O. de Jong, Maastricht). Interview O. de Jong, City of Maastricht Department of Town Development, city coordinator of the A2 project, Maastricht, May 7, 1998.
17   Interview Jamin.
18   Interview Jamin.

19 See an internal memo for the alderman public works and sports, December 30, 1974 (Archive SOG).

20 Letter Rijkswaterstaat to director of public works, city of Maastricht, February 1, 1955; see also letter Rijkswaterstaat to director of public works, March 16, 1955 (Archive Gemeentearchief Maastricht).

21 The system concept is used in the Large Technical Systems (LTS) approach, introduced by Hughes (1983). The metaphor of network is introduced in the Actor-Network Theory (ANT), developed by Callon (1986), Latour (1987) and Law (1991). Bijker introduced the notion of 'sociotechnical ensemble' (see e.g. Bijker 1995a, 1995b).

22 See for example work by Callon (1986, 1987, 1991).

23 This aspect of the SCOT approach has recently been criticized by Klein and Kleinman (2002). See also Hamlett (2003) for a discussion of the critique on social constructivists (as for instance expressed by philosopher Hans Radder) that they disregard the importance of 'non-local norms' in the development of technology. The category of persistent traditions makes clear how norms that 'transcend' local contexts play a role in the construction of urban obduracy. This discussion relates to a more general debate in social theory about models that put an emphasis on agency, locality and contingency, versus other models that foreground structure, institutions and persistence.

24 See Hommels (2000, 2005b) for an analysis of how this shift influenced public opinion on the huge urban reconstruction project in the city center of Utrecht: Plan Hoog Catharijne. The protests against the implementation of the subway in Amsterdam are also notorious in this respect, as well as the fierce actions against the construction of a highway through a nature conservation area near Utrecht: Amelisweerd.

25 See the structural plan of the city of Maastricht: 'Structuurplan Maastricht 1979', Report no. 168 (Personal archive O. de Jong, Maastricht).

26 For example, the 'declaration for the construction of international main traffic routes' (i.e. the E-routes, including the E-9) that was signed in Geneva, September 16, 1950, stated that E-roads should bypass built areas when they are led through cities and cause inconvenience and dangers, that E-roads should not have level road junctions, and that intersections and traffic lights should be avoided. These rules, though not compulsory, should be observed as closely as possible. See internal paper Rijkswaterstaat, 'Notitie: E-wegen', Rijkswaterstaat directie Limburg, July 13, 1978 (Archive SOG Maastricht).

27 See the city's structural plan 'Structuurplan Maastricht 1979', Report no. 168 (Personal archive O. de Jong, Maastricht).

28 See the final report of the trajectory study by Rijkswaterstaat, 'A2/E9 om en in Maastricht: Tracé studie', Report no. 169, Rijkswaterstaat, Maastricht, 1979 (Archive Rijkswaterstaat Limburg Maastricht, afdeling Integraal Verkeer en Vervoer).

29 Letter P.J.G. Groot to the city council of Maastricht, February 3, 1975 (Archive SOG Maastricht).

30 See Appendix, 'Central Action Committee E-9 Underground', August 1975 (Archive SOG Maastricht).

31 See proposal to the city council of Maastricht, no. 66, January 7, 1981 (p. 9) and the document that summarizes the opinion of the city board of Maastricht about the trajectory study, 'Standpunt van de gemeente Maastricht op de hoorzitting gehouden door de Raad van de Waterstaat over het rapport A2/E9 in/om Maastricht', April 29, 1981 (Archive SOG Maastricht).

32 Interview Cremers.

33 See also 'Standpunt van de gemeente Maastricht op de hoorzitting gehouden door de Raad van de Waterstaat over het rapport A2/E9 in/om Maastricht', April 29, 1981 (Archive SOG Maastricht).

34 Interview Jansen.

35 See proposal to the city council of Maastricht, no. 66, January 7, 1981 (Archive SOG Maastricht).

36 See memo 'Gespreksnotitie t.b.v. periodiek overleg RWS en gemeentebestuur Maastricht op 21 augustus 1979 betreffende studie E-9 en Maastricht' (Archive SOG Maastricht).

37 Interview A. Lutters, city manager between 1977 and 1998, Maastricht, July 21, 1999.

38 Interview J. Kroon, Bouwdienst Rijkswaterstaat, highway engineer involved in trajectory/EIS study, Apeldoorn, August 18, 1999.

39 See memo 'Notitie E-9 en Maastricht ten behoeve van het overleg RWS-Maastricht op 10 april 1978', Rijkswaterstaat Limburg (Archive SOG Maastricht).

40 Letter Rijkswaterstaat to city board member Dols, January 27, 1978 (Archive SOG Maastricht).

41 Interview Jamin. I studied Jacques Jamin's personal archive, which included numerous highway designs that were made for the A2 by Rijkswaterstaat in the period 1967–93. Indeed, I found no plans for a tunnel until the 'rough' 1981 tunnel designs. It is thus likely that closed tunnel variants were first investigated by Rijkswaterstaat in detail in the Working Group Tunnel Design (established in 1989) and the trajectory/EIS study (started in 1995).

42 See 'Inspraakreacties inzake de nota A2/E9 om en in Maastricht', November/December 1980 (Archive SOG Maastricht, 'Rijksweg 75', code 1.811.111/1).

43 Letter of the Council for Water Works to the minister of transportation, November 11, 1981, my translation (Archive SOG Maastricht).

44 See 'Vaststelling van het tracé van de rijksweg A2/E9 in Maastricht', July 23, 1982 (HW/WWO 39296) (Archive Rijkswaterstaat Limburg Maastricht).

45 Interview Jansen.

46 Interview Jansen. For an analysis of shifts in postwar Dutch governmental policies on traffic and mobility, see Peters (1998).

47 See a report by G. van Heusden, 'Plan van Aanpak Tracé/Mer-procedure
Rijksweg 2 Passage Maastricht', Projectbureau MER, Maastricht,
February 16, 1995 (Archive Rijkswaterstaat Limburg Maastricht).

48 Interview Cremers.

49 See a concept internal memo by P. Jansen, 'Concept: Stand van zaken
met betrekking tot de ondertunneling van de A2 te Maastricht',
Gemeente Maastricht, Maastricht, 1991 (Archive SOG Maastricht).

50 Interview Cremers.

51 Interview R. Daniëls, town planner Buro 5, Maastricht, April 21, 1999.

52 Interview Cremers.

53 Letter P.J.G. Groot to the city council of Maastricht, February 3, 1975
(Archive SOG Maastricht).

54 See Appendix, 'Central Action Committee E-9 Underground', August
1975 (Archive SOG Maastricht).

55 See for example 'Maastricht, stad in evenwicht, balans in beweging:
Hoofdpunten van het ruimtelijk en economisch beleid 1990–2000',
March 1992 (p. 28) and 'A2-Traverse: Voorlopige uitgangspunten en
randvoorwaarden Gemeente Maastricht', Confirmed by city board
November 19, 1996 (Personal archive O. de Jong, Maastricht).

56 See Williams and Edge (1996) for a similar line of argumentation linked
to Actor-Network Theory. It should be noted however that in other
interpretations of the notion of embeddedness, as in Harvey's work
(Harvey 1985), there *is* attention to structure and power factors such as
the forces of capitalism.

REFERENCES

Aibar, E. and Bijker, W.E. (1997) 'Constructing a City: The Cerdà Plan for the
Extension of Barcelona', *Science, Technology, & Human Values*, 22(1): 3–30.

Berggren, H. (1956) *Maastricht in 1956*, Maastricht: Leiter-Nypels.

Bijker, W.E. (1995a) *Of Bicycles, Bakelites and Bulbs: Toward a Theory of Sociotechnical Change*, Inside Technology, Cambridge, MA: MIT Press.

Bijker, W.E. (1995b) 'Sociohistorical Technology Studies', in S. Jasanoff, G.E. Markle,
J.C. Petersen and T. Pinch (eds), *Handbook of Science and Technology Studies*,
Thousand Oaks, CA: Sage Publications, pp. 229–56.

Bisscheroux, N. and Minis, S. (1997) *Architectuurgids Maastricht 1895–1995*,
Maastricht: Stichting Topos, Dienst SOG Gemeente Maastricht.

Bruijnzeels, K.V.H. (1960) *Maastricht in 1960*, Maastricht: Leiter-Nypels.

Callon, M. (1986) 'Some Elements of a Sociology of Translation: Domestication of
the Scallops and the Fishermen of St Brieuc Bay', in J. Law (ed.), *Power, Action and
Belief: A New Sociology of Knowledge?*, London: Routledge & Kegan Paul,
pp. 196–233.

Callon, M. (1987) 'Society in the Making: The Study of Technology as a Tool for Sociological Analysis', in W.E. Bijker, T.P. Hughes and T.J. Pinch (eds), *The Social Construction of Technological Systems: New Directions in the Sociology and History of Technology*, Cambridge, MA: MIT Press, pp. 83–103.

Callon, M. (1991) 'Techno-economic Networks and Irreversibility', in J. Law (ed.), *A Sociology of Monsters: Essays on Power, Technology and Domination*, New York: Routledge, pp. 132–61.

Callon, M. (1995) 'Technological Conception and Adoption Network: Lessons for the CTA Practitioner', in A. Rip, T.J. Misa and J. Schot (eds), *Managing Technology in Society: The Approach of Constructive Technology Assessment*, London and New York: Pinter Publishers, pp. 307–30.

Damoiseaux, R. (1981) *Maastricht 1981*, Maastricht: Leiter-Nypels.

Ellis, C. (1996) 'Professional Conflict over Urban Form: The Case of Urban Freeways, 1930 to 1970', in M.C. Sies and C. Silver (eds), *Planning the Twentieth-Century American City*, Baltimore, MD: Johns Hopkins University Press, pp. 262–79.

Garud, R. and Karnøe, P. (eds) (2001) *Path Dependence and Creation*, Organization and Management Series, Mahwah, NJ: Lawrence Erlbaum.

Gorman, M.E. and Carlson, W.B. (1990) 'Interpreting Invention as a Cognitive Process: The Case of Alexander Graham Bell, Thomas Edison and the Telephone', *Science, Technology, & Human Values*, 15(2): 131–64.

Graham, S. and Marvin, S. (1996) *Telecommunications and the City: Electronic Spaces, Urban Places*, London: Routledge.

Graham, S. and Marvin, S. (2001) *Splintering Urbanism: Networked Infrastructures, Technological Mobilities and the Urban Condition*, London: Routledge.

Hamlett, P.W. (2003) 'Technology Theory and Deliberative Democracy', *Science, Technology, & Human Values*, 28(1): 112–40.

Hård, M. and Misa, T.J. (eds) (2008) *Urban Machinery: Inside Modern European Cities*, Cambridge, MA: MIT Press.

Harvey, D. (1985) *The Urbanization of Capital*, Oxford: Blackwell.

Hommels, A. (2000) 'Obduracy and Urban Sociotechnical Change: Changing Plan Hoog Catharijne', *Urban Affairs Review*, 35(5): 649–76.

Hommels, A. (2005a) 'Studying Obduracy in the City: Toward a Productive Fusion between Technology Studies and Urban Studies', *Science, Technology, & Human Values*, 30(3): 323–51.

Hommels, A. (2005b) *Unbuilding Cities: Obduracy in Urban Sociotechnical Change*, Inside Technology, Cambridge, MA: MIT Press.

Hughes, T.P. (1983) *Networks of Power: Electrification in Western Society, 1880–1930*, Baltimore, MD: Johns Hopkins University Press.

Hughes, T.P. (1994) 'Technological Momentum', in M.R. Smith and L. Marx (eds), *Does Technology Drive History? The Dilemma of Technological Determinism*, 2nd edition, Cambridge, MA: MIT Press, pp. 101–13.

Kitt Chappell, S.A. (1989) 'Urban Ideals and the Design of Railroad Stations', *Technology and Culture*, 30: 354–75.

Klein, H.K. and Kleinman, D.L. (2002) 'The Social Construction of Technology: Structural Considerations', *Science, Technology, & Human Values*, 27(1): 28–52.

Konvitz, J.W. (1985) *The Urban Millennium: The City-Building Process from the Early Middle Ages to the Present*, Carbondale: Southern Illinois University Press.

Latour, B. (1987) *Science in Action: How to Follow Scientists and Engineers through Society*, Cambridge, MA: Harvard University Press.

Law, J. (1991) 'Power, Discretion and Strategy', in J. Law (ed.), *A Sociology of Monsters: Essays on Power, Technology and Domination*, London: Routledge, pp. 165–91.

Moore, S.A. (2001) *Technology and Place: Sustainable Architecture and the Blueprint Farm*, Austin: University of Texas Press.

Peters, P. (1998) 'De smalle marges van de politiek', in H. Achterhuis and B. Elzen (eds), *Cultuur en Mobiliteit*, Den Haag: Rathenau Instituut, pp. 39–63.

Thewissen, M.A.F.C. (1958) *Maastricht in 1958*, Maastricht: Leiter-Nypels.

Venne, J.J.J. van de (1962) 'Maastricht: stad in een keurslijf', *Publieke Werken*, April.

Venne, J.J.J. van de (1964) *Maastricht, een visie op de toekomst*, Maastricht: Gemeente Maastricht.

Williams, R. (1990) *Notes on the Underground: An Essay on Technology, Society, and the Imagination*, Cambridge, MA: MIT Press.

Williams, R. and Edge, D. (1996) 'The Social Shaping of Technology', *Research Policy*, 25: 865–99.

7 Mutable immobiles

Building conversion as a problem of quasi-technologies

Michael Guggenheim

INTRODUCTION

In 1992 the administrative court of the canton of Zürich in Switzerland had to decide whether a former industrial building, now rezoned in a zone for habitation, should be allowed to be turned into offices (Verwaltungsgericht des Kantons Zürich 1992b). As such, the case is a variation of a typical case of gentrification and urban renewal turned into a legal problem. A city changes and thus in certain city parts the structure of habitation, uses, buildings and residents changes too. Owners of buildings lose their tenants or force them out and have to find different kinds of tenants to replace them. In the above case, as in many others, an industrial area suffered from de-industrialization and the authorities subsequently changed the designation of the zone from industrial to habitation. The owners of the building sought to make a few changes such as inserting internal partitioning to turn the building into offices. The lower building authorities deemed these changes illegal, because they did not conform to the changed zoning. The administrative court of the canton ruled that a change into flats could not be enforced because such a change would require 'considerable construction expenditures' that would not even lead 'to satisfying solutions' (Verwaltungsgericht des Kantons Zürich 1992b: 21). The court also denied the argument of the authorities that various examples of former factories that were converted into flats would prove the feasibility of a conversion. The court stated that these were always made with 'the voluntary cooperation . . . of the owners within an integrated concept', neither of which were given in the case (Verwaltungsgericht des Kantons Zürich 1992b: 21).

This case highlights the precarious role of buildings in processes of urban renewal and gentrification. For the municipal authorities the building is a blank slate that can be used for anything and should thus be used as flats according to the zone. For the owners and the administrative court, the building is a technology that allows only specific uses. It resists being used for habitation and can only be made to do so by the unlikely combination of huge expenditures, an integrated concept and the cooperation of the owners. The case is not only about a specific building but has become a setting to negotiate what buildings can do.

162   *Michael Guggenheim*

In this chapter, I wish to analyse this precarious role of buildings from a theoretical and empirical point of view, by analysing change of use as a test case. As we have seen in the above example, change of use is a test case because it challenges two opposing but prevailing notions of buildings. On the one hand, many in urban studies assume the identification of buildings and uses. Ethnographies and historical studies of specific building types routinely describe use patterns and building forms as being interrelated (Foucault 1979). By doing so they assume that building forms affect or even control the uses. Thus they describe the uses as shaping the building types. On the other hand, studies on gentrification processes, housing renewal and urban development often do not consider buildings to be a factor in these processes (Smith 1996). They often simply ignore them and treat them as mere material strata for social processes.[1] However, as I will show in this chapter, the relationship between buildings and uses cannot be described in either of these ways, either on a theoretical level or on an empirical one.

The problem for a theory of buildings is thus to explain the specific notion of a building type in terms of a more general theory of objects and technologies. I would like to show in this chapter that buildings can be considered as technologies in a double sense: they are technologies as buildings, but they are also considered to be technologies on a second level, as building types. But, as building types, they are unstable technologies, or what I call quasi-technologies. More specifically, since buildings occupy a stable location and are singulars they are open to different uses at the same time, which turns them into what I call mutable immobiles. This feature of being quasi-technologies and mutable immobiles creates the problem of accounting for change of use – as we have seen in the opening paragraph. Change of use presupposes building types as a form of classification that conflates the use of buildings with their materiality. Change of use, then, demonstrates that a building type is not a proper technology, since it can be easily circumvented and turned into another building type, even by keeping the building intact.

In this chapter I establish the theoretical basis by first discussing whether buildings can be considered as technologies, and I highlight the classification as building types as a specific feature of buildings. I introduce the notion of quasi-technology to account for the fast changes of buildings, and I explain how change of use can serve as a test case to analyse the problem of buildings as quasi-technologies. I then move on to use several court cases related to issues of zoning and change of use of buildings to demonstrate the theoretical perspective previously laid out. I first discuss cases of the privilege of continued existence. These refer to buildings that were built before a change in zoning took place and should be adapted to the new zoning. Second, I discuss cases of changing uses that violate existing zoning. In all of these cases the courts struggle with conceiving of buildings as technologies only to find out that they are not, and vice versa.

BUILDINGS AS TECHNOLOGIES?

For an analysis of the design, production and use of buildings it seems fruitful to me to ask in what sense buildings can be considered to be technologies. Bruno Latour's early work established against the mainstream of sociology the idea that 'technology is society made durable' (Latour 1991). He showed these stabilizing effects of technologies in several case studies, such as doors or sleeping policemen, some of them related to architecture. Whereas Actor-Network Theory has not shown much interest in differentiating different *types of objects*, my task here is to look at the specificity of buildings. I am interested here in *whole buildings as technologies* and therefore it is important to clarify whether buildings are comparable to other technologies, such as door openers.[2]

A working definition of technology would be that technology is a black-boxed actant-network (Latour 1987: 81) or a strict coupling (Luhmann 2000: 370).[3] The important feature of these notions that sets them apart from other definitions of technologies is that they do not necessarily understand technologies as material or high-tech, but as *procedures* with specific features. Parts are assembled in such a way that they are not changed during use and cannot be changed, or only by dint of considerable work. Such technology speeds processes up, because it always accomplishes the same output with the same input. This is so because no consensus is needed to make technology work: the consensus is already part of the black box. Second, it is possible to calculate the resources needed to operate technology and, third, failures become visible as irregularities that can be diagnosed and repaired. A door opener is a technology, because he/she/it always opens the door when somebody approaches, no matter whether the door opener is a human being or a machine (Latour 1992).

Following such a definition, we can ask whether buildings are technologies. Clearly, they are composed of technologies, such as door openers and walls and roofs. But, as I would like to claim, they are also technologies with regard to specific uses. As such, they are technologies that locate specific interactions. The emphasis here is on 'locate': buildings are technologies for the location of interactions because buildings occupy *permanently* a specific location and are designed with respect to functional aspects (with respect to society). These locational technologies work as classifications, as building types, such as bank, church or office building. A type is a classifying word such as 'bank' that is tied to a use and a form. A building type is a classification on a secondary functional level and specific to buildings. Such secondary-use classifications do not exist for other objects.[4] Buildings as building types thus coordinate interactions by providing interactions with a location and resources. In other words, buildings as building types are generic black boxes with respect to specified uses. For example, banks permanently locate and facilitate money transfers and credit lending by assembling black boxes such as teller's windows, bullet-proof glass, prestigious façades, sign systems, separate rooms for meetings and back offices.

So far, I have established buildings as technologies in a way compatible with ANT. The perspective of ANT is to explain how things hold together and how they have the capacity to delegate actions. Buildings, specifically when seen as building types, are technologies to locate interactions. That is, they are assumed to be stabilizers of society. But, as I would like to show, building types are a very special kind of technology and their role in society poses specific problems for the theory.

As I said above, buildings seen as being part of a building type are technologies as wholes. The notion of forming 'a whole' is important, since building types cannot be defined by providing a list of requisite parts, as is possible, for example, for an air pump.[5] None of the aforementioned individual black boxes (teller's windows and so on) are necessary to build a bank. It is not even possible to give a minimal list of items necessary; however, without any of these black boxes, a bank-building becomes unlikely.

Note that all of the above refers to the building type 'bank', not to the organization called 'bank'. For analytical reasons it is thus important to separate the building type from the organization or type as form from type as use. Certainly, it is possible to declare a container to be a 'bank' and operate a bank-business therein. Thus an actor-network 'bank' may consist of none of the above-mentioned black boxes. But, when we see a building which we attribute to the building type 'bank', we immediately attribute all its parts to be part of the building type. The formation of a whole is thus a process in which a process of attribution and of Gestalt perception or rather Gestalt construction takes place, rather than a process of seeing and naming its precise constituents. In a building type, the semiotic aspect is thus much more important than in most of the actor-networks known from ANT studies. A building classified as type works as technology because we recognize it as belonging to a type.

Historically, the codification of types and therefore the fact that buildings as types became technologies dates back to the spreading of a huge variety of building types in the early nineteenth century and the need to teach architects how to build these types in design books (Markus 1993). Thus historically, building types had to be made technological. The codification became increasingly standardized and technologized with building types as a very complex assemblage of technologies, as can be seen in the transformation from early design manuals such as Durand (1821) through the different editions of the best-selling 'Neufert' now in its 38th edition (Neufert 2005).[6] Not only architects but also historians and sociologists have interpreted buildings classified as building types as technologies, when they termed hospitals 'machines' (Foucault 1976) or endowed prisons with the faculty to 'fabricate virtue' (Evans 1982).[7] Building types became thus extremely complex actor-networks, assemblages of material-semiotic parts forming a whole. However, once buildings were made technologies, their status was contested, and they were unmade as technologies again, because they did not always perform what they were supposed to do.[8]

BUILDINGS ARE QUASI-TECHNOLOGIES

The historical construction and deconstruction of buildings as technologies are due to the fact that they are not always technologies. This is because buildings are composed of many parts, some of them technologies and some not. As building types, they are themselves black boxes, containing other black boxes. They are manifold on their 'inside'. With their inside, I would like to describe everything that is not normally accessible by the user, the black-boxed part. Just like CD players, and unlike for example stones, buildings have an inside that is manifold in the sense that it contains too many black boxes, independent of each other, to make it predictable.

But like stones, and unlike CD players, buildings are also manifold on their outside, the parts that are accessible to users. They lack a clearly specified interface, such as a play button, but contain a scattered array of interfaces that specify neither an order nor a hierarchy of use. The windowsill is here to prevent the rain from dripping on to the façade, but you can use it to sit on and drink a coffee in the sun or to dry your wet clothes.

This double manifold makes the use of buildings unpredictable. This is further aggravated by the fact that, once buildings are built, the creators of these networks, the architects, usually lose control to the users. 'Nobody is really in charge' (Star 1999: 382) and nobody can really be in charge, because the manifold interfaces simply allow too many starting points for different uses by different people at the same time.

The double manifold of buildings turns them into what I would like to call quasi-technologies. The term 'quasi-technology' derives from Michel Serres's concept of the 'quasi-object', a mixture between a subject and an object (Serres 1987: 352). Quasi-technologies, as I define them, are objects that are sometimes real technologies, functioning as black boxes, but at other times they lose this quality. They are turned from technologies, in the sense of black-boxed procedures, into 'mere' masses of materials. They become *materialized*, as I would like to call it. To materialize in this sense means that an object is freed from its actor-network and reduced to its material qualities.

'Quasi-technologies' are not objects that prefigure actions, but objects that are sometimes technologies and sometimes not, depending on who is using them and how. At some point in time and under certain circumstances, buildings as building types work like proper black-boxed actor-networks, where an actor (the architect) controls the network, but this can shift quickly and the building loses its properties of a black box. Quasi-technologies depend thus much less on the pre-programming of designers and inventors than other technologies do.

The concept of quasi-technologies allows us to look at objects not only from the perspective of how they are turned into technologies or black boxes, but how they are made *not to act* at the same time. The concept of quasi-technologies asks us to look at those procedures and circumstances that turn objects into technologies and those which turn them into mere material.

Other than what classical ANT would propose, such a turning of an object into a technology or merely material is not necessarily related to creating long networks involving many actants and a lot of work. It is sometimes a matter of a single sentence.[9]

ANT stresses that we are surrounded by mixtures of humans and non-humans and that a proper description of the world necessarily and always requires us to include and mix social and material aspects. In contrast, the notion of quasi-technologies requires that their descriptions vary according to their state. While some descriptions may be ANT descriptions, others may be much more physicalist or socialist. ANT descriptions rest on an algebra where the addition of further actants necessarily results in a strengthening of the network. With quasi-technologies such an algebra does not work. The outcome of a procedure cannot be calculated from the number of actants, because it relies very much on situational categorizations that can easily override existing networks. In the case of building types too, as I have shown, the number of actants that constitutes a type is unknown and situational.

WHY BUILDINGS ARE MUTABLE IMMOBILES: SINGULARITY AND LOCALITY

We may now ask why buildings are specifically likely to be quasi-technologies as opposed to other objects, since being a quasi-technology is not a matter of ontology. The answer I propose is found in two specific and interrelated features of their *use* – and not of their production. The use of buildings differs from that of most other objects in two respects: they occupy a fixed location and they are singulars, and for this reason I call them *mutable immobiles*. These two features turn buildings into the opposite of what Bruno Latour has called immutable mobiles (Latour 1987: 226–7). Immutable mobiles are objects that are stabilized as technologies to perform the same actions in different locations. They are technologies that operate independently of their context, because their creators (often scientists) can shield the objects from interfering user groups. Buildings as mutable immobiles do quite the opposite.

First, occupying a fixed location – being *immobile* – exposes them to many different user groups. Buildings cannot be shielded away like other objects into private homes, laboratories, courtrooms or museums, where the respective constituencies can control them. Even if access can be denied to certain user groups, the outside of buildings is open to interpretation and definition by anybody. Furthermore, since the location is fixed, buildings are inevitably bound to their local contexts. This is why buildings are the only type of object for which norms and laws exist as to how they should relate to neighbouring objects.

Second, as singulars, buildings cannot be standardized, but, like biological organisms, each one has its own form. The singularity of a building links its local stability and its openness to users and makes it changeable: parts that were once considered to be necessary for the whole to operate are exchanged, disposed of or simply ignored in interactions. Rather than being an immutable and stable 'technology across contexts, buildings are unstable' and *mutable*. Once a building is built, by being used in specific ways and by being locally stable and thus connecting to its changing environment, it inevitably acquires a biography that makes it distinct from all other buildings. This even applies to seemingly identical and standardized buildings, as is nicely shown in Philippe Boudon's pioneering study about the changes of Corbusier's houses in Pessac (Boudon 1969).

BUILDING CONVERSION: A TEST CASE

So far I have explained the concept of quasi-technologies as objects that are sometimes real technologies and sometimes not. Furthermore I have explained that type is a specific secondary level of classification that links uses to forms. In a third step I have explained that buildings are mutable immobiles, because they occupy a stable location and are singulars; this is what makes them more likely to be quasi-technologies.

Taken together, these issues amount to the widespread phenomenon of change of use of buildings – the change of building types. Not only are buildings routinely changed and renovated to add certain features,[10] but some of the changes lead to a change of building type. There can be no change of use if there has never been a predefined and inbuilt use. If buildings were not classified as types, the problem of change of use could never occur. However, change of use does not necessarily imply many changes of the building itself, since the category of a building type is, as explained above, a category that cannot be linked to any specific building part. But change of uses presupposes that the assumed technicality of buildings can be muted, thereby materializing the building. The change of a building type is thus a Gestalt switch.

Therefore change of use as a concept already presupposes the concept of quasi-technologies. Change of use shows that buildings are mutable immobiles. For a reconstruction of change of use we have to find out, first, how buildings are made to be building types and then how they are unmade and remade as other types. The empirical questions here are thus: What is it that changes the building type of a given building? How many little black boxes have to be changed to get another building type? Are there black boxes that prevent the change to a specific type? Can specific building types overrule the resistance of black boxes that define other types?

THE LAW OF BUILDINGS

For an empirical discussion of the problems of change of use, let us now turn to the law. The following discussion of law cases will provide one glimpse of the ambiguity, complexity and contradictions that building types pose for society. The field at stake here is cases where change of use leads to a conflict with zoning regulations. My aim is to show with these cases that the very fact of buildings being quasi-technologies leads to ever-ongoing attempts by the law to relate buildings and their uses. This is not to show the incapacity of courts to unambiguously define types, but to show that it is impossible to define the technicity of building types, because they are quasi-technologies. The descriptions of the twists and turns of the courts merely reflect this general ambiguity of building types. Analysing court decisions has the following advantage: the courts have to perform exactly those operations of deciding if and why a specific building belongs to a type, by relating forms to uses.

For the following analysis I use as empirical material several decisions regarding change of use made by the building appeal commission and the superior administrative court, both of the canton of Zürich, Switzerland.[11] Zoning cases are useful because they allow us to focus again on the specificity of buildings as objects – as mutable mobiles. Since buildings are locationally stable, building codes not only regulate security issues, as for other objects, but also regulate the use of buildings via zoning laws – a kind of law that exists only for buildings. Zoning laws tie uses to buildings by attaching uses to defined patches of land called 'zones'. Changing a building is thus likely to run into a conflict with zoning law if a change of a building is considered to be a change of use. Zoning laws define thus the technicality of buildings with respect to building types. They can keep the technicality by stating that a specific building cannot be changed, or they can turn buildings into quasi-technologies by allowing change. The cases also show that *minimal* changes of buildings lead to decisions that *recategorize the whole building* to a new building type.

Before I look at specific cases, let me analyse the definition of zones in the law. The written law in Switzerland is ambiguous because it does not separate materiality and use and therefore already assumes that buildings are quasi-technologies. The federal law says: 'zoning plans regulate the admissible *uses* of the land' (Schweizerische Eidgenossenschaft 1979, Art. 14, emphasis added). The regulation is defined as concerning the uses, not the buildings. One could infer that the law does not hold a technological view of buildings! However, as we will see over and over, this is only partially true. The confusion already starts at the level of written law when it defines specific zones: the building code of the canton of Zürich, in section 60, reserves a zone for 'public *buildings*' – and not public *uses* (Baudirektion des Kantons Zürich 2005, section 60, p. 15, emphasis added). Thus, for the law, buildings are real technologies by assuming that buildings and uses *merge*. Because there are no further definitions of 'public buildings' in the law, we can infer that the law knows what a public building is and that this 'being a public building' is a use of a certain surface area of land and a building that is recognizable as a public building at once.

Zoning law restricts change of use, because once a house has been built in a specific zone the owner is not allowed to use the house for another use. Since, in cases of change of use, building form and use are no longer co-extensive, it provides a test case to understand how the law deals with the quasi-technicality of buildings. To explore this in detail, I will first start with the question of stability and continuity of the structure of buildings and then move to the other side, changing uses as modifiers of buildings.

THE IDENTITY AND STABILITY OF BUILDINGS: HOW BUILDING PARTS BECOME REAL TECHNOLOGIES

In this first part I show by going through a succession of cases that the courts indeed hold a technological view of buildings, but this view refers not to whole buildings as building types, but to building parts. They thereby 'solve' the problem of quasi-technologies by moving it to a lower level. The stability of buildings is important in cases of change of use because of the so-called *Bestandsprivileg*, the privilege of continued existence. If a building existed before the zoning law was invented, or before the zoning law changed the zone in which the building is located, the building's use may not conform to the zone. In such a case, the privilege of continued existence permits keeping the building and its uses as before. However, once an owner changes her building, the privilege of continued existence requires her to prove that her building and its use are still 'the same' as before. The problem here is thus that the law first assumes that building types are not real technologies but mutable immobiles. The privilege of continued existence stabilizes the technicality of the building. But how can the law define that a building is still the same as before?

According to the law, a new building has to meet the following basic requirements to be considered the 'same' as the old. It has to be built on exactly the same location, and it has to be equivalent in volume and use – the defining criteria of mutable immobiles as I described them above. However, to keep the 'identity' of a building, as we will see, this is not enough. Three problems for identity emerge. First, the workings of time in the form of decay, repairs and replacements make the whole disintegrate. Second, if it is not the whole building, classified as a type, that grants the privilege of continued existence, then the question becomes how the law defines those parts that do so. Third, the question is whether identity means that a building really enforces its use or whether this is rather an unwillingness to accept other uses.

The first aspect, the problem of time, can be illustrated with a case where a claimant wanted to use a building outside a building zone for a car-repair shop. The court denied this use, because the building had first served as a carpenter's workshop, then during the 1960s as a warehouse and, finally, during the recession in the mid-1970s it remained empty, until it was used as a warehouse again. The court denied the privilege of continued existence on the grounds that it refers only to 'existing facts' (Baurekurskommission des Kanton Zürich 1983: 56). The court stated: 'If only a ruin is left, no rights follow from the privilege of continued existence' (Baurekurskommission des Kanton Zürich 1983: 56). For the law, a building is therefore unstable over time. As soon as it is not used for a certain period and the network of the building vanishes, this constitutes a break in its congruence between use and materiality. Under these circumstances, the building ceases to belong to *any* building type. In the view of the court, the old uses were thus tied to the material building. It assumed that, as a technology, the uses can only persist if they are backed by an intact material base and, once this material base vanishes, the uses vanish too.

This is further elaborated in another case, where somebody wanted to change a sauna and an electrical repair shop situated in a house in a residential zone into offices. The court denied a privilege of continued existence because the law only protects 'existing facts', defined as 'built parts of a certain size serving the prohibited use'. If, as in the case under consideration, 'all parts that served the forbidden use are replaced', the proprietor waives the privilege of continued existence (Baurekurskommission des Kanton Zürich 1990: 38). The court justified this claim because the privilege of continued existence does not protect 'the use as such, but the material assets that serve the use' (Baurekurskommission des Kanton Zürich 1990: 38). The court defines the classification 'office building' as an assemblage of parts that enable the use as office. Again, the building is understood as being a technology proper as a building type, an assemblage of various parts that stabilize a use. If *all* these parts are gone, then the classification is disassembled and has lost its technicality to stabilize the uses.

But not *every* material part of the building grants the privilege. Therefore, the question is: how can the law define the parts that do grant the privilege and what do they really do? As the court further elaborated in the same case, the privilege of continued existence rests on the idea that 'existing structures prohibit the use of the building according to the law'. In some cases, it reasoned, 'the whole building is constructed in such a way that no use as ordered is possible at all. But in most cases, it is only some rooms, whose structure negatively set a precedent for a use contrary to regulations' (Baurekurskommission des Kanton Zürich 1990: 38). It turns out that, for the court, only in a few cases does the *whole* building necessitate the classification as a given type. In most cases the defining power is delegated to specific rooms. These rooms do not so much define a type as resist being integrated into a new type.

The privilege of continued existence is thus deeply technological. In the view of the court, uses can be easily changed, but not the buildings. The court starts with the assumption that building types are technologies (as in the first example) and, when it turns out that building types are not coherent, the capacity to define a use is shifted to specific rooms, the next lower level. These parts at the lower level can be identified, because they would *force* the owner to break the law. The court defines the technological parts as black boxes that cannot be opened, not even by the force of law. The privilege of continued existence is therefore not so much a privilege, but a burden: it proves that building *parts* can be, at least for the court, *real* technologies.

However, this leads to the third question, namely whether building parts are positive technologies in the sense that they enforce certain uses, or whether they are negative technologies in the sense that they simply prevent specific other uses. The case mentioned in the introduction concerning the transformation of industrial spaces into offices elucidates this issue. As mentioned, the court ruled that, even though a conversion to offices would be possible in principle, it would force the owner into 'considerable construction expenditures' that would not even lead 'to satisfying solutions' (Verwaltungs-gericht des Kantons Zürich 1992b: 21). Conversion would only be possible with 'voluntary cooperation by the clients in the context of integrated concepts' (Verwaltungsgericht des Kantons Zürich 1992b: 21).

As becomes clear, the technology here is negative: it is not a technology that enforces a use, but a resistance to acquire specific new uses. Furthermore the resistance is by no means total. Conversion in principle is always possible. But the conversion would require a new network to be put in place, consisting of cooperative owners, a lot of money and an 'integrated concept'. The integrated concept, I infer, would be a replacement of the resistance by the technological parts with a new building type, that is, a new 'whole' that turns the building from an industrial building into flats. The fact that the building parts are technological is a negative attribute: it cannot take on new uses. We deal here with what we could call a *white box*, which contains many actants that neither belong to a specific network nor link to another network. But the building is not totally materialized or muted either. The white box can be muted in principle, but only at a very high cost: namely, the insertion of a new network. But, for the law, buildings are not always technological, but more often they are simply material and without defence against foreign and pro-hibited uses. To analyse this more closely, we turn to those cases where change of use itself is under consideration.

PROBLEMS OF IDENTIFYING AND SEPARATING USES: TECHNOLOGY VERSUS OBSERVATION

As we have seen, the legal definition of zones implies a notion of building type that conflates use and material building. But in cases of change of use, use and material building fall apart. For the law the problem emerges whether the building or the use defines the type, since change of use can occur without materially changing the building, and materially changing a building does not constitute necessarily a change of use. The two mirror problems are thus: When and how do changing *uses* redefine a building type? And when and how do *material* changes redefine a building type? I look at the two problems in turn, each time looking at a specific case that exemplifies the problem.

EPHEMERAL USES AND MATERIALIZED BUILDINGS

Starting with the uses, the law is confronted by the fact that uses are ephemeral. A use is an interaction between humans and a building. It takes place once or twice or ten times, and neither its time-scale nor its scope is known beforehand. Change of use, then, must be defined as a *quantification* of repeated (mis-)uses that ultimately sums up to a reclassification of the building type. The idea here is that uses can or cannot conform to the building and, if they repeatedly do not conform, then the building itself becomes transformed. So when does repeated (mis-)use indicate a change of use?

The problem of repetition is exemplified in a case where a community has asked the owners of a nursery to file a building proposal for the use of the nursery as a party space, which the owner rents out every second week or so. The appeal court states that change of use according to federal law requires a building proposal only if it is likely to have 'localized impacts on the order of uses' (Verwaltungsgericht des Kantons Zürich 1992a: 6). After an inspection, the court found out that for the following six months no more than ten events would take place, each with 30–60 participants. The court reasoned that the above-mentioned 'impacts' would include the additional traffic by the guests, unspecified 'emissions', probably noise, and the additional load of the sewage system. Then it stated that, in case of doubt, the community should ask for a permit, only to counter this statement by adding that this should not lead to 'a farmer or a gardener having to apply for permission, if he wants to use his building in the rural zone from time to time for a convivial evening with family members, employees or a third party' (Verwaltungsgericht des Kantons Zürich 1992a: 7). The court added that in the present case 'the use under consideration in no way displaces the authorized nursery' (Verwaltungsgericht des Kantons Zürich 1992a: 7). The court finally left it open as to whether a permit is required.

As we see, it is impossible for the court to define the exact number of parties required for a change of use. For the court, the party is not technological and thus difficult to grasp, especially since it does not mute the other uses. The building is muted only for short instances of time, which for the court are just repeated events and do not add up to permanence. The ephemerality of use is key to understanding that uses do not redefine building types. As long as the normal uses and the material building remain intact, then repeated other uses do not constitute a change in building type. The view of the court here is that a type is material only (and not technological) and that uses are ephemeral for definitions of type.

INDIFFERENT BUILDING PARTS AS MULTIPLE TECHNOLOGIES

In the last case material and interactional changes occur together. Here the question is how the interactional changes are linked to the material ones. In this case, a farmer used a wooden barn 25 kilometres away from his farm as a refuge when taking care of his sheep. Because the farmer needed protection from the cold, wind and rain, he replaced a weathered façade. He also filled three openings used for throwing hay in and out of the barn with windows and installed a wood stove. The community asked in respect of the replacement of the windows and the stove, claiming he had thereby turned the barn into a summer cottage. The court denied the request by the community, with the following reasons. First, it stated that the outer appearance of the building remained the same and that the owner showed no intention to change the use of the building, nor would the changes imply such uses. The court held that staying overnight seemed suitable only 'in good weather conditions in summer' (Verwaltungsgericht des Kantons Zürich 1985: 9). It described the situation as follows: 'The thin timber wall fails to protect from the cold; many chinks do not even protect from draught' (Verwaltungsgericht des Kantons Zürich 1985: 8). Furthermore, since the 'sanitary conditions needed for habitation are lacking entirely' it would be an unlikely summer cottage even for somebody with 'the most primitive demands' (Verwaltungsgericht des Kantons Zürich 1985: 9). On the other hand, the court wrote that the 'cattle need a barn to spend the time until the onset of winter in these rough regions' and that 'minimal comfort' for the farmer should exist 'to warm up during breaks in the cold season' (Verwaltungsgericht des Kantons Zürich 1985: 9). Furthermore, 'a sudden change in weather or an accident might force the owner to stay overnight', making the changes acceptable (Verwaltungsgericht des Kantons Zürich 1985: 10). Finally, the court added, if the owner would not adhere to these proposed uses, 'the community administration would quickly notice it' (Verwaltungsgericht des Kantons Zürich 1985: 12).

More than in the case before, we can see here the close connections between building parts and uses, with *both* being unstable. The building is unsuited to use as a summer cottage, and at the same time it has to allow the farmer to stay overnight, just in those adverse circumstances that the building is not made for. The reasoning rests on the assumption that using a barn as a summer cottage and using a barn for emergencies in adverse circumstances are not only socially different uses but also black-boxed in technology. The two uses represent two different building types: a barn and a summer cottage. Parts of the building, the timber wall and the chinks, are technologies to prove that it is unsuitable for holidays; other parts, such as the stove, are technologies to prove that it is suitable for staying overnight. But neither of them is a proper technology to *discern* the two uses and define a building type. The chinks do not prohibit the use as a summer cottage. How could they, if they even allow the farmer to stay overnight in winter? Conversely, the stove cannot be prevented from cooking a meal on mild summer nights either. Even if the changes of parts of the buildings can be observed, and even if these changes are considered to be technologies, change of use cannot be read from that. The problem here is that all these building parts are technological, but they are technological for *both* of the uses. They can be inserted into two different networks at the same time without muting the other one. Creating either of these types does not require other actants than those that are already there. The only solution, on which the court relies, is to constantly observe the actual uses. Real and actual use is decisive, but cannot be technologized; thus it has to be monitored. The cottage proves to be a typical case of a quasi-technology: in some instances it is a technology, in others it is not, and all possibilities to stabilize the separation between the two fail. Furthermore, the court assumes that a supervision of the uses by the village is easily accomplished. However, reflecting on the case on parties I assume that it would have become an even more difficult task to prove that the uses – cooking a meal, for example – turn the barn into a cottage. Luckily, the records indicate that none of the parties involved attempted such a proof.

CONCLUSION

The goal of this chapter was to show the specificity of buildings as objects. The specificity of buildings relates to their secondary classification as building types. As building types, buildings act as technologies to locate specific interactions. However, as I have shown, buildings are not real technologies, but quasi-technologies – they operate as technologies only under very specific instances and for specific users. Furthermore, as mutable mobiles, they are located and singulars. They are impossible to standardize and they are likely to be changed and used differently by different constituencies. I have also shown how this status as mutable immobiles creates a problem for the law. Zoning law is already on the level of the written law undecided about whether it defines zones as relating to uses or to building types, thereby mingling buildings and uses. The problem continues on the level of individual cases. In cases of continued existence, the law tends to deny the technicality of whole buildings but grants it to building parts or rooms. In cases of change of use the courts decide that ongoing differing uses do not constitute a change of use. On the other hand, they decide that, even though minor material changes of the buildings may constitute change of use, the material changes are not proof enough to stabilize a change of type. These varying decisions of the courts do not signify that the court is absent-minded. It merely hints at the fact that building types are not proper technologies and can never properly stabilize uses. What zoning law and the courts accomplish is a short fixation in the ongoing puzzle that the classification of buildings as types poses. By defining the relationship between buildings and use for a given moment in time, they allow or forbid the owners upcoming specific changes and uses. They do so with the power of the law, because buildings cannot accomplish this feat on their own.

This conclusion also leads us back to how to understand buildings with regard to the general notion of urban assemblages. We can now see more clearly the specific location of buildings in a more general theory of the urban. The location of buildings, as indicated with their designation as mutable immobiles, is in between a classical sociological conception of 'urbanism as a way of life' (Wirth 1938) and the technological notion of the urban as backed by infrastructure (Graham and Marvin 2001; Star 1999). Whereas the former understands the urban via interactions only and misses the technical part, the latter conception refers to an understanding of cities as composed of mostly invisible real technologies that either work or don't. The focus on mutable immobiles not only highlights buildings as primary objects of cities, but also enables us to see why they create a specific problem for society: they are immobile and used and therefore pose a constant problem of defining and categorizing them. The very idea of a city consists in an assemblage of mutable immobiles. Our very joy in rambling through a city derives from this fact: we orient ourselves with types and we enjoy being surprised by the failing of our own classification of types.

NOTES

1 For an exemplary proof to the contrary, see the pioneering study of lofts by Sharon Zukin (1982).
2 For some studies in the tradition of STS relating to particular technological aspects and thus different parts of buildings see Slaton (2001) and Thompson (2002).
3 Although ANT and Social Systems Theory seem to cover opposing terrain, their concept of technology is remarkably similar because it refers to stabilizing processes and not to a materialist definition.
4 As a kind of classification it is maybe comparable to the classification of gender among humans, a similar classification on a secondary level that deeply affects the definition of the whole. For other objects, such secondary classifications do not exist, because classifications usually refer to defined features. A car may have a four-wheel drive or it may be a diesel or a cabriolet, but none of these classifications affects the car as such. The classification of a building as a bank, however, is not limited to any specific part.
5 Quatremère de Quincy, who provides us with the most well-known definition of type in architectural theory, put it like this: 'All is precise and given in the model; all is more or less vague in the type' (Quincy 1977: 148). Because of the vagueness of the idea of type, no boundary work with regard to types exists. There are no architectural texts where an author could state that a given building is in fact *not* a bank. We can give a positive description of the general features of specific building types (Pevsner 1976), but no negative criteria for the exclusion of a building from a type. The indexical empirical assertion ('This is a bank!') always overrides analytical definitions.

6   Neufert has sold more than 1 million copies and is estimated to be the world's best-selling architecture book. On the history and influence of Neufert on standardizing building types see Prigge (1999).
7   For an overview of current discussions and analyses of the problem of type see Franck and Schneekloth (1994).
8   For an historical account of how buildings were made and unmade as technologies see Guggenheim (forthcoming) and Vanderburgh and Ellis (2001).
9   ANT puts heavy emphasis on the work used to create networks and make them work. However, this is mostly due to its preoccupation with controversies in scientific contexts where documentation, experiment and proof are the prerequisites for winning a controversy.
10   For other works on changing buildings in the context of STS see Gieryn (2002) and Hommels (2005).
11   The following material is taken from the publications of the courts. These publications contain only the ruling of the court and no pleadings or texts of lawsuits. Switzerland has a national space-planning law which was instituted in 1979 (Schweizerische Eidgenossenschaft 1979). According to this law each canton is obliged to regulate basic zoning, whereas each community is obliged to divide the land into zones, which have to conform to the regulations of the cantons. Conflicts are first dealt with by the local administration; from there they can go to the building appeal commission of the canton and next to the superior administrative court of the canton.

REFERENCES

Baudirektion des Kantons Zürich (2005) Planungs-und Baugesetz des Kantons Zürich (PBG). Entwurf für die Vernehmlassung 19. August bis 19. Dezember. Kanton Zürich.
Baurekurskommission des Kanton Zürich (1983) 'BRKE I, Nr. 358/1982', *Baurechtsentscheide Kanton Zürich*, 2/3(33): 55–6.
Baurekurskommission des Kanton Zürich (1990) 'BRKE I, Nr. 836, 837/1989', *Baurechtsentscheide Kanton Zürich*, 1(9): 36–40.
Boudon, P. (1969) *Pessac de Le Corbusier*, Paris: Dunod.
Durand, J.-N.-L. (1821) *Précis des leçons d'architecture: Données a l'école royale polytechnique*, Paris.
Evans, R. (1982) *The Fabrication of Virtue: English Prison Architecture, 1750–1840*, Cambridge: Cambridge University Press.
Foucault, M. (1976) *Les machines à guérir: Aux origines de l'hôpital moderne*, Paris: Institut de l'Environnement.
Foucault, M. (1979) *Discipline and Punish: The Birth of the Prison*, Harmondsworth: Penguin.
Franck, K.A. and Schneekloth, L.H. (eds) (1994) *Ordering Space: Types in Architecture and Design*, New York: Van Nostrand Reinhold.
Gieryn, T. (2002) 'What Buildings Do', *Theory and Society*, 31: 35–74.

Graham, S. and Marvin, S. (2001) *Splintering Urbanism: Networked Infrastructures, Technological Mobilities and the Urban Condition*, London: Routledge.

Guggenheim, M. (forthcoming) '(Un-)building Social Systems: The Concrete Foundations of Functional Differentiation', in I. Farias and J. Ossandon (eds), *Observando Systemas*, Vol. 2, Mexico City: Universidad Iberoamericana.

Hommels, A. (2005) *Unbuilding Cities: Obduracy in Urban Socio-Technical Change*, Cambridge, MA: MIT Press.

Latour, B. (1987) *Science in Action: How to Follow Scientists and Engineers through Society*, Cambridge, MA: Harvard University Press.

Latour, B. (1991) 'Technology Is Society Made Durable', in J. Law (ed.), *A Sociology of Monsters: Essays on Power, Technology and Domination*, London: Routledge, pp. 103–31.

Latour, B. (1992) 'Where Are the Missing Masses? Sociology of a Few Mundane Artefacts', in W. Bijker and J. Law (eds), *Shaping Technology – Building Society: Studies in Sociotechnical Change*, Cambridge, MA: MIT Press, pp. 225–59.

Luhmann, N. (2000) *Organisation und Entscheidung*, Opladen: Westdeutscher Verlag.

Markus, T.A. (1993) *Buildings and Power: Freedom and Control in the Origin of Modern Building Types*, London: Routledge.

Neufert, E. (2005) *Bauentwurfslehre: Grundlagen, Normen, Vorschriften über Anlage, Bau, Gestaltung, Raumbedarf, Raumbeziehungen, Masse für Gebäude, Räume, Einrichtungen, Geräte mit dem Menschen als Mass und Ziel. Handbuch für den Baufachmann, Bauherrn, Lehrenden und Lernenden*, 38th edition, Braunschweig: Vieweg.

Pevsner, N. (1976) *A History of Building Types*, Princeton, NJ: Princeton University Press.

Prigge, W. (ed.) (1999) *Ernst Neufert: Normierte Baukultur im 20. Jahrhundert*, Frankfurt am Main: Campus Verlag.

Quincy, Q. de (1977) 'Type', *Oppositions*, 8: 147–50.

Schweizerische Eidgenossenschaft (1979) Bundesgesetz vom 22. Juni 1979 über die Raumplanung (Raumplanungsgesetz, RPG), Schweizerische Eidgenossenschaft, Online, available at: http://www.admin.ch/ch/d/sr/c700.html.

Serres, M. (1987) *Der Parasit*, Frankfurt am Main: Suhrkamp.

Slaton, A.E. (2001) *Reinforced Concrete and the Modernization of American Building, 1900–1930*, Baltimore, MD: Johns Hopkins University Press.

Smith, N. (1996) *The New Urban Frontier: Gentrification and the Revanchist City*, London: Routledge.

Star, S.L. (1999) 'The Ethnography of Infrastructure', *American Behavioral Scientist*, 43(3): 377–91.

Thompson, E. (2002) *The Soundscape of Modernity: Architectural Acoustics and the Culture of Listening in America, 1900–1933*, Cambridge, MA: MIT Press.

Vanderburgh, D.J.T. and Ellis, W.R. (2001) 'A Dialectics of Determination: Social Truth-Claims in Architectural Writing, 1970–95', in A. Piotrowski and J.W. Robinson (eds), *The Discipline of Architecture*, Minneapolis: University of Minnesota Press, pp. 103–26.

Verwaltungsgericht des Kantons Zürich (1985) 'VB 128/1984', *Baurechtsentscheide Kanton Zürich*, 1(2): 8–12.

Verwaltungsgericht des Kantons Zürich (1992a) 'VB 91/0156', *Baurechtsentscheide Kanton Zürich*, 1(1): 5–8.

Verwaltungsgericht des Kantons Zürich (1992b) 'VB 91/0004', *Baurechtsentscheide Kanton Zürich*, 4(30): 20–2.

Wirth, L. (1938) 'Urbanism as a Way of Life', *American Journal of Sociology*, 44(1): 1–24.

Zukin, S. (1982) *Loft Living: Culture and Capital in Urban Change*, Baltimore, MD: Johns Hopkins University Press.

8 Conviction and commotion

On soundspheres, technopolitics and urban spaces

Israel Rodríguez Giralt, Daniel López Gómez
and Noel García López

> Much more than colours and shapes, sounds and their arrangement shape
> societies.
>
> Jacques Attali (1977: 15)

NOISE IN THE DARK: SOUND IN SOCIAL SCIENCES

Barcelona, 23 July 2007. At around 10 p.m., the residents of different areas of the city began a loud and spontaneous protest. A day earlier, the fall of a high-tension cable left around 350,000 people without an electric supply, causing considerable chaos in the city's traffic, commercial and hospital activity. On the following day, almost 110,000 residents still did not have an electric supply. As the evening fell, when the blackout clearly darkened the city, many of them decided to take to the streets because of the lack of solutions provided. They called for a quick response to the situation. This is how one of the largest and loudest *caceroladas*[1] (a word that is apter than ever) that has taken place in the city began. It was thus, spontaneously, that many residents affected by the blackout, visibly angered, started to make noise with saucepans and whistles, and even bangers, as a form of protest. In the street, in front of the police, at home, from balconies, indignation was expressed to the metre of a metallic sound. Little by little, the noise took over the darkness. We want light! Out, out! This is a robbery! This is a circus! It's a disgrace! They're laughing at us! . . . shouts and other chants that joined the rhythm produced by the metallic banging against the saucepans. Invisible but noisy, increasingly numerous, the protest spread around many zones of the city. A mosaic of spontaneous and interconnected protests filled Barcelona with a single sonority.

This event reshaped the city by taking advantage of the exceptional primacy of the ear over the eye, a change that helps us to bring the daily sonorous practices of the city to the foreground as political practices through which collectivities and spaces are set up. Nevertheless, a sonic practice like that is traditionally treated as noise, according to physical and psychological measures, as a form of expression with meaning, normally a message. Thus, what is usually neglected is the sound as a social practice. Since there is no clear message or rhythm (music), the *cacerolada* appears as the collateral effect of a protest (just its sound, the physical accident of its existence). In this case, only its meaning but not its sonority would be treated as important. In effect, as noted by Augoyard, Grosjean and Thibaud (2001), usually the urban researcher hears badly in two ways. On the one hand, he suffers from bad hearing owing to an excess of noise. But he is also a victim of hearing badly owing to not listening enough. For him, what is sound related occurs only as excess or silence. As a consequence of this bad hearing or hearing badly, the study of sound in urban space has basically been limited to noise, as environmental pollution,[2] or to music, as significant noise, and, moreover, with the setting of a Manichean and ideological scale between the two poles (Shepherd and Wicke 1997).

It is surely for this reason that sound has traditionally been considered something potentially harmful for the city, almost as if it were an obstacle to civility. The cleansing campaigns against acoustic pollution are a paradigmatic example of what we are commenting on. The civilized city is attained in a silenced city, where the organization of life in common is executed through an oculocentric architectural design (see Mumford 1937; Wirth 1938). The traffic of people, goods and vehicles must be primarily regulated on the basis of a design of the urban framework that is geometric and clear, and through unequivocal visual signals. On the other hand, to live in the city in a civil manner, the self-containment of its sonorousness is required. This being the case, both the bad hearing or hearing badly of social scientists and the cultural and historic identification of sound in public spaces as a sign of disorder and incivility have resulted in sound being considered a matter of secondary importance in urban studies – as an anomaly, as a dimension that is external to urban reality or as something improper against which the only available option is to defend oneself (Bijsterveld 2001; Iges 1998). For this reason, when we study urban space, we always tend to silence the conflict between the physical space of the city and the space that is used. We think of the block and the stroll, without considering their sonorousness. We systematically overlook the implications that considering sound as a constitutive element of urban space, and not merely an insubstantial wrapping, would have. Consequently, as claimed by Jean-François Augoyard (2004), our social sciences spend most of their time looking – watching and reading. It seems as if they have not yet learned, as Attali (1977) adds, that, more than seeing it, one hears the world. More than reading it, you listen to it.

As a result of this primacy of what is visual, we face an obvious difficulty when treating sound as a constitutive element of urban space. Although auditive perception has been an important object of research for psychology, for sociology it has been its tool for analysis, its instrumental method for interpreting social behaviour, in only a very few cases. In this sense, the sensation and the perception of shapes – not only in art – have not progressed from being mere 'objects' for sociology or social anthropology; they have hardly been viewed as immanent modes of the social bond (Bull 2000; Eyerman and Jamison 1998). Beyond considering sonorousness in terms of its excess or absence, trivial listening and the perception of the sounds of everyday life have not been worthy of careful analysis, and even less a general theory of the sonorous experience. This trend has only started to shift somewhat in the last few decades.[3] Thanks to the work of different authors, it has been possible to highlight that the analysis of the urban sound space cannot be restricted to acoustic evaluation in a strict sense, nor merely to combating noise (see Atkinson 2007). On the contrary, it has been clearly shown that it is increasingly more necessary to be able to interpret the sound space of a place in a complex and meaningful way, to be able to relate the sounds and spaces of a place with its communicative practices, with its social relations and with the manner of conceiving urban spaces and their cultural, political and social dynamics (Augoyard and Torgue 1995).

Thus, following Serres (1969, 1996), we could well say that in the *cacerolada* the sound turns into the operator that sets up a space and a collective through its continuous displacement and spreading and transformation. Like a shock, a metallic sound alters the street's attention. Suddenly, the energetic rhythm that a load of old saucepans creates, the beating with spoons, ladles and skimmers on all sorts of pans and aluminium lids, invades everything. Little by little, the din enters every home. People join the protest: from the balcony, from an open-air area outside a bar, from the pavement, on the roadside. The cars also stop and honk their horns. Employees from a grocery store come out, striking metal objects against façades. Gradually, more neighbours join in: in the attics, basements, inner courtyards, hitting with strength, whistling and shouting in unison, filling all the space in a deafening manner. In a few seconds, the streets, houses, shops and squares, the whole neighbourhood, have turned into a deafening racket, into an improvised chorus of shouting in unison, into a sonorous body. Through a simple metal timbre, a harsh and accelerated rhythm, a *soundscape* is created, an atmosphere that establishes an acoustic bond between the neighbours (Schafer 1977; Smith 1994, 2000). Those affected thus find a way of recognizing each other, *making themselves heard*, displaying their discontent and protecting their interests.

So, even though the power of this apparently intangible domain has generally been under-examined in urban studies, it is our opinion that sound is a central element in the definition and understanding of urban space. It is an important sensory departure point that provides a means of exploring the more ephemeral and shifting elements of urban life. More specifically, we are going to show how the urban space is made up of different competing sonorous practices that must be addressed as technopolitics owing to their ability to give rise to different spatialities and collectivities. To better develop this argument, we will structure the following text in three sections. Drawing on sound and urban studies, we will firstly focus on the argument that the production of a specific *soundsphere* is not solely a 'formal' question, but rather that it entails the emergence of different spatialities and forms of sociality. In this regard, secondly, we will employ Actor-Network Theory and two concepts drawn from Sloterdijk (2003), practices of commotion and conviction, to discuss the political character of the use of sound in a small dispute concerning the use of *soundsystems* and *dance music* in a demonstration. Finally, we will see how the tension between commotion and conviction-based sonorous technopolitics actually expresses a redefinition of the very urban space in which they are expressed.

CITY SOUNDSPHERES

That ours is a culture that is dominated by what is visual is a platitude that, although it is easily assumable, avoids the important fact that we rarely perceive, understand or feel an image without sound. That is why we agree with Michael Bull and Les Back (2003) when they state that sound is tremendously important to (re-)thinking sense, nature and the meaning of our social experience; to thinking about our relationship with the community and our relationship with power; and to revisiting our relational experiences, the way in which we relate to others, with ourselves and with the spaces and places that we inhabit. As we are reminded by Attali (1977), sound is the audible band of the manufacturing of society, the sonorous vibration of the production and transformation of social life.

   But it is probably in urban space that one can see more clearly that this auditive experience plays a fundamental role. In effect, urban space has a sound; sound is a permanent protagonist in our relationship with it, an essential element in the narrative and understanding of its space and inhabitants. Noise, just like music, forms part of the *sound of the city's everyday life*, of its very expression; it is a characteristic feature of the ways of residing that are a place's own, and it is necessarily integrated into urban lifestyles. Each flat makes up a sonorous space with elements and characteristics that are specific and recognizable: the specific sound of the alarm clock or of the toaster; the creaking of a door that is waiting to be lubricated; the voices of the people who dwell in it, their accents and exclamations, as well as the rhythm of their steps; the way in which the bell is rung; the preferred radio station; or the unique crooning in the shower. However, this immediate sonority, this immediate link, inevitably coexists with the noise of neighbours and the street; with the drains from the flat above and the arguments or nighttime parties of the family that has just moved in, with the horn of the taxi passing by, the greetings that cross the street, the noise of engines or the unexpected demonstration. Together, the sounds shape that distinctive soundsphere[4] of the urban, the one that we do not even listen to but, rather, find ourselves shrouded in. Thus, the murmur becomes the very psychoacoustic link of urban life, the element that receives us and introduces us to urban evolution. We are part of it, and at the same time we constantly shape it.[5]

In this context, we should not be surprised that the sonorous experience also plays a tremendously relevant role in the reshaping of urban limits and collectives. In the midst of urban effervescence, the production of one's own space, of an identity, is intimately related to the production of a specific acoustic link. Acting upon the urban murmur, directing it or even reducing it, in sum creating a specific *refrain* (Deleuze and Guattari, 1980), turns into something fundamental and indispensable in this constant dispute to be heard that takes place in urban life. Thus, in the urban space, the production of a specific soundsphere is not merely a 'formal' question, but rather it entails the emergence of different spatialities and forms of sociality. In this sense, the *cacerolada* is a perfect example for understanding how certain sonorous practices manage to enact and differentiate spatialities and collectivities. Through the racket, we see how a momentary community (a new soundsphere) is established, a space of shared listening, how the murmur is heightened until it sets itself up as an unavoidable centre of attention that brings together a heterogeneous combination of people into the same demonstration and shared complaint.

It is not an isolated case. As highlighted by several authors, we find many other ways of operating sonorously upon the murmur of what is urban and of setting up differentiated spaces and collectivities.[6] Walking down the street or walking around the city while listening to music on an MP3 player is another example of a soundsphere, another way of re-creating one's own space, of experiencing urban space in an intimate and personal way (Bull 2000).[7] The regulations aimed at managing and controlling noise in the city constitute another instance of these operations intervening upon the urban murmur that we are referring to. Likewise, the presence of sirens and alarms in our daily life also exemplifies this importance of what is sonorous in what is urban, the social relevance that the production of specific soundspheres to establish and limit spaces and collectives possesses (Garcia 2005). At any hour, sound and sonorous practices reveal themselves to us as a valuable means for ordering, attracting, advertising, complaining, limiting, affecting, silencing or producing a breaking point within urban life, which is already booming on its own.

THE STRUGGLE TO BE HEARD: TECHNOPOLITICS OF CONVICTION AND COMMOTION

There are many ways to compose a soundsphere. However, as we will explain below, these sonorous operations are not merely aesthetic operations; they are also political, because each of them displays a concrete spatiality and collectivity. Within them, we see the composition of a habitat at the same time as that of an inhabitant. Establishing which sounds are its own and which are from elsewhere, which sounds are significant and which ones are mere noise, what it is important to hear and what is best silenced, becomes, as we will see, not just a technical and aesthetic matter, but rather a political issue of importance that has direct implications for the way in which we define the city.

There are numerous examples of this throughout history. When the Roman emperor Nero came across the hydraulic organ invented by Ctesibius of Alexandria on one of his trips to Naples, he imagined at first that, by just playing it, he would convert all his enemies to his cause. Through this new sonorous technology, Nero could turn them into soldiers who would obey his orders at the same time as his empire expanded at the speed of sound, something that, as Virilio (1980) explains, would not be an exaggeration at all. The very development of symphonic music shows how the orchestra conductor has reached the privileged rank of a sole leader, when he controls not only his company but also the body of his listeners, whom he immobilizes in their seats and strives to rid of any ability to generate noise. What are the clapping, whispers and coughing that can be heard at the end of a piece, if not the recovery of this ability by the body? Thus, a power without the means for making itself heard is inconceivable. Transformations of power have always been accompanied by transformations in sound technologies (Attali 1977). In fact, the Greeks were already conscious of the power of the word, first as a voice and then as an idea. For this reason, their organization was already a political technology, a way of governing the *polis*. Participative democracy was founded on a space that was flat and polyphonic, the *agora*, where citizens moved erratically from one group to another discussing different themes; administrative democracy, on the other hand, was exercised in the theatre, a space that guaranteed governability, as it ensured that while someone talked the others were facing him, sitting and listening.[8]

Nonetheless, a dispute that happened as a result of the May Day demonstrations of 2005 will help us to delve deeper into the comprehension of this productive and political character of sound. For the last few years, a parallel organization to the one organized by the large trade unions has transformed the nature of the first of May in Barcelona. In the first place, it has entailed the emergence of a new political tonality. It is not just a matter of a new claim by workers, but rather it is a medium that seeks to bring together local and heterogeneous struggles (students, precarious workers, immigrants, etc.) to widen, open and extend the mobilization to different spheres and attain a global dimension. On the other hand, this political shift has also been accompanied by a change, which may well be more important, in the *sonorous* tonality of the event. The demonstrators no longer have only their throats with which to shout slogans, or loudspeakers to extend and direct them; now we find other sonorous landscapes: protest *batucadas* (a form stemming from Brazilian percussion), bands playing live, and mobile soundsystems that seek to increase the repercussion and strength of the demonstration through the pleasure of the collective party.

These changes gave rise to an important dispute at the time. The cause of the disagreement, and this is not by chance, was precisely the change in the sonorous tonality. After the celebration of the May Day event in 2005, on the www.indymedia.org website[9] a number of messages appeared that complained about the music at the demonstration because it made it impossible to hear the addresses and slogans. The discussion derived from an opposition between two ways of demonstrating: on the one hand, the demonstration-with-a-message, in which demonstrators, harangued by the organizers' loudspeakers, shouted slogans, chanted, and listened to manifestos; and, on the other hand, a demonstration with lots of music and partying, in which people joined the mobilization, but where it seemed impossible to put on any unitary organized action. Some said:

> The May Day was a failure. It was like a disco. Those of us who wanted to fight for the freedom of the CNT's[10] prisoners were either denied the right to speak or the pill-poppers of the May Day had the music on at full blast.
>
> (Puta Bofia, 1 May 2005, 5.16)

> The people who were in Plaza Universitat did not decide anything, simply because, when it was said over the PA system that there were people arrested, it had already been decided that the demo would continue moving ahead without responding in any particular way, and moreover the message was unintelligible for those of us who were behind the lorry with the techno music because they didn't turn down the volume when this was being said.
>
> (Ei, això de les 10.000 . . ., 1 May 2005, 6.35)

Others, on the other hand, supported the view that the use of sonorous technologies such as the soundsystems with music on at full blast implied a different form of political action that was equally respectable:

> I think that radicality can be expressed in many ways, and the May Day's gamble on creating a truly open assembly, of breaking away from the forms of representation that are group- and identity-based, of experimenting with other graphic codes and other languages, is an interesting challenge – the attempt to avoid that people who possibly would never go to a demo that is 'overrepresented' by codes that are sometimes self-referential and internal for people who already mobilize can be dismissed through a simple charge of tele-tubbyism[11] . . .
>
> (Una precaria, 1 May 2005, 4.49)

In one case, there is a reflexive action, with a speech and clear objective; in the other, there is an emotive action that seeks to widen the mobilization. Evidently, both images are used rhetorically as a way to make stances attain visibility in a context of highly polarized discussion. Nonetheless, they make it possible to focus on what, being that which is most evident, paradoxically tends to be less noticed: the sound itself. So it is that the sound is not just the medium for the expression of meanings, but also, and very especially, a way of producing affections and, hence, of shaping subjects, both collective and individual. Thus it is that we already find something in the Greek *polis* that can be heard in the May Day argument. The issue is that practising politics necessarily entails building up sonorous technologies. For this reason, in practice, what gives rise to anger in the May Day demonstration is the technical and pragmatic dimension of sound, because it is the means to build up a collectivity as well as specific spatiality.

However, the struggle to be heard on May Day 2005 would normally be addressed as a technical problem, but not a political one. There is nothing wrong with the soundsystem in itself. The problem is how it was used and which goals were pursued through its use. This would seem to be the real political problem. Our point is quite different. Regardless of its goals and intentions, the music of the soundsystem was a political practice, because politics is not just the goals that we pursue, and nor is technology just the means we use to accomplish them. In this regard, Actor-Network Theory could be helpful in understanding the May Day dispute without splitting sound, technology and politics into different realities. According to Latour (2002) technology must first be understood as an adjective rather than as a noun. There is no ontological region that we may classify as technology which is external to others, as is the case for ones involving science, art, religion and so on. Technology is not an object, but rather a way of relating to each other, which Latour classifies as a *fold*. 'Technology is the art of the curvature' (Latour 2002: 251). In fact, if the soundsystem makes it possible to mix and play music at a great volume in a demonstration, this is because, within it, a number of unforeseen paths are invented allowing different elements to be articulated: demonstrators, passers-by, sound engineers, cables, loudspeakers, computers, synthesizers, records, ways of dressing and so on. Thus, the *mobile soundsystem playing music* is the very fold that makes them relate to each other and defines them. Moreover, articulating these elements implies setting up the limits of the demonstration, what is included and excluded. What is at stake is the spatiality of the demonstration (Mol and Law 1994): the perimeter of the demonstration, established by a planned tour through the main three streets of the centre of Barcelona and the police cordon, is blurred and redefined because of the music spread out to the alleys of the old city. The boundaries of the demonstration turn fluid and quite uncontrollable.

In fact, this is the idea that is drawn in the sense that the ancient Greeks gave to technology, *metis*. What allows us to reach goals is not a straight line but, rather, the work of curvature, namely the creation of unexpected and unforeseen relations among multiple elements.[12] Hence, the soundsystem, like the loudspeaker or other sound technologies, is something more than an instrument that pursues different ends – to seduce through rhythm or convince through words. We are looking at unique folds and articulations between heterogeneous elements – symbolic, material, social – that produce these effects. Consequently, to talk of the 'art of the curvature' and to talk of the 'fold' are different ways of granting technological means an adequate ontological dignity to be questioned and analysed as political agents, something that cannot be done if politics is identified only in the human goals that are pursued by technical means. Therefore, the implications of this argument go beyond a political claim on behalf of sonorous technologies (Winner 2000). What is truly interesting is that not only do sonorous technologies turn into a political issue but politics itself becomes redefined. Drawing on Gabriel Tarde, Bruno Latour (2005) defines the duty of putting together associations as the main political duty. If, in the technical mediation, courses of action open up or are shut down, that is, if life projects are created, from this point of view politics cannot simply be understood as a merely human and deliberative matter. To put this otherwise, politics is not just a deliberative matter of reaching, through discussion or argument among citizens, some kind of conflictive or agreed collective life project together. Politics is this duty of building up spatialities and collectivities by setting up new associations among heterogeneous elements, making them durable or just removing them (see Mol 1999). This is why we think that these practices of using soundsystems and loudspeakers in a demonstration are forms of technopolitics.

But what kind of technopolitics or what kind of spatialities and collectivities was being established in the May Day 2005 demonstration? This case shows us at least two different, competing sonorous technopolitics that have, as we will see, important implications in the definition of the urban space: on the one hand, a technopolitics of *commotion* and, on the other hand, a technopolitics of *conviction*. As Peter Sloterdijk (2003) has explained in a very suggestive way, both kinds of technopolitics must be understood as strategies for building up collectives that struggle to be heard. The sound is their modus operandi. Sloterdijk criticizes the excessive prominence that technologies of vision, linked to writing, illustrated culture and the capacity for rational thought, have had in social sciences. He considers that the modern individual is, as Foucault (1975, 1978) demonstrated marvellously, the result of technopolitics of super-vision, of disciplines, although there are also technopolitics of sound that are equally important and have not been adequately studied. In his view, modernity can be characterized as a struggle between com-*motion* and con-*viction*. Modernization is indeed the step from the former to the latter: 'critical subjectivation (a product of visual technologies such as writing and geometry) is based upon de-fascination, alongside restraint in being moved' (Sloterdijk 2003: 434). This means that the conviction of the critical subject is achieved starting from a calculated adhesion of the sounds that move us towards preconceived ideas: for example, national anthems, the populist *speeches* during election campaigns, and so on. These sonorous expressions would appear as redundant expressions of discernible, and hence real, ideas. Thus, sound is and should be just the transparent vehicle of thinking.

If we return to the initial question and interpret the acoustic disputes of May Day 2005, what we have sought to show here is that there are different *sonorous technopolitics* in which a collectivity and a spatiality emerge. On the one hand, we have the sonorous technopolitics of commotion, in which the soundsystem takes part as its ace, operating by producing *herds*, that is, dispersed collective subjects that, however, cannot be de-composed, which transform qualitatively and are crossed by multiple and changing differences, although none of these is sufficiently stable to produce a hierarchy (see Canetti 2002; Deleuze and Guattari 1980). For this reason, the use of these means falls within a strategy of rupture with identitary codes, with the blooming of heterogeneity and the widening of the movement. The soundsystem amplifies the aggregative capacity of sound to produce the demonstration as a space fully occupied that contains multiple differences and where the boundaries are hard to know. And, on the other hand, we have the sonorous technopolitics of conviction, where the loudspeaker acts as the main catalyser, which operates by producing a *mass*, that is, organized collective subjects, with clear internal and external limits, and whose functioning depends on the capitalization of its forces. Through the loudspeaker and the slogans that are shouted through it, a collectivity is formed that gathers its forces and projects them towards a single goal. But, moreover, the space of the demonstration gets bordered and organized through a specific distribution of silences. Thus, the loudspeaker opens up a collective space of hearing organized by a hidden message that remains behind the amplified voice and appeals to beliefs and ideas, that is, to the intellect as the stem of the mobilization.

BETWEEN COMMOTION AND CONVICTION: THE SONOROUS ONTOLOGY OF URBAN SPACE

> Perhaps history itself may be a titanic struggle to be heard by the human ear; in it, voices that are close by fight with those that come from far away for privileged access to move, to make themselves heard, the voices of ancestors with those of the living, the voices of those who govern with those of people opposing power.
>
> Peter Sloterdijk (2003: 433)

As we understand it, these sonorous forms of political action, the technopolitics of commotion and conviction, constitute an essential hinge for work on the analysis of the city and the constitution and transformation of urban space. Hence, if we translate this scheme of sonorous intelligibility suggested by Sloterdijk to urban studies, we find an interesting and productive way of redefining the study of cities. As we have striven to demonstrate in the analysed dispute, sonorous practices are forms of political action that are involved in the production of specific subjectivities and forms of government, and in this sense they are also useful in understanding how the city is produced and changes, how it relates to the functioning of these technologies and their products. The struggle to be heard that the analysed dispute embodies allows us to understand not only the formation of collectivities and social actions, but also urban transformation itself. The distinction between commotion and conviction may well be presented as a tension that defines and articulates the very constitution of the city and all things urban. What is the city, other than the effect of a relationship with what is sonorous, with hearing? Does it not perhaps reflect this dispute to be heard that opposes or is co-produced through different sonorous technopolitics?

To delve deeper into this argument, it may be important to recall the distinction between the city and the urban that we owe to Lefebvre (1972). While the city is equivalent to a site, a morphology or infrastructure, a plan or design, the urban would be more like an ephemeral city, which is lived, the 'perpetual oeuvre of its dwellers who, in turn, are mobile and mobilized for and towards this oeuvre' (1978: 158). Following this argument, we can understand the constitution of the city, the modern city-concept, as a project based on the transition of sonorous technopolitics aimed at commotion to State technopolitics based on conviction. The conceived city, associated to the structuring of urban territories, would thus be the effect of sonorous technopolitics essentially based on contention or control of sound's power to move – in conviction. This being the case, good governance requires maintaining the city in a good state. To do so, it has to determine the meaning of the city by using mechanisms that may lend coherence to spatial compounds that are extremely complex. Working on and administrating spaces that are essentially represented, conceived, for them to be opposed to other forms of spatiality that are a feature of the practice of urbanism as a way of life – spaces that are perceived, lived, used, etc. – the control of sound and hearing is fundamental. In fact, conviction, the calculated adhesion of moving sounds to preconceived ideas, becomes indispensable for the good governance of the city – for its legibility, to change something dark for something lighter, to embody an ideology that aspires to become operationally efficient and achieve the miracle of absolute intelligibility, thus moulding the political pipe dream of an organic and ordered, stabilized city.

This is certainly where the importance and obsession of the forms of administration of the city to control sound, to convert it into noise or silence, stem from. For the sake of progress, of the constitution of an abstract and global sphere, the organization of the modern city is unequivocally established upon the control of subversive noises, because they announce requirements of cultural autonomy, claims regarding differences or marginalization.[13] That is also the source of the Kantian republican ideal of a *soundless* public space, alien to the moving power of sound, full of critical and reflecting subjects, of equal individuals, insofar as their free exercise of reason and freedom is concerned. For conviction, the public space is a condition of possibility at the same time as it is a consequence of free and transparent communication between equal individuals, which is indispensable for the discursive exchange between reasonable positions when faced with problems of general interest and in which the public comes together to frame a public opinion (Habermas 1989). And it is probably due to this very project of an ordered and legible *polis* that we also find, in spite of its anthropological importance, this stigma, this lack of consideration, even this marginalization, on the matter of sound – this forgetfulness in the theoretical sphere and this fear in the political one.

Nonetheless, as shown by the *cacerolada* and the May Day demonstration, this is an unfeasible task, an impossible utopia. The modern functionalization of the city is not exempt from resistance, conflicts or unexpected appropriations. The city may be ordered, but urban life is not (Delgado 1999). Even if it is made possible by this very project, or as its consequence, the transformation or rejection of certain practices is expressed through actions that are unexpected and innovative, thus opening up new fields for experience. In effect, practices of commotion not only carry us to a demotic multilingualism that is laden with unresolved hostilities, but also bring us closer to the production of ephemeral cities that are socially radical, the settings and products of collectivity, constantly creating and re-creating itself – a de-territorialized territory that is always new, in which there are no objects or substances, but rather social relations and bonds that are full of countless heterogeneous actions and actors. With these, the urban space turns into pure potentiality, into an open possibility of uniting, an immanence that is only bound to specific actions and practices and that is only recognizable in the very moment in which it records the social articulations that enable it. Through practices of commotion urban space becomes the product of an articulation of sensory qualities that result from the practical operations and time-spatial schematizations of its inhabitants.

CONCLUSIONS

In order to conclude, firstly, we would like to reaffirm something that is as obvious as it is forgotten: that the city has a sound. In this sense, underlining the political importance of soundspheres makes it possible to reject the traditional forgetfulness of urban studies for what concerns sound. Far from continuing to relegate these practices as an insubstantial wrapping for our social relations, what has been said thus far highlights the need to invest in social sciences that are more sensitive, capable of studying sound as a means of constructing the city. It is not just a matter of theorizing about sonorous matters; it has to do with theorizing for matters pertaining to sound, from sound-related matters. Consequently, urban studies cannot and must not continue to restrict the study of sound to mere acoustic evaluation or a fight against noise in the city and public spaces.

However, our main aim has not just been to highlight the importance of sound in urban studies but to put forward an approach centred on sonorous practices, their heterogeneous ontology and their political dimension. By focusing on technologies such as the soundsystems in demonstrations, or the making of noise with saucepans in *caceroladas*, but also mundane practices like listening to music with an iPod while walking around or alarms and sirens howling, we have shown that urban space is not static or unitary. On the contrary, urban space thus appears as a heterogeneous multiplicity composed by a myriad of *soundspheres* that are overlapping, repelling or reinforcing each other. All sonorous practices, thus, turn out to be technopolitical practices, as they set up a specific spatiality and collectivity. But, moreover, the ontological multiplicity of urban space reveals a sonorous-based political struggle to define the city. As we have tried to show, the city can be understood as the result of a dispute between sonorous technopolitics of conviction and commotion that enacts the city as an abstract and ordered infrastructure, a site, or as an immanent and de-territorialized space, respectively.

NOTES

1   *Cacerolada* (also known as *cacerolazo, caceroleada* or *caceroleo*) is a form of demonstration, of citizens' protest. Its most distinctive feature, which sets it apart from other kinds of protest, lies in the fact that demonstrators express their discontent through a cadenced noise, at an agreed time or spontaneously, rhythmically striking the objects that they have within their reach (generally saucepans and other domestic items, which is where the name comes from). The origin of this form of protest leads us back to the early 1980s in Chile, although its geographical presence covers practically the whole of South America. Nonetheless, in Barcelona, *caceroladas* became popular as part of the repertoire of action and protest thanks to alter-globalization movements, reaching a point where they became one of the symbols of the great protest against the invasion of Iraq in March 2004.
2   In the study of relations between human beings and their environment, noise draws an increasing interest from disciplines such as acoustics, architecture, town planning, psychology and social sciences in general. In concrete terms, what stands out is what has come to be known as acoustic pollution. This approach treats acoustic contamination as a theme for analysis and intervention, particularly from the point of view of the quality of urban life (Baranzini and Ramírez 2005; Chan 1988; Imrie 2000; Staples 1996).
3   For several years, different theoretical and intervening approaches have presented a complex of theories and methodologies of analysis for understanding sound beyond noise and music. In this sense, one may highlight the conceptual contributions (the *sonorous object* and *sonorous*

landscape) of Augoyard and Torgue (1995) or the work of Pierre Schaeffer (1966) to balance academic classifications between noise, sounds and music. It is also worth highlighting *The Tuning of the World*, a work with which Murray Schafer (1977) develops the concept of 'soundscape' by representing the sonorous environment as a musical composition and opening a field of studies about listening in the world and the integration of sound with human beings.

4   As Sloterdijk suggests, being in your own home, with your relatives, among family, essentially means participating in a given *soundsphere*. For example, just as a student explained in a classroom, turning a house into a home requires, albeit not solely, a gesture that is as trivial as it is essential: ringing the bell in a certain way, that is to say, to compose a specific soundsphere through which it is possible to separate an outsider from a family member (Sloterdijk 2003).

5   Some have referred to this murmur of the urban as 'sonic ecology' (Atkinson 2007) or 'urban aether' (Scanner 1994).

6   There have been many authors in the last few years who have tried to map out this invisible city that is erected through sound. Thus, Hopkins (1994), for example, tells us of the subtle control or scripting of consumption that is involved in a particular urban roar. Others tell us of the importance of specific sonorities when it comes to designating work rhythms and spaces (Lanza, 1994; Packard 1957). From a more anthropological perspective, Rice (2003) has explored the 'acoustemology' of institutional contexts and Pink (2004) has described some domestic soundscapes. And some who study space and the city, for example, have stressed the importance of the new acoustic semiotics of space in the transformation of contemporary public space and of its conditions of access (Smith, 1996; Sorkin 1992). For example, the use of 'monster stereos' in cars (Muir 2005) creates an increasingly constant and critical presence, challenging notions of public use and access. Or the use of functional music, or *muzak*, which is seen as a strategy of pacification, scripts public spaces and frames the range of behaviours deemed acceptable by the co-ordinators of these spaces (Atkinson 2006; Jones and Schumacher 1992).

7   In a study of personal stereo users, Bull (2000, 2007) uncovers how such users employ these devices as a way of escaping the urban soundscape in which these aural sanctuaries create 'bright' experiences which can be contrasted with the mundane world that lacks this personal soundtrack. The experiences related by Bull highlight the ways in which the dominance of city soundscapes is seen as something intrusive and to be blocked out through the substitution of a personal soundtrack.

8   Just as Sennett explains in *Flesh and Stone* (1994), in the simultaneous and changing activities of the *agora* the chattering of voices easily dispersed the words, and the mass of bodies in motion experienced only fragments of continued meaning. In the theatre, the individual voice

established itself as a work of art through the techniques of rhetoric. The spaces in which people listened were so organized that spectators often became victims of rhetoric, paralysed and dishonoured by its flow (Sennett 1994: 56).

9 Until recently (the link is no longer active) this dispute could be viewed on the Indymedia Barcelona site: http://barcelona.indymedia.org/ newswire/display/174861/index.php.

10 The CNT acronym refers to the Confederación Nacional del Trabajo, a Spanish confederated union of autonomous trade unions with an anarcho-syndicalist ideology. Founded in 1910 in Barcelona, the CNT is an organization that has played a very significant role within social movements related to anarchism.

11 Here, 'teletubbyism' is a metaphoric reference to the popular BBC children's programme *Teletubbies*. Thus used, it refers ironically to the markedly colourful, festive, naive and pleasant character of this form of protest, which also contrasts with the more serious and formal character of other repertoires of action that are more orthodox.

12 Daedalus is the father of engineering owing to his ingenuity, that is, his ability to invent these kinds of strange relations. For this reason, in Greek, a *daedalion* is defined as something curved, a deviation from a straight line, ingenious but false, beautiful and artificial (Latour 1999a, 1999b).

13 Because sound is a source of power, power has always been fascinated by listening to it. In a text that is not well known, Leibniz meticulously describes the ideal political organization, 'the Palace of Marvels', a harmonious automatic body exhibiting all the sciences of the time and the instruments of power: 'These buildings will be built in such a way as to allow the home-owner to hear and funnel what is said and done without anyone noticing through mirrors and tubes, which would be something very important for the State, and a sort of political confessional box' (Attali 1977: 16).

ACKNOWLEDGEMENTS

We would like to give our thanks to Yasha Macanico for his excellent job in the translation of this text from Spanish into English.

REFERENCES

Atkinson, R. (2006) 'The Aural Ecology of the City: Sound, Noise and Exclusion in the City', Occasional Paper No. 5, Housing and Community Research Unit, University of Tasmania.

Atkinson, R. (2007) 'Ecology of Sound: The Sonic Order of Urban Space', *Urban Studies*, 44(10): 1905–17.

Attali, J. ([1977] 1995) *Ruidos: Ensayo sobre la economía política de la música*, México: Siglo XXI. Trans. Brian Massumi (1985) *Noise: The Political Economy of Music*, Minneapolis, MN: University of Minnesota Press.

Augoyard, J.-F. ([1979] 2004) *Pas à pas: Essai sur le cheminement quotidien en milieu urbain*, Paris: Seuil.

Augoyard, J.-F. and Torgue, H. (1995) *A l'écoute de l'environnement: Répertoire des effets sonores*, Marseille: Parenthèses.

Augoyard, J.-F., Grosjean, M. and Thibaud, J.P. (eds) (2001) *L'espace urbain en méthodes*, Marseille: Parenthèses.

Baranzini, A. and Ramírez, J. (2005) 'Paying for Quietness: The Impact of Noise on Geneva Rents', *Urban Studies*, 42(4): 633–46.

Bijsterveld, K. (2001) 'The Diabolical Symphony of the Mechanical Age: Technology and Symbolism of Sound in European and North American Noise Abatement Campaigns, 1900–1940', *Social Studies of Science*, 34(1): 37–70.

Bull, M. (2000) *Sounding out the City: Personal Stereos and the Management of Everyday Life*, Oxford: Berg.

Bull, M. (2007) *Sound Moves: iPod Culture and Urban Experience*, London: Routledge.

Bull, M. and Back, L. (eds) (2003) *The Auditory Culture Reader*, Oxford: Berg.

Canetti, E. (2002) *Masa y poder*, Barcelona: Galaxia Gutenberg. Trans. Carol Stewart (1987) *Crowds and Power*, Harmondsworth: Penguin Books.

Chan, M. (1988) 'Specific Effects of Urban Noise', *Ekistics*, 331/332: 208–14.

Deleuze, G. and Guattari, F. (1980) *Mil Mesetas: Capitalismo y esquizofrenia*, Valencia: Pre-textos. Trans. Brian Massumi (1988) *A Thousand Plateaus: Capitalism and Schizophrenia*, London: Athlone Press.

Delgado, M. (1999) *El animal público*, Barcelona: Anagrama.

Eyerman, R. and Jamison, A. (1998) *Music and Social Movements: Mobilizing Traditions in the Twentieth Century*, Cambridge: Cambridge University Press.

Foucault, M. (1975) *Vigilar y Castigar*, Madrid: Siglo XXI. Trans. Alan Sheridan (1987) *Discipline and Punish: The Birth of the Prison*, London: Penguin Books.

Foucault, M. (1978) 'Nacimiento de la biopolítica', in *Estética, Ética y Hermenéutica*, Barcelona: Paidós.

Garcia, N. (2005) 'Alarmas y sirenas: sonotopías de la conmoción cotidiana', in VV.AA. *Espais sonors, tecnopolítica i vida quotidiana: Aproximacions per a una antropologia Sonora*, Barcelona: Quaderns-e de l'ICA.

Habermas, J. (1989) *The Structural Transformation of Public Space*, Cambridge, MA: MIT Press.

Hopkins, J. (1994) 'Orchestrating an Indoor City: Ambient Noise inside a Mega-mall', *Environment and Behaviour*, 26: 785–812.

Iges, J. (1998) *La ciudad resonante*, Madrid: Ministerio de Fomento.

Imrie, R. (2000) 'Disabling Environments and the Geography of Access Policies and Practices', *Disability and Society*, 15(1): 5–24.

Jones, S. and Schumacher, T. (1992) 'Muzak: On Functional Music and Power', *Critical Studies in Mass Communication*, 9: 156–69.

Lanza, J. (1994) *Elevator Music: A Surreal History of Muzak, Easy-Listening, and Other Moodsong*, New York: Picador.

Latour, B. (1999a) *La Esperanza de Pandora*, Barcelona: Gedisa.

Latour, B. (1999b) *Pandora's Hope: Essays on the Reality of Science Studies*, Cambridge, MA: Harvard University Press.

Latour, B. (2002) 'Morality and Technology: The End of the Means', *Theory, Culture & Society*, 19(5–6): 247–60.

Latour, B. (2005) *Reassembling the Social: An Introduction to Actor-Network Theory*, New York: Oxford University Press.

Lefebvre, H. (1972) *Espacio y política*, Barcelona: Península.

Lefebvre, H. (1978) *El derecho a la ciudad*, Barcelona: Península.

Mol, A. (1999) 'Ontological Politics: A Word and Some Questions', in J. Law and J. Hassard (eds), *Actor Network Theory and After*, Oxford: Blackwell.

Mol, A. and Law, J. (1994) 'Regions, Networks and Fluids: Anaemia and Social Topology', *Social Studies of Science*, 24(4): 641–71.

Muir, H. (2005) 'Monster Stereos Are Hot Ticket for Clamp Down', *Guardian Weekly*, 19 August.

Mumford, L. (1937) 'What Is a City?', *Architectural Record*, 82. Reprinted in R. LeGates (ed.) (1996) *The City Reader*, New York: Routledge.

Packard, V. (1957) *The Hidden Persuaders*, Harmondsworth: Penguin Books.

Pink, S. (2004) *Home Truths: Gender, Domestic Objects and Everyday Life*, Oxford: Berg.

Rice, T. (2003) 'Soundselves: An Acoustemology of Sound and Self in the Edinburgh Royal Infirmary', *Anthropology Today*, 19(4): 4–9.

Scanner (1994) *Scanner 1*, London: Ash International.

Schaeffer, P. (1966) *Traité des Objets Musicaux*, Paris: Editions du Seuil.

Schafer, M. (1977) *The Tuning of the World*, New York: A.A. Knopf.

Sennett, R. (1994) *Flesh and Stone: The Body and the City in Western Civilization*, New York: Norton.

Serres, M. (1969) *Hermes I: La communication*, Paris: Minuit.

Serres, M. (1996) *La comunicación: Hermes I*, Barcelona: Anthropos Editorial.

Shepherd, J. and Wicke, P. (1997) *Music and Cultural Theory*, Cambridge: Polity Press.

Sloterdijk, P. ([1998] 2003) *Esferas I*, Madrid: Siruela.

Smith, N. (1996) *The New Urban Frontier: Gentrification and the Revanchist City*, London: Routledge.

Smith, S. (1994) 'Soundscape', *Area*, 26: 232–40.

Smith, S. (2000) 'Performing the (Sound)world', *Environment and Planning D*, 18: 615–37.

Sorkin, M. (ed.) (1992) *Variations on a Theme Park: The New American City and the End of Public Space*, New York: Hill and Wang.

Staples, S. (1996) 'Human Response to Environmental Noise: Psychological Research and Public Policy', *American Psychologist*, 51: 143–50.

Virilio, P. (1980) *Estética de la desaparición*, Barcelona: Anagrama. Trans. P. Beitchman (1991) *The Aesthetics of Disappearance*, New York: Semiotext(e).

Winner, L. (2000) 'Do Artifacts Have Politics?', *Daedalus*, 109(1): 121–36.

Wirth, L. (1938) 'Urbanism as a Way of Life', *American Journal of Sociology*, 44(1): 1–24.

Interview with Stephen Graham

Ignacio Farías

Ignacio Farías: I still remember very clearly the first of your papers I read. It was the Spanish translation of 'Telecommunications and the Future of the Cities: Debunking the Myths', a programmatic article, where you critically addressed myths of technological determinism, of universal access, etc. Still, I wonder whether these myths, as they highlighted the role of new technologies of information and communication, contributed in a way to open up the field of urban studies towards the role played by network infrastructures?

Stephen Graham: Let me first say that those myths still function. Every new layer of communication technologies is always surrounded by a big discursive push, as far back as the telegraph actually, the telephone, videotext, the Internet, Wi-Fi and everything else. There's a sense of a big rush of myths of transcendence, myths by which we will finally be able to do away with materiality, with physical movement, sometimes with the city itself, which in a way is a communications device, and to overcome the time constraints of interaction. But I think the myths actually deflect attention from network technologies rather than bring attention to network technologies. For example, the myth of cyberspace as a sort of separate realm that exists in a parallel world tends to totally neglect, obscure and cover up the huge materiality of communications systems. These are powered by vast amounts of electricity; they are materialized through unimaginably complex systems of fibers and servers and satellite dishes and radio-systems and so on, which are sedimented into the landscape and city as a means to overcome spaces and times. So the consequence of these ideas about the end of geography tends to be that we don't pay attention to these hidden substrata of technologized, materialized infrastructure.

IF: In this major book you wrote with Simon Marvin, *Splintering Urbanism*, you speak of networked infrastructures as socio-technical assemblies and about the city as a socio-technical process. Can you explain what you mean by these and how do they relate to each other?

SG: The idea of socio-technical assemblages obviously comes from some of the sort of material semiotics work of actor-network theorists and the post-structural philosophies of Gilles Deleuze and his colleagues. Clearly the key point there is that technologies are not merely material artifacts and that the whole tradition in social sciences of rendering the world into this binary of the social, which is the subject of social science, and the technical, which is perhaps the subject of engineering, is radically unhelpful. So our starting point is that the world of human life and the worlds of urban life are simultaneously imbued with all manner of relationships that are socio-technical (simultaneously social and technical). It is impossible to live a modern, urban life that is not profoundly based on the whole multiplicity of assemblages that blur the social and the technical, on cyborg-like assemblages that continually connect human life with distant times and spaces through the metabolism of the body, through the metabolisms of excretion, consumption, food, energy production, transportation and so on. This is the notion of socio-technical assemblage we are starting with.

The second point is process. We criticize the tradition of environmental determinism, which suggests the forms of cities are simply the results of the imprints of a technology. This whole discourse of nuclear age city, information age city, automobile age city, this idea that we move into a linear series of discrete ages of human history based on a defining technics, is profoundly unhelpful. What we argue is that, in order to capture the way assemblages are meshed into the politics of space, the politics of the city, you have to look at this sense of a dynamic process of an urban life continually coming into being, in a profound sense of ongoing process and dynamism. Thus, what this process opens up is a sort of politics of interruption and disruption. What happens when those metabolisms are interrupted through accidents, blackouts, infrastructure collapse or political violence? Ironically what often happens is that when the assemblages stop working they become a great deal more visible politically and culturally. In New York there is a blackout and therefore everybody is absolutely preoccupied by the arcane technicalities of the geography of electric systems.

IF: In a recent article with Nigel Thrift you call attention to the fact that networked infrastructures constantly fail, that their services are interrupted, and that they are sustained by constant practices and systems of maintenance and repair. In the article you briefly mention the tendency to entropy as a reason for this, but I am not sure whether you want to give a thermodynamical explanation of technological failure and decay, do you? So, why do we have decay? Is it a technological question or a question of urban political economy?

SG: Certainly we wouldn't reduce decay to a purely thermodynamic process, but there is no doubt that decay happens all the time as built structures, infrastructures, move in more entropic directions. There is therefore a sense of continual need for work, huge amounts of very hidden work in order to reverse, continually work against, the inbuilt tendencies of systems to fail, against what Charles Perrow would call normal accidents. All systems will fail normally; it's just that the timing of the failure is not known. And this becomes more complicated, as systems become more complex, more distanti-ated, more coupled together. Many systems in a city run, for example, on electricity – everything, from information systems to logistic chains, transportation, heating and water supplies and so on. Once electricity is interrupted, you have cascading effects as multiple systems fail. This is the material vulnerability that underlies the so-called 'virtual' world. In addition, software codes are fragile, complex and vulnerable, and the infrastructures of the city are now organized through software complexes. This has a very complex series of impacts on urbanism and urban life in cities.

IF: This is interesting because, even though they might constantly fail, at the same time they are very difficult to change, precisely because they are coupled together. So, once sunk in the city, they gain, technologies gain, systems gain, in obduracy. It is as though there were no other alternatives than maintenance and repair.

SG: You are absolutely right. There is this sense of dependency. Once a system becomes stable, it is translated in the actor-network sense of becoming a black box that becomes relied upon continuously, for example on electricity systems or communication systems. It is radically difficult to inscribe a new structure or a new architecture into that system. Sometimes it is easier to simply start again and build something new.

There is an interesting contrast, for example, between the French and the British railway systems. The British rail system has been based on an attempt to continually upgrade very, very old infrastructure rail systems, sig-nal systems, at vast expense with massive disruptions, whilst in France they have decided to follow their grand-project tradition of state monopoly to just parallel a completely new TGV system with new tracks and new signals. This is a great deal easier because you don't have to face the sense of the obduracy of old sunk infrastructures which are very, very costly and very difficult to restructure. Think of the contrast between the process of building a new freeway and the 'Big Dig' in Boston. This was a vast effort to remodel one single highway going through the city core. It cost well over 12 billion dollars to put one highway beneath the city's surface within a very tightly built, congested and complex environment.

IF: At the same time you observe that maintenance and repair involve much more than keeping infrastructures working. You speak of a 'politics of main-tenance and repair' and beyond this of a 'multiscaled environmental politics'. Can you illuminate us on these connections? On these political consequences of these technical issues?

SG: There are some very dramatic political and environmental consequences of repair or the lack of repair and the way in which it is deeply invisible in contemporary society. Repair tends to be hidden and subdued. Particularly in information systems and in many other consumer products we are seeing a shift to a rapid turnover, as we move towards continual upgrading. Artifacts, and technical artifacts especially, become radically temporary, become stop-gaps owned for a very, very short time. Mobile phones in the UK, for example, now have an average life of six months: six months before they are upgraded because of the whole structures of contracts and the increasing capabilities of the device. Now, very few people ever question what happens to the hundreds and millions of VCRs that no one ever wants now because they have DVDs, what happens to the hundreds and millions of computers that are a few years old but deemed as completely obsolete. This is in fact a whole global geography of trade and supply in e-waste or electronic waste, and very often these things are transplanted to a region or country like China to be taken apart by children and women in terrifyingly bad environmental conditions. People recycle the trace elements, diamonds, gold or whatever. There are very serious health crises and problems and there are very catastrophic environmental problems too. So there is this sort of hidden circuitry of waste which is the direct result of built-in obsolescence and failure to repair.

IF: Just as cities depend on practices of maintenance and repair of infra-structures, you have shown that they can be literally switched off through warfare strategies of destruction and disruption of precisely these infra-structures. Are contemporary cities easier than ever to switch off, given their major dependency on technologies and the greater interdependency of networked infrastructures?

SG: This goes back to as long as cities have existed. We shouldn't forget the centrality of the city siege in pre-modern warfare and the efforts to contaminate water supplies and food supplies and all of that history. There is the key question of what happens when a city and its population become completely reliant on a series of material assemblages and infrastructures. This clearly means that there is a very profound vulnerability, even when systems may not fail catastrophically and while failure might be a routine thing. We shouldn't forget that, in many parts of the world, infrastructure is a process of continual failure, particularly in many of the cities of the global South, where improvisation and failure is an absolute basic reality of administration. Nonetheless, when infrastructures are targeted in political violence, this has radical implications.

Most of the emphasis has been on the way in which non-state terrorists and insurgent groups are using infrastructure as a form of projecting violence. So, obviously, 9/11 was the result of exploiting the potential, catastrophic impact of aircraft used with an Internet backup of coordination. In Madrid the detonations were operated through mobile phones. The same in Iraq, where there is a lot of insurgent activity to disrupt infrastructural supply: the electricity system, the communications system, the oil supply. There is much less emphasis on the activities of states which I think are much more dangerous in the long run, much more catastrophic even than terrorist attacks. The US and Israeli militaries, in particular, have a very sophisticated doctrine of the infrastructural warfare. On one end this involves simply bulldozing infrastructures, as in the Israeli case, and on the other end using special bombs which are made out of graphite crystals and which simply short-circuit entire electricity systems. And there are elaborate models of how these switching-off processes are purported to bring political pressures on leaders, to bring moral pressures on civilians as well. This goes back to World War II, when there was a fairly crude effort to target infrastructure systems, particularly in Germany.

In a world of growing urbanization, by targeting infrastructures you might inhibit the potential for military activity. But this is legitimizing a war on public health in urban society. In the case of Iraq after 1991 when all of the Iraqi electricity systems were systematically destroyed, there was a catastrophic health crisis, because the water and sewage systems are electrically powered and repair was not possible. The UN and UNESCO have estimated over half a million early fatalities. So this was very much a question of the mass killing of civilians. But it was a question of mass killing distanced in time and space from the initial impact of the bomb. This seems less pernicious to the global media, which move on and look at other things. Militaries have even justified destroying urban infrastructure as a 'less lethal' form of warfare.

IF: I have been somehow puzzled by your work on the imaginative geography of 'war on terror', because it seemed to me that you were giving up the networked part of what you called critical networked urbanism. On the other hand, we can certainly not leave such powerful discourse regimes out of the picture. My question has a practical and methodological character. How are you dealing in your current work with this gap between representations and infrastructures?

SG: It's a good question. One needs to struggle with that divide. It is important to look at the discourses, representations and sort of geopolitical work that they do, whilst addressing the materialization of this new security politics. What is interesting, what is striking, are the sorts of contradictions that are emerging. On the one hand the imaginative geographies and geopolitical imaginaries of the neo-conservatives, for example, render the world as a simple two-sided Manichean binary, amounting to them and us: you are either with us or against us; we are the good societies of freedom and democracy and you are the demonized 'Other', racialized, Orientalized, terroristic 'Other' over there. This is the classic politics and geopolitics of rendering the other as enemy, as unclean, as unfit and almost as an Agambenian bare life, but overlooking that there are indeed stark connections and crossovers between the security politics in domestic cities and the frontier war zone city.

In military doctrine both domestic and frontier war zone cities are being imagined as a sort of battle space where military forces have to somehow track and locate and surveil the adversary who blends into the mass of the population, who doesn't wear a uniform and uses civilian infrastructure and techniques either to execute terrorist attacks in London or New York or to undertake insurgent attacks in Baghdad. There is a sense of an integrated urban battle space in the military language. If you go to a conference of these military urban specialists, which is something I have done, you will see a presentation about Chechnya, followed by a presentation about the bulldozing of the Palestinian city of Jenin by the Israeli Army, followed by a presentation about the LA riots, and there will be a seamless discussion about technologies of control, architectures of control and the enemy. So these military doctrines have a very broad imagination of the enemy as organized from network politics to network technologies. The Zapatistas are very often discussed as a sort of network enemy, because of their role in using the new media. So there is a stark connection, but a contradictory one.

IF: Let's go back to *Splintering Urbanism*. There you were mainly focusing on the co-evolution of processes of unbundling networked infrastructures and splintering urbanism. For me it was sometimes hard to follow how you were tracing the limit between the two processes. So let me put it this way, what do you mean by splintering urbanism that is not said by unbundling?

SG: Well, I think a starting point is that, if you render the divide between the technical and the social as unhelpful, you render the divide between infrastructure and the city as unhelpful. Therefore you need to look at reconfigurations of mobility and infrastructure systems and spaces in a profoundly integrated way. And if you make that your starting point it is no wonder that the line is blurred. That's deliberate. The line should be blurred. For example, if you look at the ways in which the gated communities emerge, it is very often not adequate to look at how gating operates in terms of the definition of the gating area, the secession of the governance process. The same with the production of privatized public space, of privatized mall spaces or the production of techno-polls and techno-parks. Very often these material productions of urban space are linked to reconfigurations of mobility and infrastructure systems, whether it would be about private security systems or private energy or water supplies or private highways which have tolls regulated for premium users. What we are trying to capture there are the seamless interconnections between configurations of network spaces and infrastructures and urban spaces. That's the need to maintain this parallel perspective.

IF: I have the impression that you avoid thinking, or at least writing, in terms of scale. My question is whether working with empirical material necessitates introducing or avoiding scalar distinctions?

SG: Well, I don't deliberately avoid issues of scale or the politics of scale. I think they are probably more implicit in a lot of the work I have done with Simon Marvin. We do try and look at issues of rescaling in terms of the new economies of premium infrastructure and the way they connect valued enclaves in a sort of tunnel effect. So I think scale issues are permeated through some of our work, but we deliberately haven't tried to engage with the ongoing debate, particularly in geography, on scale. Those issues are clearly dramatically relevant to all of these things we are talking about here, and in many cases the neglect of a sort of critical politics of infrastructure has gone on in strangely parallel worlds with this massive theoretical discussion about scale.

We have to remember that the processes through which human life transcends scale, connects scale, rescales, are not just about imagination and imaginary geography, but about new production systems, about new governance systems, about how spaces and times are remade through technical infrastructure. So this is always about the politics of scale, the politics of space. A lot of the material I am looking at now in terms of cities and securities dramatically represents new discussions of scale in terms of state security. For example, the whole idea of borders is being reimagined based on new technologies, surveillance, border and technologies, as well as being the classic sort of linear separations of territory. Borders are now being exteriorized beyond nation states as calculative assemblages, which are continually paralleling flows of passengers, goods and containers linked to massive database management, data-mining, profiling, tracking, anticipation, which are continually trying to render these systems visible, in the very near future. There is a big politics of temporality. On the other hand, bordering technology is becoming a mobile and ubiquitous assemblage which operates within the territorial borders of nations as well as beyond them. It is profoundly about the notion of citizenship, based on certain populations being problematized and deemed to be threatening and rendered under the gaze of intensified surveillance.

IF: There is a sort of a gap in urban studies between an established political economy tradition and this interest for technology and infrastructure. How should we deal with this gap? And what is the way to connect both?

SG: The problems are clear. The study of infrastructure is deemed to be the work of engineers with a technical, rational paradigm based on very old-fashioned notions of public good and instrumental rationality and so on. Not surprisingly those traditions have no space for that political economy whatsoever. They are completely blind to questions of political economy. So it's up to critical urbanism and critical urban research to render these things a political economic project. I think there are some traditions that are very useful to draw upon, a lot of the Marxist geography, for example David Harvey and Eric Swyngedouw, specifically looking at how infrastructures are kind of fixed in space to facilitate the mobility of cities. But I believe that those traditions can be combined with some of the insights of STS and Actor-Network Theory. I am not saying there aren't some issues of contradiction in those theoretic traditions, but I do think there is a need to bring several perspectives to bear, and not be obsessively concerned with merely constructional or merely historical materialist and structuralist traditions.

IF: On the other hand, sometimes STS or ANT-inspired perspectives on the city focus on very small technical and sophisticated issues, which are miles away from these political issues . . .

SG: I think that's true. I think there's a danger in that tradition that you become preoccupied with minute, minute details which can be very elegant and very enlightening, but which make it a struggle to bring out the political and politicized nature of technological assemblages. Some of the ANT work has been criticized for being depoliticized, in the sense that you are rendering technologies with agencies, and you are looking at these complex processes of formation and translation of text, but there is the question of what does this mean for the right of the city. Can we bring in a specific political language here and can we possibly generalize in a world where you are looking at such inherently individualistic moments and case studies? Some of the work by Bruno Latour, I think, does move into political questions. I think the political ecology tradition builds on some of these notions of hybridity in terms of socio-technical, but also bio-social and socio-natural, hybridities.

IF: This is interesting. Which traditions of urban studies, which concepts and authors, should be rescued to help us think about this hybridity?

SG: Since we wrote *Splintering Urbanism* there has been a big growth in political ecology and probably before that book too. The whole literature on political ecology is very much informed by the sorts of traditions of cyborg theory and ideas of urban processes as transforming nature into culture through metabolized processes, water-based metabolisms, production and transportation of food, productions and transportations of energy. The whole very interesting world of political ecology is doing much to render the infrastructure materiality of cities visible and political in new and important ways.

But going back to the point you made about who can we rescue, there are some great theorists and writers who have done great work on looking at various aspects of urban infrastructure in a non-technocratic, non-deterministic tradition. I would mention people like Gabriel Dupuy, working on France, who has a brilliant book called *Networked Urbanism*, which is very much looking at the various technical systems and the infrastructures and traditions of it. France has a very long, powerful history of looking at infrastructures and territories, and they have longstanding research centers such as the LATTS research center in Paris, where Jean-Marc Offner is working. Lewis Mumford clearly had a major preoccupation with technics of the city and had a specific politicized agenda. In the communications scene we can probably look at people like Ron Abler, a geographer of communication. So there are some outstanding scholars, but I think what we were saying in *Splintering Urbanism* was that there is a failure to have a systematic, critical and sophisticated approach to network infrastructures as a whole, to look at their multiple interconnective importance in urban life.

IF: And in that sense too at the end of *Splintering Urbanism* you plead for a fully networked urbanism, which should be more about relations and less about objects. To what extent do you think are contemporary urban studies moving in this direction?

SG: I think there are real signs of progress. The growth of these discussions about flat ontologies and scalar distinctions is paralleled by a really serious engagement with assemblages of various sorts and the way they are instantiated often very reliant on these urbanistic, network urbanistic relationships. Similarly in theoretical terms there is a big shift towards relational theory which transcends any simple divisions between scales and social structures. Finally there is a big engagement with the politics of new technology in urban debates. There is a massive growth of engagement by artists, activists, social scientists and researchers with the sort of informational cities, emerging and brought into being through digital technologies. I still think there's a need to work at the politicization of network space. There is still a long way to go in looking at the more banal and familiar infrastructural systems of water and electricity and sanitation, street spaces as well. But again there's some progress there in terms of politics of resistance against neoliberalized privatization. There's some very good research in places like South America, South Africa, on post-neoliberal reimagination of the right of the city and the network systems that make the city possible. So there's been real progress.

New York City, 2007

Part 3

The multiple city

9 The reality of urban tourism

Framed activity and virtual ontology

Ignacio Farías

Here, there, everywhere, tourism is a well-established urban reality. There was a time when one should open articles on urban tourism by featuring tourism as a growing urban industry (Judd and Fainstein 1999), highlighting its capacity to lead processes of urban resurgence (Law, C.M. 1992), and discussing its reshaping of urban spaces, landscapes and imaginaries (Cocks 2001; Selby 2004a; Sheller and Urry 2004; Smith 2005; Urry 1999). Similarly the socioeconomic and political relevance of studying tourism was often stressed by discussing issues of commoditization, entrepreneurialism, gentrification and forms of social exclusion in urban locales (Holcomb 1999; Huning and Novy 2006; Shaw *et al.* 2004; Zukin 1996). Such problems and questions have been at the core of most of the thinking, researching and writing about urban tourism leading to important findings and profound insights about the interconnections between tourism and urban change and development. Still one could fairly argue that none of this literature says much about what tourism is or, better, about what it means to say that tourism has become an urban reality.

This chapter provides a double answer to that question. On the one hand, the reality of tourism is described in terms of the actual practices of touring a city, understanding that tourism is enacted in concrete situations. On the other hand, tourism is understood as a virtual plane of activity along which tourist situations occur. The point I would like to make is that, for tourism to become an urban reality, visitors, attractions and tourist devices per se do not suffice. A double movement, rather, is necessary: tourism needs to become territorialized in the streets of a city, needs to become enacted and actualized in concrete situations, and simultaneously it needs to become de-territorialized, that is, it needs to become a virtual urban plane.

The larger part of this chapter is about the first of these movements. The first two sections offer an original analysis of the situational conditions in which tourist activity is actualized in urban spaces. Thereby these two sections problematize the common assumption that tourism embraces all situations and activities that occur while travelling for leisure away from home (Bruner 2005; Cohen 1984; Graburn 1995; Urry 2001). The focus is rather posed on the fragile and fluid frame that sustains tourist activity by separating it from the banal and totalizing overflows of the urban every-day. This analysis is based on the ethnographic study of sociotechnical arrangements of sightseeing bus tours that circulate in the city of Berlin. The second movement necessary to render tourism into an urban reality, de-territorialization or virtualization, is only briefly discussed in the third, concluding section. However, the thesis presented there is strong. It is a form of tourist sense-making through communication that creates this virtual plane for tourist activity, this urban ontology.

THE FRAMING AND OVERFLOWING OF TOURIST ACTIVITY IN THE URBAN EVERYDAY

A brief look at the classical and contemporary literature on tourism shows that notions of everyday life are constantly invoked, not just to explain people's need to leave home, but also to describe the experience of being a tourist. Tourists are said to be escaping their everyday lives (Boorstin 1987; Enzensberger 1996), but fascinated with the everyday lives of 'Others' (Graburn 1995; MacCannell 1973, 1999), even though their own everyday lives are already mixed with the everyday lives of 'Others' (Franklin and Crang 2001; Urry 2001) and their travels are nothing but heightened exten-sions of their own everyday lives (Franklin and Crang 2001; Obrador Pons 2003). It is true: the everyday has become a heuristic device lacking much conceptual clarity. But, still, everyday life constitutes such a crucial aspect of city life that it can certainly not be ignored when analysing the reality of urban tourism. The urgent question is then: how do we understand the relationship between tourism and everyday life?

A good departure point completely disregarded in tourism studies is Lefebvre's emphatic assertion that 'the everyday is not a synonym for *praxis*' (2002: 45), but a *level* of social practice. Thus, 'everyday life is profoundly related to all activities, and encompasses them with all their differences and their conflicts; it is their meeting ground, their bond, their common ground' (Lefebvre 1991: 97). Lefebvre's theory of everyday life does not designate a region, type of activity or system of knowledge, but seeks rather to describe a level of society that crosses through different regions and systems of activ-ities. Above all, the everyday involves banal experiences that slip in through the interstices of specific types of activities, filling up its gaps, 'technical vacuums' and blank moments. From this perspective, one could start arguing that everyday life is present in urban tourism as a *level* of practices. The question is then how this intromission of the everyday in tourist practices should be imagined and understood.

The work of Michel de Certeau supplies here the key insight, as he speaks of an overflowing (*débordement*) of the everyday. In the general introduction to *The Practice of Everyday Life* (1988) de Certeau explains that the book has a twofold aim. It involves, on the one hand, the careful description and res-toration of cultural legitimacy to everyday practices such as reading, talking, walking or dwelling and, on the other hand, investigating 'the extension of these everyday operations to scientific fields apparently governed by another kind of logic' (de Certeau 1988: xvii–xviii). De Certeau speaks here of an overflowing of the everyday: 'the work of overflowing operates by insinuation of the ordinary into scientific fields' (1988: 5). Morris (1990) and Seigworth (2000) have also observed that the all-encompassing character and pervasiveness of the everyday is to be understood above all in terms of these banal overflows.

One way of grasping the challenge posed by banal overflows to tourist practices in urban spaces is to revise the description of situations of 'radical urbanity' developed by the Spanish anthropologist Manuel Delgado (1999, 2007). He argues that the specificity of the urban experience lies with the immanent instability and effervescence of forms of minimal sociality based on constant movement, ambiguity and transitivity and on the principles (and political rights!) of anonymity and indifference. A crucial feature of streets, sidewalks, boulevards, squares, transit spaces and other sites is the radical minimization of users' signs of identity or authenticity. In these spaces, a new *a*-social figure emerges, the transient passer-by, who does not just represent, but literally is, '*someone* or *anyone in general*, or, if you prefer, *everyone in particular*' (Delgado 2007: 188, transl. IF). This is precisely the challenge that the urban everyday poses to tourism, namely, overflowing tourists with its pull towards anonymity and minimal sociality – transforming or, more properly, reducing all of them, tourists and non-tourists, to transient shadows of themselves. Considering this is crucial, for, if one aims at understanding the urban reality of tourism, one needs to first understand which are the most pervasive urban forces destabilizing that tourist reality.

There is yet a second phenomenon that undermines the possibility of holding a tourist stance towards the world. Besides 'banal overflows', tourist activity is also entangled with multiple symbolic, cognitive and affective non-tourist practices and experiences based on moral values, biographies, social relations, group pressure and so on. The reality of tourism is thus undermined by 'totalizing overflows'. The work of the British anthropologist Daniel Miller (1998, 2004) on economic practices in market societies is here particularly helpful, as he shows that market transactions often are highly entangled social practices. His hypothetical ethnographic example of a recently divorced French woman named Sophie, mother of two children, with a stylish circle of friends, who is about to buy a new Renault car, shows the multiple elements and considerations, including aesthetics, thrift, parenting, sense of autonomy, environmentalism and nationalism, entangled with her decision, preventing her and the purchase situation from being framed in terms of economic calculation:

> we are closer to Jean Paul Sartre's (1976) conceptualization of a moment of aesthetic totalization, in which everything – from past suffering to possible future pleasure, from all her social relationships to all other

economic possibilities that are contingent upon this particularly large purchase – come together in this choice of a particular style and the weighing up of a constellation of values.

(Miller 2002: 226)

The challenge that can be extrapolated from Miller's position for the study of tourism is that looking at the practices of people in the situation of leisure travel *as* tourist practices is a tricky assumption, for a closer look reveals that specific tourist activities, such as sightseeing, are rather peripheral. Indeed, totalizing overflows are very common among tourists, who often bring into travel experiences all possible kinds of personal and socio-cultural entanglements, rendering the trip into a total experience constitutive of lives, identities and societies. This is well known among tourism researchers, but not the logical conclusion that Miller reaches, which in the case of tourism would suggest that tourist practices do not exist as such. They are so embedded and so entangled with all kinds of personal, social and cultural values, elements and practices that describing and qualifying them as being tourist practices is insufficient and inadequate.

All this poses a central question for the study of urban tourism, namely, how does tourism despite these banal and totalizing overflows of everyday life become a separate region of activity? How is tourism disentangled and reproduced as a particular form of experiencing, inhabiting and gazing at the city? Or, in other words, how does tourism become an urban reality? In what follows, I would like to explore the relevance of thinking in terms of dynamics of framing/overflowing to understand how tourism becomes real in the urban realm.

Any attempt at a frame analysis of tourist practices needs to refer to the work of Erving Goffman on the situational organization of experience (1974). At the very core of Goffman's innovative investigations was the question of how society is possible. The decision to look at interactions was certainly a central part of Goffman's answer: infinite threads of situations were understood by him to be the basic flows, which as they circulate make society, and our experience of it, possible. However, Goffman's answer to the question about social order does not point only to this circulation. At the root of his work also lies the assumption that the consistency of social bonds is organized around regions of signification with rules of pertinence that circumscribe interactions.

In *Frame Analysis* (1974), his last major work, Goffman understands such consistency through the notion of 'frame'. The frame is a 'ready-made' that conveys cognitive and practical dispositions for interpretation and involvement, i.e. providing a way to describe events and for participants to become spontaneously engrossed and caught up by the activity in which they are participating. In this sense, frames do not represent states of mind, but are 'principles of organization which govern events – at least social ones – and our subjective involvement in them' (Goffman 1974: 10–11). Rather than a schema producing interpretations, frames constitute moments and sites of situated activity. As boundaries, or 'partitions, hideaways, fire-doors' as Latour has called them (1996), frames provide an answer to the question 'What is it that's going on here?' (Goffman 1974: 25). In thinking of tourism in these terms, it becomes clear that it can't just equal travel, leisure, being away from home, global mobility or anything like that. Tourism is primarily a situation or, more precisely, tourism is a ready-made frame available for defining, connecting and getting involved in situations.

Frame analysis aims at two aspects: first, to 'isolate some of the basic frameworks of understanding available in our society for making sense out of events and [second] to analyze the special vulnerabilities to which these frames of reference are subject' (Goffman 1974: 10). Regarding the first aspect, I would suggest that tourist frames involve a transcription, a transposition, a transformation of primary frameworks (leisure travel) into a patterned activity (tourism). Correspondingly, they should be grasped as what Goffman (1974) called a keying; this is a 'set of conventions by which a given activity, one already meaningful in terms of some primary framework, is transformed into something patterned on this activity but seen by the participants to be something quite else' (Goffman 1974: 43–4). Tourism implies thus a sited process of keying and transformation in which individuals engage, rather than a condition associated with a primary framework, such as leisure travel.

Regarding the second aspect, the vulnerabilities of frames, one should state that in the case of urban tourism it is not so much the occurrence of the unmanageable – events that cannot be ignored and to which the frame cannot be applied – that leads to a frame breaking (indeed, what can't become a tourist attraction?), but above all the fact that tourist activity always, almost inevitably, coexists with other modes and lines of activity, from pedestrian and transit regulations to commercial activity and outdoor advertising. However, despite these 'totalizing overflows', Goffman suggests that it is always possible to perceive and describe fields of activity as constituted in terms of one chiefly relevant frame which is pursued by participants across a range of events, elements and storylines. Correspondingly, tourist practices in urban spaces should be understood as highly vulnerable framed activity *pursued across* the banal overflows and totalizing entanglements of everyday life, *across* the parasites, non-authorized activities, ambiguity and complexity that make evident the improbable and 'expensive' character of tourist frames.

This simultaneity of framing and overflowing dynamics has recently been emphasized by Michel Callon (1998). His basic argument is that, in situations where overflowings are the rule, framing or disentangling is possible by means of sociotechnical networks of devices. This involves a double innovation with respect to Goffman. First, Callon suggests that processes of framing do not occur despite, but as a consequence of, overflows. While a wholly hermetic frame would be condemned to sterile reiteration, Callon observes that the very same resources invested in the process of framing constitute intermediaries to wider networks, enabling a bi-directional flow between the inside and the outside of the frame. Second, Callon emphasizes the role played by material, textual and technological devices for the maintenance of the frame. Even though Callon recognizes that actors must agree upon the rules and frame within which they are interacting, he inscribes the analysis of frames in a socio-material ecology.

In the next section, I empirically explore the thesis that tourist situations are sustained by sociotechnical frames by presenting the main findings of my ethnographic research in two types of bus tours. It could be argued that bus tours are not a good example of the process of framing, because they correspond to standardized situations, being over-controlled and detached from everyday life. I shall make the opposite argument. The detachment of bus tours means that multiple sociotechnical elements have been arranged in complex ways, in order to hold the frame together. Therefore, if one is able to recognize the dis/entanglement of bus tours with the urban everyday, it follows then that similar framing/overflowing dynamics apply to other tourist situations.

THE SOCIOTECHNICAL ARRANGEMENTS OF BUS TOURS

Sightseeing bus tours are hybrid machines. A large set of heterogeneous elements – such as pathways, tourists, tour guides, double-decker buses, stories, anecdotes, jokes, headphones, drivers, video cameras, images, photos, TV screens, audio recordings, pedestrian and transit regulations, and urban rhythms – must be organized in specific ways in order to produce a sightseeing bus tour. However, and this is crucial, there is no one unique form of structuring these elements; nor should all these elements be simultaneously present in all bus tours.

A bus tour is defined by the organization of its elements, not by its structure. I follow here the distinction introduced by Francisco Varela and Humberto Maturana (1973) between the *mode* of assembling elements together (organization) and sets of *concrete relations* among the elements (structure) of any complex unit. The organization of sightseeing bus tours can be analysed in terms of four basic arrangements that have to be made in order to frame tourist activity. The first of these can be termed a 'spatial arrangement'. This involves the production of destination space by means of some basic operations, including the production of routes, the introduction of spatial distinctions, and the distribution of objects in space. Second, bus tours need to produce a particular 'visual arrangement', which involves enacting a moving glance. Third, I shall distinguish a 'narrative arrangement', which functions as a source of themes and topics available for the reproduction of tourist communication. Fourth, bus tours entail particular 'performative arrangements', in the sense that they organize the performance of certain tourist roles, such as the tourist and the tour guide, that are conditions for framing tourist activity.

While sharing a common mode of assembling elements, the structure of bus tours varies more than people would expect. In this sense, bus tours can be understood as 'fluid technologies' (de Laet and Mol 2000) that hold their shape in a fluid manner (Law, J. 2002). Looking at bus tours as fluid technologies involves thus uncovering the variety of sociotechnical arrangements that enable the enactment of tourist situations. I do this by focusing on two extreme forms of bus tours that can be found in the city of Berlin: the so-called 'Videobustour' and the public transport bus route 100. Even though both exhibit singular ecologies calling for more comprehensive ethnographies, they are used here in a more analytical way to discuss particular aspects of the dynamics of framing/overflowing tourist situations.

Hermetic frames and blind spots: the Videobustour

In 2004 the company Zeit-Reisen: Erlebnisagentur (Time-Travel: Agency of Experience) launched the Videobustour (from now on, VBT), a new invention aimed at bringing together a historical approach to Berlin and tourism. The VBT consists of a private single-decker bus, not a public bus, not a double-decker, equipped with monitors and speakers for playing numerous videos, animations, photos, images, graphics and audio recordings relating to the urban sites and attractions visited during the tour. The VBT's speciality is the provision of thematic tours on topics such as the history of the Wall, the Golden Years, Hitler's Berlin, urban planning in Berlin, 1945 or the End of the War.

During recent years the VBT has attracted some media attention and, indeed, almost all the main German newspapers have written positively about it. Feedback from customers has similarly been quite positive, remarkably from Berlin residents, who represent approximately two-thirds of the passengers. However, in terms of broad public appeal the VBT is far from being a hit. Whenever I did participant observation in 2005, the tour group was never larger than 20 persons. For the director, Arne Krasting, low numbers are a problem of limited advertising and the consequence of the fact that his company cannot compete with the large sightseeing bus tour companies operating in Berlin. He particularly regrets not reaching young people, who he believes can react much faster to the multiple inputs and different media of the VBT. Despite this, the VBT has been on the streets now for almost four years and it has been particularly well received by groups, such as firms, foundations and school classes, that privately book it. On the basis of these good results and experiences, 2007 marked the beginning of a new phase when this tour format was exported to Arne's home town, Hamburg.

The VBT is a very extreme case of a hermetic tourist frame built upon very tight sociotechnical arrangements. Indeed, the first striking element of the VBT is its very unique *spatial arrangements*. The elaboration of particular routes is a central aspect, as the VBT specializes in thematic tours. In the case of the 'Berlin Divided City' tour, for example, the VBT goes north to reach the Wall Memorial in Bernauerstraße, and east to reach the East Side Gallery, and crosses right through Kreuzberg to visit the Springel building, places which are usually not visited by standard sightseeing bus tours. What is particularly striking about some of the VBT routes is that they are chronologically structured, as though history would be chronologically distributed over urban space. In the case of the VBT focusing on the history of the Wall, this means that first the tour jumps back in history and space to the period preceding 1961, the year that the Wall was built. As the bus moves through the city, tourists see how the Wall was constructed, modified and amplified, and other relevant sites, including the place where the first victim of the Wall was killed. The tour is structured in order to witness chronologically all the main historical events until the fall of the Wall. After the VBT passes the Palast der Republik where German reunification was signed by the RDA parliament, it goes to its terminus, close to the Brandenburg Gate, the main symbol of the German reunification, where people can get off in a new reunified Germany.

Such historical narrative is closely intertwined with a second spatial arrangement. On the VBT on the history of the Wall, as in other of its routes, the number of places mentioned is significantly lower than on standard sightseeing bus tours. While in the latter tour guides point to, name and refer to around 80 sites, in the VBT the history of the Wall is told by means of no more than 35 places. The VBT functions as a highly selective device, filtering out places and sites that do not fit into its chronological narration. When I asked whether this 'blinding out' involved much work, Arne explained that people seldom were interested in or asked about buildings or sites not thematically connected with the tour. However, instead of explaining this in terms of an intersubjective agreement between tour guide and tourists about what the tour is about, I would argue that this blinding out of the urban environment is enabled by the most basic technology at play, the bus, which produces a very particular *visual arrangement*.

Standard bus tours use open-roof double-decker buses and function on the premise that tourists sit on the upper deck. The so-enabled peripheral perspective upon the city constitutes one of the central goods that tourists pay for when buying a ticket for standard sightseeing bus tours. The VBT, however, does not use double-decker buses. Its visual arrangement relies on a mixture of monitors and normal side windows, which allow a constrained visual interface with the city. In this context, touring with the VBT is mainly structured as a combination of two main modes of visual engagement: places are seen through the side windows (as the VBT stops for a few seconds in front of them or passes by slowly) and places are shown on the monitors (primarily as the VBT goes from one place to the next).

Indeed, the most characteristic moment of the VBT sightseeing experience occurs right after gazing upon a sight, when at least five different types of videos are shown. Some films show what once stood and people cannot see any more. Other films show buildings that still exist and relate them to historical events or small stories that took place there. A third kind of films makes it possible to get inside the buildings, showing the different rooms of the now unbuilt Palast der Republik or the Nordic embassies. Excerpts of movies filmed at places passed by the tour are also shown. Last but not least, graphics, maps, animations and 3D images are also presented.

> Arne explained that the idea is always to keep a tension between the place and the films, so that people can look at both, compare past and present and understand the city and its history. Otherwise, he said, the films could be shown anywhere. He then explains that the VBT is about finding the right mixture between media and reality.
>
> (Fieldnotes, 12 October 2005)

The videos and images shown on the VBT perform a very important task, namely, the elimination of the space between sites and places. Sightseeing with the VBT becomes a very extreme form of sightseeing, in which the city is not just reduced to a collection of sights, but, more radically, stripped of the ambiguity and fluidity of everyday life in urban spaces. Thus, the VBT renders the urban environment into a radically ductile material for tourist narrations. Such transformation of the city allows very extreme treatments, such as the chronological ordering of the urban environment. Rather than being embedded in the urban context, the sightseeing experience enabled by the VBT is thus radically self-referential and enclosed. The air-conditioned bus, and not the city, constitutes the medium in which practices of sightseeing are embedded. Stopping at particular places, getting off the bus and inhabiting the city are treated mainly as interruptions, not as intensifications, of the tour: 'We get off at Checkpoint Charlie to see the polemical private Memorial of the Wall constructed by Mrs Hildebrandt.[1] After some brief comments, the guide invites us to come back to the bus: "in order to understand this place" ' (Fieldnotes, 26 March 2005). The idea that the city can be better understood – or even experienced – in the bus than *in situ* overtly contradicts the basic logic of most forms of touring and it makes evident the uniqueness of the frame enacted by the VBT.

Indeed, a rather constrained visual and embodied experience of the city is crucial to support VBT's *narrative arrangements*, which are based on the claim that the VBT discloses historical backstages and reconstitutes authentic historical atmospheres that couldn't otherwise be seen or felt. Arne explained that they have different maxims for the VBT, such as 'We show what you otherwise couldn't see' or 'We see not only what we can see from the bus, but we see more; we make the history of the city visible.' At the same time, even though the VBT is 'really about history', about making multiple city pasts visible, 'boring' academic narratives are explicitly avoided. Arne sees the VBT as a form of making people curious about the city and its history rather than teaching people. Thus, while side windows, monitors and short stops are necessary for detaching people from the city and thus making history visible, images and films are expected to perform a complementary operation, namely, making people curious and reattaching them to the city history.

The VBT frame functions for most of the time as an all-encompassing translation mechanism detaching tourists, filtering overflowings, reorganizing the relationship between urban space and city history, and prescribing performances. However, the VBT does not simply contain individuals in a hermetic frame. Framing is a process enabled by constant overflowings, sources of indeterminacy and transformation, and in the VBT there are indeed important overflows which need to be acknowledged and accounted for. The first and most obvious source of overflows can be located at the few stops when tourists get off the bus.

> We get off at the Günter Litfin Memorial [the first victim of the Wall] and by chance we meet the brother of Günter Liftin, the promoter of the memorial . . . The old man speaks for at least 10 minutes to the group about the lack of support from the city government and suggests that Berlin is not taking charge of its history. Once in the bus, the guide says to the group, 'well, as the saying of German historians goes, there is no worse enemy for history than the witnesses' and starts to explain 'objectively' the efforts of Berlin Senate.
>
> (Fieldnotes, 26 March 2005)

As this situation shows, a casual encounter can have disruptive effects. Narrative elements suddenly no longer fit together, and the possibility of an interpretive conflict is foreseen. At the same time, however, such unexpected encounters open up the frame for new connections, new stories and new memories.

Even within the 'protected' environment of the bus, overflowings are also common. The sociotechnical arrangements I have described have indeed a blind spot, which, although it is a condition for the well-functioning of the VBT, remains an uncontrollable and unpredictable variable. This is what I would call the *performative arrangements*, i.e. the set of performances that humans need to enact in order to hold together the tour frame. It might seem that VBT passengers would just need to sit there and let themselves be detached from the city, guided through historical ambiences and reattached to the city. But this is not the case. The whole process requires tourists to be very active and dynamic. This is precisely what **Arne** is saying when he regrets not reaching a younger public more interested in and more used to multimedia. Even though at the front of the bus he has no access to the small conversations among the tourists during the tour, he probably has the sense that the following situations are not unusual:

The bus stops in the Chaussestrasse for two or three seconds. Our guide mentions an important tower. The woman sitting in the row in front of me looking through the window asks her companion about the location of the tower. He answers rapidly, 'No! On the video! It does not exist any more.'

(Fieldnotes, 26 March 2005)

Indeed, the VBT poses a challenge for tourists, who have to discriminate, switch and make connections between the films shown on the monitors and the urban environment beyond the windows. In that context, it doesn't matter how tightly spatial, visual and narrative arrangements might be assembled if the corresponding tourists' capacities are not performed.

This example shows that, rather than defining a space of tourism as opposed to the space of the city, tour frames constitute boundaries regulating the flows between the two sides. As such, frames are very vulnerable fragile technologies. In some cases, complex sociotechnical arrangements are predisposed to strengthen and reinforce them. However, large and uncontrollable arrays of factors, such as casual encounters, performative arrangements, traffic jams, technical problems with the videos, off-tour sights or events that people see through the windows, make the frame tremble, revealing how it hangs by a very thin thread.

The overflowed frame: bus route 100

It is difficult to think of a tour frame more radically different from the one described above than the one being enacted in the Berliner bus route 100. Bus route 100 was the first line that connected the two halves of the city after the fall of the Berlin Wall. During the 1990s, as tourism in Berlin started to increase, it became a 'public secret' among tourists, because its route passed the major tourist attractions, such as Alexanderplatz, Museum Island, Unter den Linden, the State Opera, the Brandenburg Gate, the Reichstag, Schloss Bellevue, Tiergarten, the residence of Germany's president, the Friedrich Wilhelm Memorial Church and many other sights. Bus 100 even passed through the Brandenburg Gate until 2003, when it was closed to traffic. In 2000 Berlin's society for public transport, the BVG, opened a second line, bus 200, which takes a slightly different route and covers another important tourist area which includes Friedrichstrasse, Gendarmenmarkt and Potsdamer Platz.

Taken together, buses 100 and 200 constitute one of the most important means of transport for touring the city. On-line travelogues and guidebooks give some very good reasons for this being so:

The best way to see the whole of Berlin very quickly is to spend an hour or so on a No. 100 bus, which passes by all the main sights . . . It is one of the quickest and most cost effective ways of touring Berlin and one ticket allows you to get on and off as often as you like within the two hours permitted.[2]

The cited description does not compare these bus routes with sightseeing bus tours, but the Berliner Tourism Information agency does: 'Public bus lines with the number 100 and 200 do pass many of the major tourist attractions between Zoological Garden and Alexanderplatz. Taking any of these buses is almost comparable with a short sight-seeing tour.'[3]

What makes the comparison feasible is the spatial and visual arrangements of these buses. In fact buses 100 and 200 pass all the main attractions and are both double-deckers. They are not open-roof double-deckers, a fact that renders the front seats into a very rare and disputed good, but they do offer the same visual angle as standard sightseeing bus tours. Thus, buses 100 and 200 have become very popular among tourists, not only as a practical means of transportation between attractions, but also as an alternative to standard sightseeing bus tours.

But the bus routes 100 and 200 provide a radically different touring experience, as they do not have tour guides who could supply the names and narratives necessary to glean the urban landscape. Indeed, most of the same on-line travelogues and guidebooks that recommend touring with these bus routes advise tourists not to forget to take some information with them. 'It's a good idea to pick up a map and information leaflet from the information kiosk at the station.'[4] In other words, to capitalize on the spatial and visual attributes of the bus, passengers need to actively produce the otherwise lacking narrative arrangements, a performance in which they engage by means of different strategies. Now, for the vast majority of unprepared passengers, the most common source of narrative arrangements is the quite uneven collection of names and symbols that can be gathered during a ride. Loudspeakers on buses 100 and 200 and the display of bus stops can become the primary sources of names.

> They don't talk much and look through the window. The child repeats loudly the names of some bus stops, as they are announced over the loudspeakers ... The loudspeakers say 'Staatsoper' and as the child asks, the mother looks for the Staatsoper to show it to her son, paying less or no attention to other buildings and monuments in the same spot.
>
> (Fieldnotes, 5 May 2006)

Another fund of names is the urban signals, streets names, signposts and traffic indications that can be caught from the bus and are used by tourists as guidelines to glean the city and its attractions.

City residents, who sometimes join tourists on buses 100 and 200 play also a crucial role. They provide tourists with rich narratives and personal insights on the sites and places visited on the bus in a much more expressive way than bus stop names and traffic signs.

We are approaching the Reichstag and what started as a very lively bus ride feels now a quite normal bus ride with people sitting, looking outside, but not talking too much . . . The only ones who talk all the time are two men in their sixties, sitting on corridor seats, mostly looking at each other and pointing now and then to some sites in the city. One of them gives the facts and tells anecdotes about the sights we pass, while the other indicates that he understands.

(Fieldnotes, 5 May 2006)

Indeed, local friends tend to be durable and consistent sources of information. After most tourists have given up gleaning the city, friends continue to provide their visitors with stories, anecdotes and information.

From observing the dynamics of tourist groups on these buses, it is also possible to recognize recurrent distributions of roles and sets of practices, which like spatial or narrative arrangements contribute to the production of a tourist frame. These performative arrangements can be described by looking at the positions and the way of engaging with the urban environment assumed by different members of groups.

I realize that in this group there is a certain division of work which prevails for the whole bus ride. The boy in the left window seat at the front of the bus is very active, taking pictures, pointing to buildings and other attractive elements of the urban environment, talking to the group, proposing activities, etc. The boy sitting next to him and the girl in the corridor seat talk to each other most of the time. The two girls in the right window seats lie back, look through the window as the city passes by, and engage in other activities. However, they listen to what their friends say and they look when the first boy points to something, but they are not very active.

(Fieldnotes, 31 May 2006)

Sightseeing on these buses can be very costly. While tickets are cheap, tourists need a lively son or a very good friend to join them and help them with names and narratives. Otherwise tourists need to make very costly investments of a cognitive, visual and narrative nature. They are required to actively glean the urban space and elaborate narratives about what they see. Some tourists try with maps and guidebooks, but they prove to be very difficult to use on the bus and after a few minutes many give up. For some it is then time to make economic investments: 'As we see a standard sightseeing bus tour coming in the other direction, the woman sitting close to me tells her partner that tickets cost about 15 euros and that they should "rather" [*lieber*] take one' (Fieldnotes, 7 May 2006).

One of the main activities tourists do while riding on buses 100 and 200 is to silently lean back in their seats and contemplate as the city passes by. Rather than a collection of sights, as in the case of standard sightseeing bus tours, the city emerges here as a continuous landscape and thereby as a whole without parts. Deleuzian philosophers would describe this city as a 'Body without Organs' (BwO). Interestingly, the idea of the BwO does not simply imply radical dedifferentiation, in this case, of the city, what Deleuze and Guattari would rather call an 'empty' BwO. The city that emerges in these buses resembles rather a 'full' BwO, which is 'what remains when you take everything away' (Deleuze and Guattari 1987: 150). The city is taken away from its attractions – organic and stratified entities of tourism – and becomes a plane of intensity – a plane on which urban everyday elements emerge as intensive singularities that overflow the tourist frame. But tourist frames are also made of such overflows.

Indeed, it is not uncommon to see tired and silent tourists looking through the windows who suddenly have some kind of revelation and start to comment to their partners or friends their general impressions of the city:

> In that moment one of the girls, who hasn't said a word in the last 5 minutes or paid much attention to the conversation of her friends, suddenly starts complaining about Berlin, expressing deception and scepticism. In her opinion, Berlin is not as great as it is supposed to be and the buildings are not very beautiful either. Her friends agree. We are approaching Brandenburg Gate and from the bus it is possible to see a big football globe in front of it. She continues making her case by pointing to this football globe and arguing that with the World Cup everything is even worse. After some moments of silence, as we turn left into the Dorotheenstraße, the other boy tries to counterbalance things by pointing to a corner building, which in his opinion is great.
>
> (Fieldnotes, 31 May 2006)

In these contexts, the tourist frame being performed is very porous. Rather than by a rigid boundary allowing only very particular aspects and sights of the city, touring with the 100 and 200 is enabled by an elastic frame that allows multiple flows between the outside and the inside. Overflows are the rule here and they are crucial, for they allow this very unique framing oriented to grasp the destination as a whole without parts, as a BwO, as a plane of immanence. Moreover, the frame enacted is so abstract that tourists are sometimes even prompted to reflect on their experiences of tourism and travel at large:

> A small bus with a licence plate from Hamburg is pointed at by the young boy sitting at the window, who comments how much he likes Hamburg and says that he would like to go to Hamburg now . . . The conversation rapidly moves to tourist destinations, and then one of them mentions Paris. They all comment how little French they know and joke about how difficult it would be to get a bottle of water . . . One of the girls, I think, says that she would like to learn Spanish. They all agree that it is easier to

learn languages when you live in the foreign country where the language is spoken.

(Fieldnotes, 31 May 2006)

Such conversations on issues not related to the visited destination are also constitutive parts of the practices of touring the city. Indeed, the tourist frame enacted in such situations embraces more than just sightseeing practices and offers, rather, a general background against which it makes perfect sense to talk about touring Paris while touring Berlin. This shows that, while tourist frames are sociotechnical arrangements, they offer an abstract basis for the unfolding of a certain type of communication, namely, tourist communication.

THE VIRTUALITY OF THE TOURIST CITY

To conclude I would like to come back to the question about the reality of the tourist city. The study of sociotechnical frames provides a good understanding of how tourist situations are actualized in the urban environment, but it falls short of grasping tourism as an urban reality or, more precisely, of understanding the reality of urban destinations as existing beyond the realm of concrete interactions. But what does 'beyond' mean? We certainly don't want to think of tourist destinations as transcendental or ideal entities, for we know that they are not prior to actual tourist activity. Destinations are enacted, put into play and transformed in tourist situations. But, at the same time, they define a tourist urban realm which is only partially contained within the frames of tourist situations. It is in this sense that I previously suggested thinking of tourism as a virtual urban reality. The crucial question we need to pose then is how tourism is virtualized or, better, counter-actualized, and the answer to that question necessarily involves identifying processes that occur at the very boundary between the actual and the virtual.

One of such processes is sense-making through communication. For Deleuze (1990) sense follows the logic of virtual events, as it brings into existence that which expresses. Sense is certainly not to be confused with signification, for it is what creates the very distinction between signifier and signified (see also Hallward 2006). Sense is thus pure virtuality, which however is constantly actualized through communication. Communication, understood as a way of processing sense, as Luhmann (1995, 1997) has put it, involves thus a permanent dynamics of actualization and virtualization. The logic of communication is indeed described by Luhmann as autopoietic, which basically means self-creative. Such self-creation involves not just the actual selection of information, means of expression and understandings, i.e. actual communication, but also sense-making processes which with Deleuze can be understood to be virtual, in the sense that they cannot be located in any of the three communicative selections, but in between, connecting and making sense of them. Following the hypothesis that communication develops at the boundary between virtual sense-making and actual communi-cations, I would like to concentrate on processes of tourist communication and argue that these processes are, first, what is framed in tourist situations and, secondly, what creates a tourist destination as a virtual urban reality (for a detailed analysis see Farías 2008).

Regarding the first aspect, one should observe that, while dynamics of fram-ing and overflowing involve the boundaries of situations and interactions, it is the communicative actualization of a specific form of sense-making that defines their quality and that renders them into tourist situations. Such a form of tourist communication cannot be reduced to the intersubjective agreements on what the situation is about or to its sociotechnical arrange-ments. Tourist, economic, political and other forms of communication can rather be thought of as virtual processes available throughout world-society to be actualized in concrete situations. But how should we think about this? In 'Communication about Law in Interaction Systems', Luhmann (1981) describes such actualizations as enabled by a crossing of 'thematization thresholds'. The transformation of everyday conflicts into legal issues, he observes, occurs 'by expressing the hitherto taken-for-granted in terms of explicit legal norms' (Luhmann 1981: 242). When these are thematized, the situation is redefined in terms of the binary code articulating legal communi-cation (legal/illegal), rendered into an objective legal issue, and consequently the capacities for conflict resolution externalized. Extrapolating Luhmann's insights to the bus tours described above would suggest that what sociotech-nical arrangements do is to facilitate the crossing of tourist thematiza-tion thresholds, enabling thus the actualization of tourist communication. Consequently, one should conclude that situational boundaries separate not just materials, objects and spaces, temporal episodes and sequences, social roles and individuals, but above all a specific form of communication. Sociotechnical frames are essentially therefore frames of communication.

Interestingly, Luhmann argues that, once law is thematized, 'even where legal issues are not explicitly discussed, communication remains latently within the legal sphere, inasmuch as one, for example, keeps in mind ques-tions of responsibility or proof' (Luhmann 1981: 248). Something similar holds true for the interface of tourist situations and tourist communication. Once the destination is thematized, there is a pull towards a tourist framing of situations. In the case of tourism, this latency or, better, virtuality might be even more lasting than in the case of communication about law, which, according to Luhmann, involves a kind of alienation and pressure that parti-cipants seek to avoid. Unlike legal communication, tourist communication is highly attractive. Multiple studies show tourists actively seeking and pushing for a crossing of thematization thresholds even before engaging in physical travel, and postponing the de-thematization of the destination and commu-nicating about it even after physically leaving it and returning home (Harrison 2001; Koshar 1998; Michalski 2004; Selby 2004b; Selwyn 1996). This makes clear that, during the stay at a given destination, tourists are most of the time involved in the thematization of the destination. This, again, involves two related things: first, that the tourist thematization of the city occurs independently of the sociotechnical arrangements making up particular tourist situations and, second, that it is in virtue of processes of tourist sense-making and communication that tourism becomes an urban reality.

We move thus to the second point, namely, how should one conceive tourist destinations if their reality cannot be derived from specific tourist situations? The argument I have made is that processes of tourist sense-making should be distinguished from its communicative actualizations at concrete situations, for what tourist sense-making does is to literally create the tourist destination or, better, render the city into a tourist destination. Tourist communication makes sense of cities as planes or horizons for tourist activities and possibilities. Urban destinations correspondingly emerge as virtual objects, planes or horizons along which tourist communication can occur, that is, can be situationally actualized. But, as virtual objects or planes, destinations are not just a result of actual communication about the city; both, actual situations and virtual planes, are rather simultaneously constituted. Situated tourist communication in urban environments occurs within the virtual plane defined by the destination, and such a virtual plane can emerge only out of concrete tourist situations.

Up to now, I have sparsely used the notion of ontology in a rather unsystematic way as a synonym for reality. Now it is possible to be more precise about this crucial notion, which is, by the way, gaining momentum in the ANT and STS worlds (Hacking 2002; Law, J. and Urry 2004; Mol 1999, 2002; Woolgar *et al.* 2008). I would like to argue that the notion of ontology is particularly helpful and adequate to refer to this virtual urban plane in which tourism occurs. Indeed, if one takes this notion seriously as referring to a reality which is beyond categories and distinctions, it can then hardly be thought of as an actuality (cf. Franklin 2004 for a use of ontology for actual tourist orderings). The notion of ontology defines rather a virtuality. My sense is that most of the current talk about ontologies in STS and ANT equates these with enactments and assemblages of practices, materialities, spaces, knowledges and so on. If ontologies are really brought into being in actual practices, then what does the notion of ontology add to the notion of assemblage or enactment and why should we need it? Perhaps the problem is that ANT has no theoretical tools to grasp the virtual, to distinguish the multiple tourist assemblages brought into being in bus tours, hostels and the infinite number of urban settings where tourism takes place from the virtual plane of activity and sense-making that enables and holds together these multiple tourist assemblages. Consequently, distinguishing between urban ontologies and urban assemblages is not a matter of principle, but one which has to do with stability and flexibility. The tourist urban destination should be stable enough to be recognized as a more or less distinguishable object across sites of tourist practices, and should be flexible enough to support multiple tourist practices, the actualization of multiple assemblages.

So what does it mean then to say that tourism is an urban reality? It certainly means that tourism creates a unique virtual ontology, but it also means more than that: it involves moving beyond the idea that the city is made of multiple assemblages to the idea that the city is made of multiple virtual ontologies. And here lies precisely the biggest complication for urban studies: since assemblages are actual and ontologies virtual, empirical urban studies can directly study only actual urban assemblages. The challenge is then to make visible through these assemblages these multiple ontologies of the city.

NOTES

1　In December 2004 Alexandra Hildebrandt, director of the private Checkpoint Charlie Museum, set up a memorial to the people shot and killed while trying to escape from East to West Germany. The memorial consisted of 1,065 large wooden crosses erected in two vacant sites on both sides of Friedrichstrasse, one of Berlin's central commercial thoroughfares. The memorial triggered much public discussion given its blatant historical and symbolic inaccuracy. It was erected on a spot where no one had been killed. It remembered all victims of the Wall with crosses, even though only a few of them had been Christian. It suggested that 1,065 victims had fallen whilst attempting to cross the Berlin Wall. In fact the number of victims is thought to be between 125 and 262 persons. The vacant land was leased by Mrs Hildebrandt from a bank, which after discussions started notified her that the deal was cancelled and obligated her to move the installation by the end of 2004. Mrs Hildebrandt refused to dismantle it. After a long legal battle, on 5 July 2005 the police tore down the illegal memorial.

2　See http://www.a2zlanguages.com/Germany/Berlin/berlin_bus100.htm (accessed 8 October 2007).

3　See http://www.berlin-tourist-information.de (accessed 15 June 2006).

4　See www.a2zlanguages.com/Germany/Berlin/berlin_bus100.htm (accessed 7 October 2007).

REFERENCES

Boorstin, D. (1987) *The Image: A Guide to Pseudo-events in America*, New York: Atheneum.

Bruner, E.M. (2005) *Culture on Tour: Ethnographies of Travel*, Chicago: University of Chicago Press.

Callon, M. (1998) 'An Essay on Framing and Overflowing: Economic Externalities Revisited by Sociology', in M. Callon (ed.), *The Laws of the Market*, Oxford: Blackwell Publishers, pp. 244–69.

Cocks, C. (2001) *Doing the Town: The Rise of Urban Tourism in the United States, 1850–1915*, Berkeley: University of California Press.

Cohen, E. (1984) 'The Sociology of Tourism: Approaches, Issues, and Findings', *Annual Review of Sociology*, 10: 373–92.

de Certeau, M. (1988) *The Practice of Everyday Life*, Berkeley: University of California Press.

de Laet, M. and Mol, A. (2000) 'The Zimbabwe Pump: Mechanics of a Fluid Technology', *Social Studies of Science*, 30: 225–63.

Deleuze, G. (1990) *The Logic of Sense*, New York: Columbia University Press.

Deleuze, G. and Guattari, F. (1987) *A Thousand Plateaus: Capitalism and Schizophrenia*, Minneapolis: University of Minnesota Press.

Delgado, M. (1999) *El animal público: Hacia una antropología de los espacios urbanos*, Barcelona: Anagrama.

Delgado, M. (2007) *Sociedades movedizas: Pasos hacia una antropología de las calles*, Barcelona: Anagrama.

Enzensberger, H.M. (1996) 'A Theory of Tourism', *New German Critique*, 68: 117–35.

Farías, I. (2008) 'Touring Berlin: Virtual Destination, Tourist Communication and the Multiple City', Unpublished Ph.D. dissertation, Humboldt Universität zu Berlin.

Franklin, A. (2004) 'Tourism as an Ordering: Towards a New Ontology of Tourism', *Tourist Studies*, 4(3): 277–301.

Franklin, A. and Crang, M. (2001) 'The Trouble with Tourism and Travel Theory?', *Tourist Studies*, 1(1): 5–22.

Goffman, E. (1974) *Frame Analysis: An Essay on the Organization of Experience*, Harmondsworth: Penguin Books.

Graburn, N. (1995) 'Tourism: The Sacred Journey', in V. Smith (ed.), *Hosts and Guests: The Anthropology of Tourism*, Philadelphia: University of Pennsylvania Press, pp. 21–36.

Hacking, I. (2002) *Historical Ontology*, Cambridge, MA: Harvard University Press.

Hallward, P. (2006) *Out of this World: Deleuze and the Philosphy of Creation*, London: Verso.

Harrison, J. (2001) 'Thinking about Tourists', *International Sociology*, 16(2): 159–72.

Holcomb, B. (1999) 'Marketing Cities for Tourism', in D.R. Judd and S.S. Fainstein (eds), *The Tourist City*, New Haven, CT: Yale University Press, pp. 54–70.

Huning, S. and Novy, J. (2006) *Tourism as an Engine of Neighborhood Regeneration? Some Remarks towards a Better Understanding of Urban Tourism beyond the 'Beaten Path'*, CMS Working Paper Series, Berlin: Technical University of Berlin.

Judd, D.R. and Fainstein, S.S. (eds) (1999) *The Tourist City*, New Haven, CT: Yale University Press.

Koshar, R. (1998) ' "What Ought to Be Seen": Tourists' Guidebooks and National Identities in Modern Germany and Europe', *Journal of Contemporary History*, 33(3): 323–40.

Latour, B. (1996) 'On Interobjectivity', *Mind, Culture, and Activity*, 3(4): 228–45.

Law, C.M. (1992) 'Urban Tourism and its Contribution to Economic Resurgence', *Urban Studies*, 29(3/4): 599–618.

Law, J. (2002) 'Spaces and Objects', *Theory, Culture & Society*, 19(5–6): 91–105.

Law, J. and Urry, J. (2004) 'Enacting the Social', *Economy and Society*, 33(3): 390–410.

Lefebvre, H. (1991) *Critique of Everyday Life*, Vol. 1, London: Verso.

Lefebvre, H. (2002) *Critique of Everyday Life*, Vol. 2: *Foundations for a Sociology of the Everyday*, London: Verso.

Luhmann, N. (1981) 'Communication about Law in Interaction Systems', in K. Knorr-Cetina and A.V. Cicourel (eds), *Advances in Social Theory and Methodology: Towards an Integration of Micro- and Macro-Sociologies*, London: Routledge.

Luhmann, N. (1995) *Social Systems*, Stanford, CA: Stanford University Press.

Luhmann, N. (1997) *Die Gesellschaft der Gesellschaft*, Frankfurt am Main: Suhrkamp.

MacCannell, D. (1973) 'Staged Authenticity: Arrangements of Social Space in Tourist Settings', *American Journal of Sociology*, 79(3): 589–603.

MacCannell, D. (1999) *The Tourist: A New Theory of the Leisure Class*, Berkeley: University of California Press.

Michalski, D. (2004) 'Portals to Metropolis: 19th-Century Guidebooks and the Assemblage of Urban Experience', *Tourist Studies*, 4(3): 187–215.

Miller, D. (1998) *The Dialectics of Shopping*, Chicago: University of Chicago Press.

Miller, D. (2002) 'Turning Callon the Right Way Up', *Economy and Society*, 31(2): 218–33.

Miller, D. (2004) 'Making Love in Supermarkets', in A. Amin and N. Thrift (eds), *The Blackwell Cultural Economy Reader*, Malden, MA: Blackwell, pp. 251–65.

Mol, A. (1999) 'Ontological Politics: A Word and Some Questions', in J. Law and J. Hassard (eds), *Actor Network Theory and After*, Malden, MA: Blackwell Publishing, pp. 74–89.

Mol, A. (2002) *The Body Multiple: Ontology in Medical Practice*, Durham, NC: Duke University Press.

Morris, M. (1990) 'Banality in Cultural Studies', in P. Mellencamp (ed.), *Logics of Television*, Bloomington: Indiana University Press, pp. 14–43.

Obrador Pons, P. (2003) 'Being-on-Holiday: Tourist Dwelling, Bodies and Place', *Tourist Studies*, 3(1): 47–66.

Seigworth, G. (2000) 'Banality for Cultural Studies', *Cultural Studies*, 14(2): 227–68.

Selby, M. (2004a) *Understanding Urban Tourism: Image, Culture and Experience*, Tourism, Retailing and Consumption Series, London: I.B. Tauris.

Selby, M. (2004b) 'Consuming the City: Conceptualizing and Researching Urban Tourist Knowledges', *Tourism Geographies*, 6(2): 186–207.

Selwyn, T. (ed.) (1996) *The Tourist Image: Myths and Myth Making in Tourism*, New York: John Wiley.

Shaw, S., Bagwell, S. and Karmowska, J. (2004) 'Ethnoscapes as Spectacle: Reimagining Multicultural Districts as New Destinations for Leisure and Tourism Consumption', *Urban Studies*, 41(10): 1983–2000.

Sheller, M. and Urry, J. (2004) *Tourism Mobilities: Places to Play, Places in Play*, London: Routledge.

Smith, A. (2005) 'Conceptualizing City Image Change: The "Re-imagining" of Barcelona', *Tourism Geographies*, 7(4): 398–423.

Urry, J. (1999) 'Sensing the City', in D.R. Judd and S.S. Fainstein (eds), *The Tourist City*, New Haven, CT: Yale University Press, pp. 71–88.

Urry, J. (2001) *The Tourist Gaze*, 2nd edition, London: Sage.

Varela, Francisco and Maturana, Humberto (1973) *De Máquinas y Seres Vivos: Una teoría sobre la organización biológica*, Santiago de Chile: Editorial Universitaria.

Woolgar, S., Neyland, D., Lezaun, J., Cheniti, T. and Sugden, C. (2008) 'A Turn to Ontology in STS? Ambivalence, Multiplicity and Deferral', Session in 'Acting with Science, Technology and Medicine', Four-Yearly Joint Conference of the Society for Social Studies of Science (4S) and European Association for the Study of Science and Technology (EASST), Rotterdam.

Zukin, S. (1996) *The Cultures of Cities*, Cambridge, MA: Blackwell Publishers.

10 Assembling money and the senses

Revisiting Georg Simmel and the city

Michael Schillmeier

> Knock out
> the chocks of light:
> Adrift, the word
> Belongs to dusk.
>   Paul Celan, *Glottal Stop*

ASSEMBLING GEORG SIMMEL AND ACTOR-NETWORK THEORY

Towards a molecular sociology [1]

This chapter introduces Georg Simmel's work on (urban) modernity as an early form of Actor-Network Theory. In the year 1907 Simmel writes in his essay 'Sociology of the Senses' that life sciences have moved from a mere macroscopic view to a microscopic perspective (Simmel 1997). From a microscopic view life sciences were able to follow the highly complex molecular associations and processes that organize macroscopic organs and systems. Social sciences, however, still describe society as if it was mere relationships of big macroscopic, organ-like objects. Hence, Simmel understood his work – similarly to the French sociologist and philosopher Gabriel Tarde[2] – as a contribution to keep up with the development of modern sciences. His sociology employs a 'psychological microscopy' to describe the molecular associations of human beings that constitute and configure the enduring 'life' of societal relations. For Simmel, modern metropolitan life is the most prominent site to follow the complex 'forms of association' (*Verbindungsformen*) of bodies, senses, minds and things that constitute the social realities of human beings (Simmel 1997).

More recently, in 2002, the French ethnographer, sociologist, philosopher and scholar of Actor-Network Theory (ANT) Bruno Latour argued in the same vein: in order to be a 'good sociologist', one 'should refuse to go up, to take the larger view, to compile huge vistas! Look down, you sociologists. Be even more blind, even more narrow, even more down to earth, even more myopic' (Latour 2002: 124). As with Simmel his plea for a good sociology sees social scientific rigour in researching the 'myopic' spaces, i.e. molecular practices that constitute and change societal life. 'The real life of society', as Simmel notes and with which Latour would agree thoroughly, 'could certainly not be constructed from those large objectivized structures that constitute the traditional objects of social sciences.' Rather, Simmel as well as ANT is interested in 'the thousands of relations from person to person, momentary or enduring, conscious or unconscious, fleeting or momentous' that 'bind us together. On every day, at every hour, such threads are spun, dropped, picked up again, replaced by others or woven together with them' (Simmel 1997: 110).

Third agents – mediating mediators

What makes Simmel's analysis of the metropolis – and for the purpose of this chapter especially his analysis of modern money economy and the senses – so interesting is that it offers a concept of societal reality that isn't based merely on abstract, logical or intellectual 'matter of fact' – argumentation that links well-defined and delimited entities. Rather, Simmel is paying attention to the different empirical ways of how relations bring 'societied existence' into being, of how these realities are stabilized or changed by them (Simmel 1997). To be sure, relations are not about relationships of fixed individual entities. Rather, they are about the very process of relating themselves that links humans in highly multiple ways, constituting emerging spatio-temporal realities. To describe the different ways we perceive each other is an important possibility. In that vein, Simmel's sociology of the senses, then, 'aims to pursue the meanings that mutual sensory perception and influencing have for social life of human beings, their coexistence, cooperation and opposition' (Simmel 1997: 110).

Moreover, Simmel's work on money can be read in the same light. It shows how modern economy and urban life are based on the very pragmatics of societal relationships that gain their historical specificities through relations mediated by money (Simmel 1990). In modern cultures and most conspicuously in modern urban cultures, the circulation of money functions as a perfect (inter-)mediator between humans *and* between humans and things. According to Simmel, money functions as a *third element*, a tool that doesn't relate given subjects and objects as if there were a-historical essences. Rather, subject and objects gain their subjectivity and objectivity as the effect of being intermediated by money. The circulation of money mediates (or values[3]) human beings by separating off calculating subjects from calculable objects. In effect, rationalistic, individualistic and intellectualistic relationships dominate societal relations as if they were mere 'matter of fact' affairs. Simmel's understanding of humans as mediated and valued by third (inter-)mediating elements into social beings is strikingly similar to ANT's prime interest in (inter-)mediated realities of human beings by circulating 'immutable mobiles' (e.g. money or technology) (cf. Callon 1991, 1998; Latour 1990).

Following an ANT perspective, in this chapter I would like first to radicalize Simmel's microscopic view on mediating/mediated realities by reassembling his work on the senses with his insights on money economy and urban life. Such a reading is thought to contribute to an understanding of how urban relations configure enabling as well as disabling scenarios of everyday life. I will argue that another third mediating/mediated agent sustains the relation between money economy and urbanity: vision. Following an ethnomethodological strategy (cf. Garfinkel 1967), I will demonstrate that the disruption and questioning of common modes of societal ordering visualize the very processes and practices that make up these modes of ordering in the first place. I will then look at how the taken-for-granted and mostly 'invisible' regimes of vision and the circulation of money and money technologies are questioned, disrupted and altered by blind money practices. I will illustrate how the clash of such highly divergent actor-networks associates different sensory practices (visual and blind) with money practices in everyday life and how it creates enabling and disabling scenarios of urban life.

The thesis of the chapter is simple: different bodies, senses and things configure actor-networks that articulate 'the city' in a new light. Obviously, scholars of urban studies have already done beautiful ethnographies of consumption practices, the consumption of urban space, shopping malls and so on. However, it seems that the very material practices and inscriptions of money and money technologies in conjunction with the sensory practices involved have been taken for granted. My re-reading of Simmel's ideas on modern urban life opens the conventional use of Simmel for urban studies. The latter suffices in highlighting the psycho-cultural effects of big-city public culture based on indifference, anonymity and the blasé character.

Ontological indecisions

Reassembling Simmel's work with insights from ANT gives a helpful heuristic to re-research urban space and revisit urban studies. It focuses on the multiple ways humans and things come into being as complex forms of associations that configure and conduct the specificities of urban life. What composes our world is actors as networks and networks as actors, since actor-networks are associating associated (i.e. heterogeneous) elements. Following ANT, the composite of our (human) world – bodies, minds and things – are *ontologically undecided* elements. As objects, bodies, senses, minds and things they differ from each other and at the same time they differ for us only as associated bodies, minds, senses and things (cf. Harman 2002; Latour 2005; Schillmeier 2008a, 2008b, 2008c, 2009). The lesson ANT draws from such an undecided ontology is to follow the actor-networks and how they bring humans and things into existence by assembling heterogeneity (Hetherington and Munro 1997; Law and Hassard 1999; Law and Mol 2002; Law and

Moser 1999; Schillmeier 2009). In that sense, the ontological question concerning 'the city' of urban studies is answered – empirically *and* conceptually – in multiple ways following the trajectories of how urban actor-networks emerge, relate, cut into, stabilize, get disrupted and change other actor-networks.

As a fairly new endeavour ANT evolved from the discourse and research agendas of 'science, technology and society' (STS). Central to ANT is an understanding of 'the social' that challenges the classical sociological view (Latour 2002, 2005). As I have argued most recently, classical sociological thought is concerned with a universal framework of 'the social' that questions the metaphysics of the single nature of objects, their territories and borders (cf. Schillmeier 2009). It is the methodology of 'the social' that is meant to add a new (i.e. social) dimension to the individual, psychological, biological and physiological realm. For classical sociology – methodologically and conceptually – the 'social' provides an explanatory space of human affairs given by the difference between societal/cultural relations and natural laws (cf. Schillmeier 2008a, 2008b, 2009). Accordingly, the interest in 'the social' unifies such a highly diverse discipline as 'social sciences'.

ANT, on the other hand, differs from such an understanding. According to ANT, 'the social' does not explain human organization or 'adding something social to the description' (Latour 2005: 107). Rather, the social has to be explained by the associated/associating (of) heterogeneous entities involved. Hence, the social does not refer to an explanatory space that explains the non-social. Rather, the social is 'nowhere in particular . . . but may circulate everywhere as a movement connecting non-social things' (ibid.). In the following, I will employ an understanding of the social as ANT does and will try to provide new insights for urban studies revisiting the work of Simmel accordingly.

The threefolded city: ExtenCities, DenCities and IntenCities

The materialities and imaginaries of urban life configure specific forms of collectivities by linking the human and non-human. Dense and extensive, urban spaces unfold territorialized as much as de-territorialized spatio-temporal configurations that assemble bodies, senses and things in highly diverse, quickly changing and idiosyncratic ways. Urban spaces connect a multiplicity of coexisting spaces concerning humans and things, the private and public, work and leisure, culture and politics, economics and aesthetics, local and global, face-to-face and virtual relations. These spaces fabricate an object of complex interconnections: the metropolis as extensions of densely situated and related bodies, minds, senses and things.

Simmel argued at the beginning of the twentieth century that:

[e]very dynamic extension becomes a preparation not only for a similar extension but rather for a larger one, and from every thread which is spun

out of it there continue, growing as out of themselves, an endless number of others . . . For the metropolis it is decisive that the inner life is extended in a wave-like motion over a broader national or international area.

(Simmel 2002: 17)

This combination of *extension* and *density* happens to be a symptomatic characteristic of what one can call *urban actor-networks*.[4] Urban spaces make up *ExtenCities* and *DenCities*, i.e. 'practised spaces' (de Certeau 1984; Schillmeier and Pohler 2006) that fabricate the vibrant and their imaginaries of urbanity and urban life where many people are linked with a multiplicity and concentration of people and a diversity of technologies, which shape urban 'objective culture' and configure related subjective cultural forms (Simmel 1997, 2002). Urban spaces double-bind seemingly non-related processes: the becoming subject and object that characterize modern life. Again Simmel lucidly says:

The most significant aspect of the metropolis lies in this functional magnitude beyond its actual physical boundaries and this effectiveness reacts upon the latter and gives to life, weight, importance and responsibility. A person does not end with the limits of his physical bodies or with the area to which his physical activity is immediately confined but embraces, rather, the totality of meaningful effects which emanates from him temporally and spatially. In the same way the city exists only in the totality of the effects which transcends their immediate sphere.

(Simmel 2002: 17)

Urban spaces visualize long, complex and multiple psycho/somatic/material extensions linking humans and non-humans. For Simmel, urban actor-networks bring to the fore the 'essential characteristic' of 'individual freedom', and 'the particularity and incomparability which ultimately every person possesses in some way is actually expressed, given form to life' (ibid.).

According to Simmel, extensive and dense urban spaces are also *intensive* space-times of intermittent circulations and translations of human and non-human configurations; they unleash the experiences of *IntenCities*. The nearness of humans and things unfolds very specific topologies of velocities and related proximities and distances, enacting what we know as urban life, constantly and directly linking or cutting, i.e. configuring, reshaping and imagining the temporalities of local and translocal material, sensory, emotional and cognitive practices (cf. Bridge and Watson 2002). Together, *ExtenCities, DenCities* and *IntenCities* effect and/or affect urban practice and experience, which obviously demand rich skills of urban life-world management. According to Simmel, urban life leads to the *intensification of nerve-stimulating affairs* (*Steigerung des Nervenlebens*), which unravels the ambiguities of desiring, provoking and maintaining *individual differences* within the mass of people, which at the same time demands attitudes and practices of *indifference* toward that very mass to keep one's individuality sheltered.

In sum, the actor-networks of urban spatialities and temporalities visualize most conspicuously, ambiguously and conflictingly the relatedness *and* non-relatedness of bodies, senses and things. With this in mind, the very processes of how in urban spaces humans and things associate and come into being provide excellent occasions to study the becoming of actor-networks. Assembling the paradox of related and non-related (heterogeneous) entities these actor-networks configure the ontological undecidability concerning the very sociality of urban life.

'Things' and the limits of the modern divisions – the tragedy of *itio in partes*

Simmel was struck by the ambivalences of modern 'change of cultural forms' (Simmel 1957b). Simmel argued that the cultural specificities and complexities of human subjective life very much depend on the ways it relates to its non-related other, the world of non-human objects (1957c), which have their own 'meaning, rules, values beyond . . . social and human life' (Simmel 1957d: 221). For Simmel it is precisely 'human culture' that becomes visible when it relates objects to 'human nature'. Objects, one may say, *object to* the very idea that human nature is self-explicatory. The more an object differs from a sub-ject, the more an object actualizes its own order, the more it gains cultural specificity, paradoxically the more it may gain connectivity to humans (1957c). To be sure, this does not mean that humans are always able to relate the cultural specificity of objects with their subjective cultural forms. Rather, Simmel diagnoses conspicuously adverse effects of modernization by which subjects and objects remain unrelated and whereby the differences between subjects and objects are levelled. He observes a 'tragedy of culture' (Simmel 1997) inasmuch as the subjective cultural forms cannot adequately deal with the 'culture of objects'. For Simmel there is a growing gap between the highly developed objective culture and the less developed subjective cultural forms.

Cultural objects are more than mere 'matter of facts'. They articulate 'matters of concern' (Latour 2005; Latour and Weibel 2005) of human organ-ization that become visible as the microscopic traces of human life. Simmel shows how urban life comes into being through the money economy, which makes us act *as if* we were merely concerned with matters of fact: the circula-tion of money is 1) speeding up, individualizing and rationalizing human life and 2) separating off calculating subjects and calculable objects, as well as 3) objectifying subject relations.

The German philosopher Martin Heidegger, another precursor of ANT, recalls the Old High German word *thing* and what makes it an actor-network as a matter of concern. He stresses:

the Old High German word *thing* means a gathering, and specifically a gathering to deliberate on a matter under discussion, a contested matter. In consequence, the Old German words *thing* and *dinc* become the names for an affair or matter of pertinence. They denote anything that in a way bears upon man, concerns them, and that accordingly is a matter for discourse. The Romans called a matter for discourse *res*. The Greek *eiro* (*rhetos, rhetra, rhema*) means to speak about something, to deliberate on it. Res publica means, not the state, but that which, known to everyone, concerns everybody and is therefore deliberated on in public.

(Heidegger 1971: 174)

Hence, things are – as Latour has argued – not mere objects or 'matters of fact' but 'matters of concern': things make things public; they bring our concerns into being (Latour 2005; Latour and Weibel 2005). Heidegger goes on to specify what a thing, namely a matter of concern, does: he refers to it as the 'thinging of things': things bring into nearness the manifold and the simple (such as, for example, the city). Things are nearing heterogeneity, so, for example, on nearness and farness Heidegger says:

Nearing is presencing of nearness. Nearness brings near – draws nigh to one another – the far and, indeed, *as* the far. Nearness preserves farness. Preserving farness, nearness presences nearness in nearing that farness. Bringing near in this way, nearness conceals its own self and remains, in its own way, nearest of all. The thing is not 'in' nearness, 'in' proximity, as if nearness were a container. Nearness is at work in bringing near, as the thinging of things.

(Heidegger 1971: 177–8)

This is precisely what urban actor-networks do. They are nearing human and non-human bodies, minds and senses. As things, urban actor-networks gather humans and non-humans. As 'matters of concern', 'things' or 'actor-networks' object to being merely 'objective' or 'subjective', 'cultural' or 'natural'. They object to being a mere effect of human freedom *or* the sole causal effect of necessities. Hence a thing is always the assemblage of *heterogeneous* entities. Simmel can show how 'thinging' assemblages of money divides subjects into objects and how the culture of objects strongly diverges from subjective cultures. It nears the heterogeneous (human and non-humans) and preserves the differences by configuring subjects and objects. At the same time – and this is Simmel's matter of concern – money levels the 'thinging of things' into relationships of matters of fact. This is the 'tragedy of culture'; things – human and non-human alike – become objects again. Owing to the tremendous development, fast distribution, dense circulation and close relation with objects, urban subjects are at risk to deal adequately with the culture of objects.

Simmel interprets the distribution and circulation of the modern 'culture of objects' in urban spaces not only as a significant 'change of cultural forms'. He also thinks that such processes of modernization demand a thoroughly new conceptual reflection concerning the relation of non-related collectivities such as subjects and objects (Simmel 1957a, 1957b). For Simmel the very 'matter of fact' of *being modern* turns into a conceptual 'matter of concern': Simmel argues that modern life cannot be properly thought of any longer from either a monistic or a dyadic point of view, which dominate the legacy of modern thought. Thus, Simmel has to acknowledge that the very fact of *being modern* cannot be explained solely with *modern* means of explanation. Monistic and dyadic concepts describe the relatedness of non-related collectivities (differences, heterogeneities) by either unifying them into 'one' or dividing them into 'many' (cf. Latour 1991). The 'one' and the 'many', to Simmel, appear as exclusive differences, i.e. as 'either–or' oppositions: they exclude the other in order to be what they are: the one *or* the many. According to Simmel, such a binary logic systematically excludes third possibilities.

This brings Simmel to the conclusion that the very processes of modern societal life cannot be properly understood if we see them as the unity of such modernist, exclusivist and binary differences that divide – *itio in partes* (Simmel 1957b: 102) – between freedom and necessity (as Kant did) or subjects and objects (as Descartes did). According to Simmel, the Kantian and the Cartesian models are the modernist possibilities of dealing with the other. The Cartesian view perceives what does not belong to the subject as 'object' (cf. Simmel 1994, 1997). As different from the subject, we imagine the existence of objects to be independent from us, and only because we perceive them as 'objects' can they have an effect upon us. Against this mechanistic idea, a Kantian tradition has centred the world on the subject. An object becomes an object not thanks to the distance to the subject but, on the contrary, through being perceived by the subject. The object is included as a moment of subjective perception regulated by human rational minds. The subjective realm is dependent on an outside world, a *Ding an sich* that offers the possibility to sense something in the first place. However, it remains a dark object, an unknown *Ding an sich*, as long as it is not thought. According to Simmel, neither the Cartesian nor the Kantian strategy is particularly useful, since both are based upon a clear divide between objects and subjects, which leaves the change of cultural forms – as it becomes visible in modern urban spaces – unexplained.

Etymologically the notion *itio in partes* refers to the separation of groups, gatherings and assemblages to achieve decisive orders – and this was Descartes's and Kant's foremost aim.[5] For Simmel, however, although human life affairs are made up of heterogeneous entities (mental, physical), they cannot be conceptualized *itio in partes*. Rather, he stresses,

[i]t seems to me . . . that their seamless integration has begun to crumble and that, out of the rupture, a demand or a vague intuition arises – it is no more than that yet – namely, that the essential form of our will is something beyond both necessity and freedom, and is some third entity that does not submit to this alternative . . . The difficulty lies in the fact that the accentuated rejection of duality logically seems to leave only unity as an alternative, but that in so doing this does not conform to the actual picture. For we do not really gain anything if we proclaim the human being to be the unity of the mental and physical . . . I would therefore like to maintain that neither duality nor unity adequately expresses their relationship and thus that we still possess no conceptual formulation at all for this relationship. And this is notable because unity and duality logically contradict one another so strongly that each relationship of elements must necessarily succumb to the one if the other is negated by it. Nonetheless, even this alternative is now precarious for us; it has done its duty, as it were, and we demand for the nature of life, as it is both physical and mental, an expression of form of which we have so far been unable to say anything other than that it will be a third entity beyond the apparent and hitherto compelling alternative.

(Simmel 1997: 106–7)

For Simmel, we moderns constantly produce *third elements* that mediate humans and non-humans into societal relations of things. Money practices for instance clearly separate subjects and objects. But at the same time money relations level subject–object relations. They make people treat each other as mere objects. Moreover, these processes of objectification enrich the possibilities of individual life by levelling and endangering subjective forms of dealing with these objectified and individualized relations. In that sense *modern (urban) culture has never been modern*, since the empirical experience of urban life (and not the universal concept of human being!) cannot be explained any longer by modern concepts that clearly divide between subjective and objective modes of explaining societal reality. This appears even more dramatic and tragic since it is the very modern hope in the human being that is endangered by objects and object relations: objects were thought to have been tamed once and for ever by the Cartesian subject or the Kantian human rationality. Thus, the third concept is to be found in how we assemble modern life itself and not constructed as a mere intellectual concept of theoretical, rational or philosophical thought. This is Simmel's message that is so important for ANT scholars: the empirical is more complex than the conceptual. And the city is the perfect place for finding out about it. It is precisely the empirical reality of the urban experiences of *tragedy of culture* which brings to the fore that mere intellectual divisions in the mode of *itio in partes* are at their limits. Consequently, Simmel (1997: 107) argues that the tragedy of culture visualizes that:

[n]owhere is it clearer that our means for mastering the content of life by intellectual expression are no longer sufficient than in the failure of the previously valid conceptual alternatives and the demand for a still unformulated third way. Nowhere is it clearer that what we wish to express no longer fits into these conceptual pairs, but rather bursts them asunder and seeks new forms which, for the moment, announce their secret presence only as an intimation or an uninterpreted facticity, as a desire or as difficult tentative trials.

(Simmel 1997: 107)

Beyond *itio in partes*

Simmel offered such a third way that transcends an *itio in partes*. For him, the *Wesen* of cultural *life*[6] refers to the immanence of limits set *and* limits transgressed. It is the concept of 'cultural life' that unfolds a logic of 'more than one and less than many' that again links very much with insights provided by Actor-Network Theory (cf. Law 2002; Mol 2002). Simmel argues that it is the culture of objects (e.g. technoscientific tools like the telescope or the microscope) that changes our ways of life by de-limiting and re-limiting the relations of our overall organization (*Gesamtorganisation*) to our world of representations/visualizations (*Vorstellungswelt*) (Simmel 1994: 4). Thus cultural life not only is a mental act, but involves sensory practices, material culture and nature as well. As cultural collectivities we transcend our mere human realm inasmuch as we live in extensions (with and through others – other humans, technologies, natural objects). Living culturally, i.e. extending or associating our minds with bodies and our associated minds/bodies with others, human and non-human alike, we humans imitate nothing but life.

Thus, cultural life is not something added to human life, but it *is* the process of, the becoming of, human life – life that according to Simmel is always *more-life* (continuity) and *more-than-life* (individuality) (Simmel 1994: 20). *More-life* refers to the intrinsic movement of life, as long as it is life it generates lively things (*Lebendiges*). Life has to make a difference to continue and fabricate life. So, for example, mental life becomes material form (words, practices) when it actualizes its 'soulful energy' (ibid.): it becomes other to life. Life means also the 'perpetual perishing' of life (Whitehead 1978: 29). *More-life* is transgressing, overflowing life into forms and life itself overflows all forms. As life it has to associate the forms it overflows. Being life, life needs form, and, being life, life needs more than form (Simmel 1994: 22). *More-than-life* refers to the creative, the individual act, the imaginary that overflows mental life in the sense that it generates a self-significant (*eigenbedeutsam*) and self-referential (*eigengesetzlich*) sphere (Simmel 1994: 24). This marks the creative act that produces novelty out of repetitions of differences. *More-than-life* transcends life. Transcendence is the immanence of life. Transcending life as *more-than-life* is immanent to physiological life as much as it is immanent to mental life: it generates a self-reliant other, heterogeneity.

It constitutes a collective that, by including its other, gains continuity and individuality.

Cultural life refers to the immanence of life, the event[7] of life itself. Cultural beings understood as eventful collectivities circumscribe the continuity of boundary transgression owing to the process of assembling other collectivities. To continue life, then, is associating/associated Otherness. This is precisely what makes up the secret presence of the tragedy of culture, as Simmel stresses (1957e). From an ANT perspective, the parlance of the secret presence of the tragedy of culture refers to the 'empirical metaphysics' (Latour 2005: 51) of everyday modern life that goes beyond the existing philosophical, namely conceptual, understanding *itio in partes*. Moreover, Simmel's understanding of life can be understood as a deconstruction of the self-closed notion of 'society' or the 'social' that explains the cultural life instead of giving cultural life a voice to disrupt, question and alter 'society' and 'the social'.

Again, Simmel's account on the (inter-)mediation of money practices in urban spaces can be read in such a way. Simmel explores very convincingly how the circulation of money extends our social circles and relations, and how we become highly individualized. At the same, we money users are all at risk of becoming mediated by objectified money relations levelling our individuality, and it is the city, as Simmel also shows, that magnifies such risk practices. From such a view, cultural life refers to human practices as the dynamics that unfold realities that extend human limits and consequently put the cultural limits (of humans, of humanity) at risk as well. Thus, the extension of humans that transgresses its own limits is nothing added to human being, but it is immanent to its ontogenesis – a risky process of becoming collective. Becoming human is becoming collective, and becoming collective is setting, questioning, disrupting and altering the boundaries involved. Again, we find a very precise description of the mediated ontology of actor-networks, and it is no wonder that the (inter-)mediation of money plays a crucial role in ANT (cf. Callon 1991; Schillmeier 2007a, 2007b).

My reading of Simmel suggests that it is the assemblages of heterogeneous (i.e. *non-social*) entities that mediate the practices of becoming social. Cultural life as *more-life* and *more-than-life* can be seen as a third concept beyond modernist (post-metaphysical) thinking of the Cartesian and/or Kantian legacy. To treat a thing *itio in partes*, to divide the very assemblage into the human or non-human nature of objects as if they were isolated objects, has been Simmel's main concern. According to Simmel, it is the tragedy of culture most conspicuous in modern metropolitan life, which – in order to make sense of it – demands new concepts that include the excluded third to describe the 'change of cultural forms' *in the making*, namely as a 'becoming'. As Deleuze and Guattari have noted:

> becoming does not occur in the imagination [but is] perfectly real [and] produces nothing other than itself . . . [A] becoming lacks a subject

distinct from itself . . . it has no term, since its term in turn exists only as taken up in another becoming of which it is the subject, and which coexists . . . with the first . . . Becoming is always of a different order than filiation. It concerns alliance . . . *symbiosis* that brings into play beings of totally different scales and kingdoms, with no possible filiation. [Becoming] does not go from something less differentiated to something more differentiated [but unfolds a movement] between heterogeneous terms . . . Becoming is a rhizome, not a classificatory or genealogical tree.
(Deleuze and Guattari 1988: 262–3; cf. Grosz 1999)

Simmel was interested in such becomings of urban life gathering human and non-human bodies, minds and senses. The notion 'actor-network' Simmel unwittingly was looking for offers a powerful post-Cartesian and post-Kantian possibility to think about third concepts. It tries to capture the idea that actors, human and non-human alike, their abilities or disabilities to do things (acting, thinking and feeling), are the *effects* and *affects* of associating, circulation and cutting of heterogeneous entities. Herein lies the very strength of reasoning along actor-networks. Actor-networks 'give voice' to the third elements (like money, technologies, etc.) that configure, for example, humans and non-humans as subjects and objects. These third elements ANT calls intermediaries and mediators. An intermediary 'transports meaning or force without transformation', whereas mediators 'transform, translate, distort, and modify the meaning or the elements they are supposed to carry' (Latour 2005: 39ff.). The description of actor-networks, then, is not about the focus on *relationships of entities as sources*; rather, it tries to follow and describe how actors come into being as the *effect or affects of associating heterogeneous forces of different collectivities* (e.g. minds, human and non-human bodies, senses). This process of eventful association can be called *translation*. Only through translation does the composite of our world come into individual being, remain the same and/or change. Thus actors are translated translators – human and non-human alike (cf. Callon 1991; Latour 2005; Schillmeier 2007a).

Such a reading of human and non-human actors and their actions gives voice to 'a new theory of action in which what counts are the mediations and not the sources' (Callon 1998: 267). Thus the concept of (inter-)mediating heterogeneity names the risky ontology of beings as collectivities *in the making*. Moreover, it acknowledges that culture transcends culture to be culture (culture is more than culture; it desires nature), life transcends life to be life (life is more than life; it moves towards death) and human relations transcend human relations to be human relations (humans are more than humans; they desire the non-human).

Following an ANT reading, Simmel's understanding of cultural life as *more-life* and *more-than-life* names the very *sociality* of cultural life. The social is the crossing (*hinübergeifen*), infringing and encroaching on its own limits by associating the non-social. The continuity of human social affairs exists precisely in the crossing of its own limits, in its reaching beyond the social. The form of the social is given as a moment of overflow of its own form (cf. Callon 1998). The related heterogeneity of the social, then, is not an autonomous sphere separated from the non-social, but an emerging entity that gains its durability and specificity by a constant movement/movability of linking the material, the sensory, the mental, the individual.[8] The sociality or socialities emerging from the extensions, densities and intensities of the city assemble most conspicuously Simmel's understanding of life, and the tragedy of culture is the first sociological attempt to name societal controversies over agency that cannot be explained away by the 'mental' or 'the material', 'the subject' or 'the object', or 'the human being' or 'society'.

THE CITY, SENSORY PRACTICES AND MONEY

In his work on (urban) modernity Simmel was concerned with the 'sociological mood' of human affairs. Here, two aspects (together with many others) played an important part in his analyses: sensory practices and money (Simmel 1990, 1992a, 1992b, 1997, 2002).

Sensory practices play a crucial role, since:

> sensory impressions, running from one person to another, in no way serve merely as the common foundation and precondition for social relationships, beyond which the contents and distinctive features of those relationships arise for quite different causes. Rather, every sense delivers contributions characteristic of its individual nature to the construction of sociated existence; peculiarities of the social relationship correspond to the nuancing of its impressions; the prevalence of one or the other sense in the contact of individuals often provides this contact with a sociological nuance that could otherwise not be produced.
>
> (Simmel 1997: 110)

Societal relationships that emerge from sensory practices are not a mere human affair but include the non-human world as well; and they are not a matter of mere social constructions of sensory practices but constitute the societal orders through the very material specificities of sensory practices (ibid.). The realm of visual perception dominates modern cultures, and the metropolis is very much an assemblage of visual regimes. Next to vision it is the circulation of money that mediates humans and non-humans into subjects and objects: calculating subjects and calculated objects. This was one of the lessons Simmel and scholars of ANT tried to unravel in their writings (cf. Callon 1991; Schillmeier 2007a, 2007b).

As mentioned above, Simmel shows how the very power of actor-networks of money changes individual and societal urban life. Money is strongly associated with the specific ambiguous ways of modern life inasmuch as it individualizes as much as it rationalizes human social life, as it brings about 'the concentration of men and things' (Simmel 1997: 179) whereby men in order to keep up their individuality act in a highly indifferent and impersonal way. Moreover, 'the self-preservation of certain personalities is bought at the price of devaluing the whole objective world, a devaluation which in the end unavoidably drags one's own personality down into a feeling of the same worthlessness' (ibid.). What is important to note is that Simmel shows how these actor-networks associate and (inter-)mediate the heterogeneous: bodies with other bodies, mind, emotions and matter, private and public, local and global, quantity and quality, senses and things, the personal and impersonal and so on.

Urban assemblages – vision and money

Simmel (1997) argues that in urban spaces the presence of (highly fleeting) face-to-face-relations mixes knowing, namely being affected by the other (*Kennen*), and recognizing, namely recognizing the other (*Erkennen*). Like money that separates off calculating subjects from calculating objects, visual relations enact highly separated though interacting entities that constitute and construct humans as being different from non-humans. According to Simmel, the social relation of humans and non-humans reinforces the difference between the spaces of subjective feeling (e.g. the beauty of the smell of the rose) and objective spaces that try to isolate the oneness of objects (e.g. *the* rose, *the* person). According to Simmel, both processes have to be taken into account. Hence, perception between humans and humans and non-humans is not simply the result of subjective or objective perception but emerges out of the *mixing* of being affected by the other (subjective) *and* knowing this other (objective) that mutually constitutes a complex weaving of social relations.

Urban spaces are very much visual spaces, spaces of 'sight-seeing' as it were. Face-to-face relations bring to the fore the double characteristic of the eye: 1) revealing the specificities of individual appearances and 2) grasping the likeness of beings. It is precisely the sense of the eye which allows the delimiting of the different (different people and things) and the generalizing of the alike (mass, the sun, stars, clouds). The sense of the eye, then, enables individualization and collectivization. It binds together individuals who see particularities but also individuals who share what they see. The time-spaces of actor-networks of sight, one may say, reveal the individual and the momentous as much as they render stable the common, general and universal.

Never before has a person been exposed so intensively to the look of the other as in the city. Simmel argues that the proximity of humans given by public transport and dense public spaces of economic and cultural consumption increases the frequency and fluctuation of face-to-face-relations (Simmel 1997; cf. Levin 1993). In order to deal with the 'intensification of nervous stimulation' (Simmel 1997: 175) of urban life, a new sociological mood of non-conversation, a *blasé character*, as Simmel would say (1957e, 1997), becomes dominant: a blank character who allows one to move around masking oneself behind a blank face in order to gain distance and preserve one's individuality within the metropolitan mass. Hence, the physical proximity of people as well as personal difference is socially endured and embodied by a moment of concealment.

Money plays a crucial co-forming role of urban practices. As I have said above, not only does money enact – like vision – the difference between humans and non-humans, but it performs human beings as calculative subjects and calculable objects. The economy of money installs the very modern division between subject and object. Money acts as a perfect 'immutable mobile that links goods and places' (Latour 1990: 58; Simmel 1990). It 'is mobile (once it is coined), combinable, and can circulate through different cultures, it is immutable (once in metal), it is countable (once it is coined), combinable, and can circulate from the things valued to the center that evaluates and back' (Latour 1990: 58). Money extends into a multiplicity of immutable mobiles (cash, credit and debit cards, paper money, cheques, electronic money, ATMs, bills, etc.) that make up a dense 'cascade of mobile inscriptions' (ibid.) translatable into, and thus made compatible with, other inscriptions. Hence, in order to function as such a highly immutable mobile, money must be materially visible in whatsoever form and connectable, i.e. readable. Hence, the mediation of money practices mediates sensory practices dominated by vision and vice versa. Money and vision are co-intermediaries of city life. The velocity of modern cultural life, as Simmel (1992a, 1992b) has discussed, is very much the effect of a network of visual sensory regimes, money and money technologies 'that makes rapid translation between one medium and another possible' (Latour 1990: 58; cf. Simmel 1990; Schillmeier 2007a).

Thus, the circulation of money needs mechanisms, practices, and places of organizing and institutionalizing potential translations. It requires the readability of the inscriptions provided by money and money technologies. Money inscriptions come in different forms, sizes, shapes and textures of different currencies. They come in letters such as bills or bank accounts, sometimes in Braille but mostly in 'normal' visually inscribed fonts, readable only by the person with 'normal' sight. Money comes in coins, in different paper forms, in plastic cards, through ATMs; it is shuffled to and fro electronically worldwide. Each mode of money demands a particular use and location, standardized procedures and time arrangements. One may say money and money technology come in remarkably complex and different material forms that all enact certain inscriptive specificities of readability. Dealing with money and money technologies requires complex processes of translation that make money connectable and usable. Most profoundly, one needs good sight! And good sight is normal sight. The mediation of money individualizes, rationalizes, intellectualizes and impersonalizes, but normalizes its users by enacting normative forms of sensory practices. However, as I will describe below, such processes of normalization also visualize 'non-normal' practices, practices that do not fit, assemblages that disable doing 'things of vision'. This reminds us that complex and global actor-networks such as these mediated by money and vision remain local and hence fragile at every moment of mediation.

Metropolitan dis/abilities – money, vision and blindness

I have shown elsewhere in detail how ordinary acts of everyday practices of blind people like 'going shopping' or 'dealing with money' configure enabling and disabling scenarios of human life when different sensory practices and related actor-networks meet. When people are enacted visually disabled these people experience social marginalization and exclusion, public vulnerability, socio-psychological stress, alienation, dependencies and so on. At the same time, 'blind' practices assemble counter-practices to vision, and disrupt, question and alter the hegemony of taken-for-granted modes of visual societal ordering (Schillmeier 2007a, 2007b, 2008a). To follow actor-networks of visual dis/abilities in urban spaces allowed me to address the strengths and fragility of urban actor-networks and how they are stabilized but also questioned and altered by the way bodies, senses and things connect or refuse to do so. Moreover, the very focus on mediated dis/abilities brings to the fore that common social modes of orderings and related actions and skills cannot be directly deduced from dis/abled subjects *or* dis/abling objects. Rather, it highlights again the very limits of *itio in partes* and draws attention to the very translation/mediation of abilities into disabilities and vice versa as a process of assembling heterogeneous entities. In this last part of the chapter I will discuss briefly some qualitative empirical work that I conducted in Manchester and Liverpool (North-West England) in order to show 1) how urban practices configure time-spaces of dis/ability and 2) how urban studies may gain new insights along an ANT heuristic.

Cash points

As we know, cash points or ATMs are 'points of access' to money networks that can easily become centres of fraud. However, they can also turn into centres of disability. Almost 100 per cent of the visually impaired people whom I have interviewed or followed around do not use ATMs. One reason is that publicly accessible technologies like ATMs 'subject individuals to vulnerability', as Goffman has already noted (1971: 300, 302). Public urban spaces create 'lurk lines' (1971: 293ff.) that conduct a rather unpleasant panoptic situation for the blind where everybody except the blind can see what they do. Blind people cannot look back or cannot react back, as sighted people can do. They feel a certain lack of control as they remain exposed to the scrutiny of the sighted others: the sighted see the blind, whereas the sighted often remain absent for the blind (French 1999; Hull 1991). Moreover, visual face-to-face practices permit a wide range of non-verbal communication that appears as rather limited for those who are blind. Accordingly, there is a strong asymmetry of communication.

The blind and the sighted do not share the same sensory practices, experiences and knowledge. In cities ATMs materialize complex spaces and their effects, which make some of their users disabled. Using a cash point is not easy for the visually disabled. If one has a valid card, keys in the right pin number, does the right sequences, and stays within the given time limits the operation demands, one might be successful in getting money out. If one misses a sequence, types in a wrong code or takes too long, one has to start anew, knowing though that the number of attempts is limited; after the third mistake in a row one is considered untrustworthy and the card is swallowed. Cash machines are good intermediaries of money practices because they are self-restricted and lack flexibility. In order to be safe and trustworthy, ATMs have to use highly standardized interfaces for a highly standardized user. They are what cybernetics calls 'trivial' machines, as they are made to accept only standardized procedures: they do what one wants perfectly well, in a secure and trustworthy manner, but only under *their* conditions; and these conditions are highly visually designed and configured. If one does not fit the 'normative packages' (Star 1991), which the ATM configures by design, one remains disconnected, disabled to make money flow. One is made blind as a consequence of not being able to read and follow the instructions of the ATM. To put it differently, ATMs often cannot put up with the complexities of different sensory practices, such as those of blind people; for the blind these machines are too 'simple' inasmuch as they lack flexibility to deal with blind practices. As a consequence, ATMs enact the normality of visually dominated practices and visualize blindness. To make the problem worse, cities have many different versions of ATMs that do not require the same sequences or share the same interface design, which means that, among other things, the blind person has difficulties finding the slot in which to insert the plastic card. This forces the person to fiddle around conspicuously with his or her e-card, which can be understood as 'egocentric preserve' (Goffman 1971: 29). Understandably one tries to avoid such situations, as nobody likes to display his or her insecurities with personal affairs publicly – the more so if it concerns money affairs.

If the person manages to insert the card, follow the right sequence and receive money, the next challenge is to check on the spot how much the machine has given. To make sure of the amount of money received, the blind person needs time for touching, comparing, measuring and feeling the money. But, again, one does not tend to do these things in public urban spaces, especially if one already feels vulnerable! As a result, many blind people are inclined to avoid cash machines, not only because they are blind, but because they are made disabled by the visual infrastructure of money technologies and how these intermediaries enact risky public spaces. In dense urban spaces, the extension and circulation of money enable the intensities of urban practices, but only as long as these practices comply with certain standards. Money and money technologies are standardized and standardizing; they normalize and configure the 'normativity' of their users as sighted and abled and blind and disabled; they gain or lose agency.

This example shows how blindness and sightedness are not inherent properties, but effects and/or affects constituted out of heterogeneous relations involving different materialities and technologies, different people, and different (sensory) skills, which make up riskiness of (urban) societal spaces. Cash machines mediate silently risky public spaces and the more so for blind people. Cash points create and visualize the vulnerability and disability of blind people and enforce the optical asymmetry with sighted people's spaces and their visual practices.

Paying cash

> Beforehand I plan how much I will use. Well, I only use five- and ten-pound notes. That's how I manage it. If in the shop they know you, they are usually very friendly. They say 'That's a ten-pound note', and I say 'That's a ten-pound note, yes, I know.' I cannot tell the difference from the size. I use a plastic template to measure. I fold them too, half and half again. . . . You give a note and you get coins as change and you are not sure what it is because there are a lot of people queuing around you. People tend to push. It's a vicious circle. I don't want coins as change but I always get it.
>
> (Elizabeth, 50 years old, totally blind)

> When you get change they will give you a pile of change and you are stuck there and people are in a rush. And you cannot do anything, you can put them in your compartment, and then when you come home you are in a real mess. You spend ages sorting them out.
>
> The problem is that people don't count out the change to you, do they? That's the problem. They hand it to you unless you have someone who is there to check it for you. You can't have someone around all the time, can you? Well, then you take it home.
>
> So say when there are a lot of people, people are in a rush, there you are, in a queue. You are stuck. You cannot do anything; you give the note, say a 20-pound note. You cannot check what you get back. You have to trust, trusting, yes.
>
> (Margaret, 62 years old, age-related macular degeneration)

Money and money technologies enmesh with blind practices as good or bad money-in-practice. They visualize some of the specificities concerning the dis/abling practices of blind people in urban configurations. As we have already seen above, blind money practices disrupt the standardized mechanisms enacted by the very normative readability of visual systems. They not only question the very limits of visual mediation and expose visual relations to their limits of flexibility, but they also stop the flow of money.

Needless to say, for the blind, money practices are time-consuming and rather restricted in the possible scope of transaction. This is more or less problematic depending on *where* they deal with money and *who* or *what* translates money designed for good vision into money that works for the blind person. For the blind the spaces and velocities of 'town practices' are often most forceful assemblages to become disabled, dependent and vulnerable. If a blind person can feel, sort out and mark money at home, sitting peacefully in the living room, the chances that the translation of bad money into good money is successful are quite high.

Consider Elizabeth's ten-pound note. It is good blind money-in-practice because she can be sure about the value of the note after having used a template to mediate the ten-pound note designed for good sight into a 'tenner' for blind Elizabeth. After having paid with good money she may get back bad money. Obviously, in Elizabeth's case the experience of bad change is not about money as a matter of fact. Rather, it is where and how the money exchange is performed. In Elizabeth's case, it is a matter of urban space where people do not necessarily share the same sensory practices but partake in performing the velocity of urban life: people 'tend to push', as Elizabeth says. Elizabeth and Margaret can tell many stories about situations where they receive change that turn good blind money into bad money of the sighted.

In such a moment, 'foggy' relations of money appear, and money becomes messy, fugitive and volatile (cf. Schillmeier 2007a). Indecisive, foggy money is like a patina over clear and distinct forms of visual inscriptions. In consequence, the blind person gets stuck in visual money relations. Foggy money visualizes blindness and brings the calculability of money to a halt. Currencies lose their trustful invariance and appear worn out. The same coin or note starts an ontological dance of vagueness: is it a one-pound coin?[9] A two-pound coin? A ten-penny coin? None of those? Metal or stone? Money stops working as *the* intermediary of economic relations that *mediates* humans and non-humans into calculative subjects and calculable objects. Rather, owing to the failure of money and the senses as unquestioned, smooth and predictable intermediaries, money practices *mediate* human and non-humans into disabled subjects and disabling objects. In visual cultures like modern urban life money and vision assemble good intermediaries that translate money into readable money; money circulates unproblematically and translates with no trouble into a unity of account, exchange and referencing value.

Blind practices question, disrupt and alter money/vision assemblages. In effect, money and vision assemble bad intermediaries that make up actor-networks that don't allow the fast and easy use of money. These assemblages mediate visual disabilities which mediate time-spaces of blind practices which unfold a variety of opportunities and costs, enabling and disabling practices.[10] Sighted money marginalizes non-visual practices, and simultaneously bad sight or blindness marginalizes money. For the blind, 'sighted money' feels like foreign currency or even like no currency at all – it remains pure matter. In the hands of the blind, 'sighted money' works as the money of the non-modern past used to – or as it did in the countryside: only travelling locally and for rather limited purposes. With sighted money the blind are situated within a non-modern context where they themselves appear non-modern, i.e. disabled from participating in modern urban culture.

Blind practices visualize that such a universal medium of translation as money is at every moment of mediation a highly specific, particular and situated agent. Blind practices show that (inter-)mediaries like money are never neutral. Rather, the circulation of (inter-)mediaries is enacting normative packages. Generally speaking, blind practices bring to the fore the very visibility of the sociality of money/vision practices. Sociality:

- mediates the materiality of inscriptions (different colours and textures, different shapes and weights, different figures and edges) by different sensory practices;
- mediates individual sensory practices with public and economic practices;
- mediates different materialities (e.g. visual and blind) that enact relations of time and space that again mediate other societal relations which enable and disable in various ways.

MYOPIC PROSPECTS

Obviously, revisiting 'the city' with Georg Simmel and ANT did not provide the big picture of what the city *is*. Rather, following Simmel and ANT 'the city' can be understood as a complex weaving and cutting of assemblages that make up the very specificities of urban *socialities* in the making. These socialities emerge as contingent and heterogeneous realities by which human and non-human bodies, minds and the senses are associated in highly multiple ways. Thus, in order to draw attention to the emerging socialities of the city one must come quite close to the local practices. In this chapter I focused on ordinary acts of everyday practice of how blind people deal with money and money technologies. I tried to show how these assemblages mediate sensory and money practices in highly complex and powerful ways, for example enabling and disabling social practices.

Revisiting the city with Simmel and ANT tries to give a voice to all the specificities of local practices that make up the complexities of urban assemblages. In that sense, it is the practices and assemblages that teach us about the reality of the city and not our often too inflexible concepts about them. It is even more complicated than this: as myopic social scientists – as Latour has put it – we are 'partially connected' and in turn partially enacted by the practices and assemblages observed (Strathern 1991). This means that the safety of observational, conceptual and personal distance of a 'spectator theory of knowledge' (Dewey [1929] 1960) is questioned and lost. In that sense we as scientific observers can learn very much from the world of the blind. It is precisely *looking down* – as Simmel and Latour propose – that is of big importance for a myopic seer like Sally French:

> The ground is where most of the danger lies; by looking down I have the greatest likelihood of seeing objects or steps in my path. True I sometimes get slapped in the face by an overgrown hedge, but ground objects are far more common and my distance vision is so restricted that looking down is much more functional than looking up. As for lagging behind, well, following other people, looking at what they do, observing their feet to see where they tread, is a sensible way to behave. It is they, not I, who are finding the steps, it is they who are finding the way.
>
> (French 1999: 22)

No doubt the practices of myopic seeing are risky practices. What we can learn from myopic seers is that, in order to gain knowledge of our world as social scientists, we have to follow and get in touch with the traces and obstacles of the assemblages of people and things. Such an account may provide insights into the becoming of extensive, dense and intensive socialities. Clearly, this may put the social scientific frame of reference and understanding at risk as well, precisely since it cannot be separated off from the things observed. This is the risky chance involved in revisiting 'the city' with Simmel and ANT.

NOTES

1　On 'molecular sociology' see also Lynch (1993).
2　Cf. Latour (2002, 2009); Schillmeier (2008b).
3　Simmel has developed a theory of value to describe the *relational* character of subjects and objects. See Simmel (1990).
4　Ways of describing actor-networks can be found in Callon (1991, 1998), Latour (1988, 2005) and Schillmeier (2006, 2007a, 2007b).
5　For example, the division within the Reichstag of the Holy Roman Empire after the Peace of Westphalia 1648 into one Corpus Evangelicorum and one Corpus Catholicorum.
6　I interpret the term 'life' in line with ANT as nothing but (inter-) mediation. The latter, like Simmel's 'life', is ontologically undecidable inasmuch as it is 'opposite of form'. Like 'life' it precedes or transcends all forms, and to succeed in giving a conceptual definition of it would be to deny its essence (Simmel 1997: 107).
7　See also Deleuze (1996: 129ff.); Schillmeier (2005).
8　For a similar, modest reading of the social see Gabriel Tarde (1962, 1969, 1999, 2000). See also Latour (2002, 2005) and Schillmeier (2008b).
9　On blindness as ontological vagueness, see Schillmeier (2005).

10   The blind figure begging for money is the most visible sign of such alteration.

REFERENCES

Bridge, G. and Watson, S. (2002) *The Blackwell City Reader*, Oxford: Blackwell.
Callon, M. (1991) 'Techno-economic Networks and Irreversibility', in J. Law (ed.), *A Sociology of Monsters: Essays on Power, Technology and Domination*, London: Routledge, pp. 132–61.
Callon, M. (1998) 'An Essay on Framing and Overflowing', in M. Callon (ed.), *The Laws of the Markets*, Oxford: Blackwell, pp. 244–69.
de Certeau, M. (1984) *The Practice of Everyday Life*, Berkeley: University of California Press.
Deleuze, G. (1996) *Die Falte: Leibniz und der Barock*, Frankfurt am Main: Suhrkamp.
Deleuze, G. and Guattari, F. (1988) *A Thousand Plateaus: Capitalism and Schizophrenia*, London: Athlone Press.
Dewey, J. ([1929] 1960) *The Quest for Certainty: A Study of the Relation of Knowledge and Action*, Gifford Lectures 1929, New York: Capricorn Books.
French, S. (1999) 'The Wind Gets in My Way', in M. Corker and S. French (eds), *Disability Discourse*, Buckingham: Open University Press, pp. 21–7.
Garfinkel, H. (1967) *Studies in Ethnomethodology*, Cambridge: Polity Press.
Goffman, E. (1971) *Relations in Public: Microstudies of the Public Order*, London: Penguin Press.
Grosz, E. (ed.) (1999) *Becomings: Explorations in Time, Memory and Futures*, Ithaca, NY: Cornell University Press.
Harman, G. (2002) *Tool-Being: Heidegger and the Metaphysics of Objects*, Chicago: Open Court.
Heidegger, M. (1971) 'The Thing', in M. Heidegger, *Poetry, Language, Thought*, transl. Albert Hofstadter, New York: Harper & Row, pp. 165–82.
Hetherington, K. and Munro, R. (eds) (1997) *Ideas of Difference: Social Spaces and the Labour of Division*, Sociological Review Monograph, Oxford: Blackwell.
Hull, J.M. (1991) *Touching the Rock: An Experience with Blindness*, London: Arrow.
Latour, B. (1988) 'Mixing Humans and Nonhumans Together: The Sociology of the Door Closer', *Social Problems*, 35(3): 298–310.
Latour, B. (1990) 'Drawing Things Together', in M. Lynch and S. Woolgar (eds), *Representation in Scientific Practice*, Cambridge, MA: MIT Press, pp. 19–68.
Latour, B. (1991) *Nous n'avons jamais été modernes*, Paris: Editions La Découverte.
Latour, B. (2002) 'Gabriel Tarde and the End of the Social', in P. Joyce (ed.), *The Social in Question: New Bearings in History and the Social Sciences*, London: Routledge, pp. 117–32.
Latour, B. (2005) *Re-assembling the Social: An Introduction to Actor-Network Theory*, Oxford: Oxford University Press.
Latour, B. (2009) 'Eine andere Wissenschaft des Sozialen?', in Gabriel Tarde, *Monadologie und Soziologie*, Frankfurt am Main: Suhrkamp, pp. 7–15.
Latour, B. and Weibel, P. (2005) *Making Things Public: Atmospheres of Democracy*, Karlsruhe/Cambridge, MA: ZKM/MIT Press.
Law, J. (2002) *Aircraft Stories: Decentering the Object in Technoscience*, Durham, NC: Duke University Press.

Law, J. and Hassard, J. (1999) *Actor Network Theory and After*, Sociological Review Monographs, Oxford: Blackwell.

Law, J. and Mol, A. (eds) (2002) *Complexities: Social Studies of Knowledge Practices*, Durham, NC: Duke University Press.

Law, J. and Moser, I. (1999) 'Good Passages, Bad Passages', in J. Law and J. Hassard (eds), *Actor Network Theory and After*, Sociological Review Monographs, Oxford: Blackwell, 196–219.

Levin, D.M. (ed.) (1993) *Modernity and the Hegemony of Vision*, London: University of California Press.

Lynch, M. (1993) *Scientific Practice and Ordinary Action: Ethnomethodology and Social Studies of Science*, Cambridge: Cambridge University Press.

Mol, A. (2002) *The Body Multiple: Ontology in Medical Practice*, Durham, NC: Duke University Press.

Schillmeier, M. (2005) 'More than Seeing: The Materiality of Blindness', Unpublished Ph.D. dissertation, Lancaster University.

Schillmeier, M. (2006) 'Othering Blindness: On Modern Epistemological Politics', *Disability & Society*, Special Issue: Lessons from History, 21(5): 471–84.

Schillmeier, M. (2007a) 'Dis/abling Spaces of Calculation: Blindness and Money in Everyday Life', *Environment and Planning D: Society and Space*, 25(4): 594–609.

Schillmeier, M. (2007b) 'Dis/abling Practices: Rethinking Disability', *Human Affairs* 17(2): 195–208.

Schillmeier, M. (2008a) 'Time-Spaces of In/dependence and Dis/ability', *Time & Society*, 17(2/3): 215–31.

Schillmeier, M. (2008b) 'Jenseits der Kritik des Sozialen: Zu Gabriel Tardes' Neo-Monadologie' [Beyond the critique of the social: toward Tarde's neo-monadology], in G. Tarde, *Monadologie und Soziologie* [Monadology and sociology], ed. and trans. M. Schillmeier and J. Sarnes, Frankfurt am Main: Suhrkamp, pp. 109–53.

Schillmeier, M. (2008c) 'Globalizing Risks: The Cosmo-politics of SARS and its Impact on Globalizing Sociology', *Mobilities*, 3(2): 179–99.

Schillmeier, M. (2009) 'The Social, Cosmopolitanism and Beyond', *History of Human Sciences*, 22(2): 87–109.

Schillmeier, M. and Pohler, W. (2006) 'Kosmo-politische Ereignisse: Zur sozialen Topologie von SARS' [Cosmo-political events: on the social topology of SARS], *Soziale Welt*, 57: 331–49.

Simmel, G. (1957a) 'Brücke und Tür' [Bridge and door], in G. Simmel, *Brücke und Tür: Essays des Philosophen zur Geschichte, Religion, Kunst und Gesellschaft*, ed. M. Landmann, Stuttgart: K.F. Koehler, pp. 1–7.

Simmel, G. (1957b) 'Wandel der Kulturformen' [Change of cultural forms], in G. Simmel, *Brücke und Tür: Essays des Philosophen zur Geschichte, Religion, Kunst und Gesellschaft*, ed. M. Landmann, Stuttgart: K.F. Koehler, pp. 98–104.

Simmel, G. (1957c) 'Vom Wesen der Kultur' [Towards the being of culture], in G. Simmel, *Brücke und Tür: Essays des Philosophen zur Geschichte, Religion, Kunst und Gesellschaft*, ed. M. Landmann, Stuttgart: K.F. Koehler, pp. 86–94.

Simmel, G. (1957d) 'Das Gebiet der Soziologie' [The field of sociology], in G. Simmel, *Brücke und Tür: Essays des Philosophen zur Geschichte, Religion, Kunst und Gesellschaft*, ed. M. Landmann, Stuttgart: K.F. Koehler, pp. 208–26.

Simmel, G. (1957e) 'Die Grosstädte und ihr Geistesleben' [The metropolis and mental life], in G. Simmel, *Brücke und Tür: Essays des Philosophen zur Geschichte,*

Religion, Kunst und Gesellschaft, ed. M. Landmann, Stuttgart: K.F. Koehler, pp. 227–42.

Simmel, G. (1990) *The Philosophy of Money*, Routledge: London.

Simmel, G. (1992a) 'Das Geld in der modernen Cultur' [Money in modern culture], in G. Simmel, *Aufsätze und Abhandlungen 1894–1900*, Collective Works, Vol. 5, Frankfurt: Suhrkamp, pp. 178–96.

Simmel, G. (1992b) 'Die Bedeutung des Geldes für das Tempo des Lebens' [The meaning of money concerning the velocity of life], in G. Simmel, *Aufsätze und Abhandlungen 1894–1900*, Collective Works, Vol. 5, Frankfurt: Suhrkamp, pp. 215–34.

Simmel, G. (1994) *Lebensanschauung: Vier metaphysische Kapitel* [Philosophy of life: four metaphysical chapters], Berlin: Duncker & Humblot.

Simmel, G. (1997) 'Sociology of the Senses', in D. Frisby and M. Featherstone (eds), *Simmel on Culture*, London: Sage, pp. 109–20.

Simmel, G. (2002) 'The Metropolis and Mental Life', in G. Bridge and S. Watson (eds), *The Blackwell City Reader*, Oxford: Blackwell, pp. 11–19.

Star, S.L. (1991) 'Power, Technology and the Phenomenology of Conventions: On Being Allergic to Onions', in J. Law (ed.), *A Sociology of Monsters: Essays on Power, Technology and Domination*, London: Routledge, pp. 26–57.

Strathern, M. (1991) *Partial Connections*, Lanham, MD: Rowman & Littlefield.

Tarde, G. (1962) *The Laws of Imitation*, Gloucester, MA: Peter Smith.

Tarde, G. (1969) *On Communication and Social Influence*, ed. Terry N. Clarke, Chicago: Chicago University Press.

Tarde, G. (1999) *Monadologie et Sociologie: Oeuvres de Gabriel Tarde* [Monadology and sociology: the works of Gabriel Tarde], Vol. 1, Le Plessis-Robinson: Institut Synthélabo.

Tarde, G. (2000) *Social Laws: An Outline of Sociology*, New York: Kitchener.

Whitehead, A.N. (1978) *Process and Reality: An Essay in Cosmology*, Corrected version, London: Free Press.

11 The city as value locus

Markets, technologies, and the problem of worth

Caitlin Zaloom

LaSalle Street, Chicago's main artery of finance, ends at the Chicago Board of Trade (CBOT). Completed in 1930, the CBOT skyscraper draws in the city. Men and women circle through its front doors and eddy out the back toward the el tracks. In the tunnel between its main edifice and newer buildings, financial traders and clerks spin out from the concrete mass of the dealing floor in their colorful cloaks, inhale a cigarette, and swirl back in. Likewise, the history of the city moves in, around, and away from this building. On its exterior, stern-faced Native Americans clutch sheaves of wheat and confront a machine-age vision of Ceres, the Roman goddess of grain, whose abstract edges and eyeless stare blaze across the metropolis below. A transparent walkway connects the blank walls of the 1990s trading floor to the elegant and showy Art Deco structure, while behind it sits a vision of 1980s corporate power: the glass tower. A more recent change – the ascent of electronic technologies that have drawn financial trading into a new, virtual space – reels invisibly away from the streetscape in cables that run through and beyond these buildings.

A steady beat pulses at the center of this visual and historical commotion: the stamping of prices onto wheat, cattle, and financial products. Chicago's futures markets, like those that lie within the CBOT complex, resolve a central question of commerce: What should things cost? The exchanges, like the CME Group that now owns and operates them, answer this query. In the terms of the trade, the mission of futures markets is 'price discovery', or assigning a monetary value to goods from corn and pigs to U.S. Treasury debt and stock indexes. This can be seen as a relatively simple process: the prices that emerge from within the CME Group's walls and electronic circuits represent the confluence of supply and demand. Such a heuristic for understanding value takes markets to be outside time and space. In the rhetoric of supply and demand, price is a product of the abstract forces of commercial desire. However, the process of assigning value defies such a parsimonious and placeless description.

Cities, and the special hubs of exchange within them, are key sites in the production of worth. The value of agricultural and financial goods is ambiguous. What are their qualities? How much is there? How sound is it?

Who and how many want it? These questions are also temporal. When will the product be needed? When will the situation change? Cities are 'value loci', places that settle this uncertainty.

The city's performance as a value locus requires a set of entwined infrastructures: of physical connection, of informational convergence, and of expertise. Together this centralized arrangement in concrete, steel, fiber optic cable, and human capacity legitimates financial transactions and creates a symbol of worth. Enhanced by the urban locus simple price becomes a more powerful, generalized value that carries far beyond the site of its production. However, the ability of value to move beyond place also, perhaps paradoxically, relies on the organization of particular places.

The value locus is urban both in its site and in the extension to the markets themselves. Historically, the city has been the location of capital markets, and the work of market agents has focused on drawing commercial space together, developing a central place or urban nexus (Carruthers 1996; Kynaston 1988; Leyshon and Thrift 1997). But the futures markets themselves also employ the qualities of cities. To resolve the problem of value, they concentrate the time and space of transaction, create dense populations of traders, and centralize traffic in information. They also create conditions that foster autonomy in financial decision making, seeking to maximize the number of individuals who compete in their markets, though the process works differently in the open-outcry pits where trading has long been conducted and in the new digital markets online. Although outsiders often believe that floor trading is chaotic, in fact prices in the pits are determined through a highly organized system of exchange. Online, however, anonymity bolsters autonomy and competition, and traders compete as individuals against an aggregate market that appears as a single number on their dealing screens. Traders, dense, centralized, and competing against each other, make prices. Expertly arranged markets legitimate these symbols of worth. Together these processes create value that dealers trade in locations around the world.

This chapter focuses on the physical organization of market competition, and on the production of urban value loci that results. Cities, as Max Weber famously argued, have been the most important sites for resolving value, but more recently online spaces have sprung from these urban sites to extend their operations into digital space. As Manuel Castells (1996) and Saskia Sassen (2001) have shown, even as markets move online, traders continue to do business from their desks in the world's financial centers. The global network does not end the importance of the city; rather, it raises new questions about how cities influence online action. The shift from face-to-face to electronic futures markets demonstrates a particular kind of urban virtualization: the city's qualities are reproduced in cyberspace markets. Digital marketplaces mimic the centralization and density of cities, where an agglomeration of money and opinion renders market prices legitimate beyond their source point.

Chicago's futures markets provide a powerful example. The business people and organizations that draw profit from resolving questions of value have worked for over a century and a half to organize and centralize place, information, and expertise in their metropolis. The daily work of managers, traders, and technology designers sets location at the heart of futures markets. These market agents work with the materials of city spaces – particularly architecture and technology – while embedding their technical ideas and locating their own expertise in the nexus they build.[1]

THE PHYSICAL CITY

In the nineteenth century, Chicago grew to be America's capital of capitalism as it drew together western agriculture and the eastern market with railroad tracks and shipments of grain and meat (Cronon 1991; Miller 1996). The city's facility in shuttling physical goods from place to place was entwined with its capacity to pull together information as well as rail ties. Shortly after Chicago's founding, the city became the most powerful among its sister cities of the West, coordinating trade and lobbying to make its urban network, from Buffalo to Milwaukee, America's thoroughfare of commerce. As William Cronon has shown, Chicago's success in centralizing an expansive network of agricultural markets helped it vault to the top of the nation's urban hierarchy. This was no mere symbolic victory. In centralizing America's agricultural commerce, Chicago gained leverage to help develop a transportation, communication, and economic infrastructure that would consolidate its market dominance in the industrial age. Centralizing commerce established Chicago's businessmen as experts in market-building expertise. Reciprocally, their expertise rendered the city's physical infrastructure capable of establishing value beyond its boundaries.

The tandem rise of city and market in Chicago offers a window onto the creation of a value locus, perhaps because the land resisted its settlers' efforts so powerfully.[2] Chicago's marshy ground dragged wooden wheels into its muck. The city's sandy and shallow harbor at the shores of Lake Michigan rendered its waterway second class, and its location on the Mississippi handed St. Louis an advantage as a shipping route. Railroads, too, initially treated Chicago poorly. The railroad companies that were crisscrossing the country with steel ties had erected hubs in other locations.

Amid this trying landscape, denizens of Chicago's nascent business community established a central body to develop the settlement into the commercial hub of their imaginations. The CBOT's founders' primary concerns lay with the transportation and banking challenges that faced businessmen in the growing metropolis. Development of the city, particularly its transportation infrastructure, was essential to the circulation of commodities and the commercial interests of themselves and their peers. The Chicago Board of Trade, at first a small gathering of boosters at a flour store, grew to become instrumental in orchestrating Chicago's physical channels of commerce.

Technologies, such as wooden boards to span the muck, dredgers to unsilt the harbor, and rail ties to connect the city with its hinterland, opened conduits for commerce and established Chicago's centrality in the business of trading the bounty of the American West.

That they would succeed in establishing Chicago as a commercial nexus was not certain. As the members of the CBOT well knew, there were other serious contenders to become America's hub. For a while, Cairo, Illinois, seemed poised at the brink of success as rail lines and federal influence converged on the city. A federal bill supplied more than two and a half million acres to the State of Illinois to construct a line of the Illinois Central Railroad from Cairo to Galena. Chicago had to fight for a 'branch' of the railroad. However, the tributary soon overtook the main trunk line in traffic (Cronon 1991: 161). With William Ogden, Chicago's first mayor and the nation's first railroad baron, at the helm, the city's combination of water and rail transport cinched the city's success. At the outbreak of the Civil War, Chicago was the world's largest railroad junction, with more lines meeting within its borders than any other city on earth (Miller 1996: 91).

Chicago had another, more established, competitor for the position of western gateway city. St. Louis's waterways supported its claim to be the great western city and transportation hub. St. Louis, situated at the junction of the Missouri and Mississippi rivers, seemed to have natural advantages for transporting grain to market. It had logged 75 years as the key western port and principal trading partner for New Orleans. St. Louis merchants cleared furs from the west and trafficked in other commodities on their way to and from the frontier. In addition, a narrow channel north of St. Louis meant that all upstream river traffic had to stop there to transfer to smaller boats. But it was St. Louis's connections that eroded its dominance; Philadelphia was the city's major trading partner, and the eastern metropolis was already losing markets to New York. St. Louis merchants began to switch their alliances to New York, but slowly and too late. New York capital had established ties to Chicago merchants, providing pricing advantages, and railroad money had already helped establish Chicago as the west's rail hub (Cronon 1991: 295–309).

The CBOT's influence was crucial in building and maintaining the city of Chicago. The organization's work on roads, waterways, and rail lines was critical in establishing the city as a value locus. Their first challenge was to create an infrastructure for the smooth circulation of commodities. Beginning in the mid-1800s, influential merchants lobbied for and funded the growth of railways, bridges, harbors, and buildings in the city of Chicago. By developing infrastructure and in coordinating northwestern commerce the CBOT overcame the physical challenges which obstructed their metropolis's rise. For instance, after a spring flood that destroyed nearly every bridge in the city, the board reestablished communication between the north and west sides of the city to keep the metropolis running. Recognizing their disadvantage for shipping, the board set out to rebuild the harbor. Linking their roles as infrastructure innovators and financial entrepreneurs, the board worked with city authorities to issue bonds for the project. The board shouldered the financial responsibility for negotiating the debt and managing the funds.

The board also regulated commerce through the region, passed tolls on canal freight to and from the Mississippi, and debated how to manage the ever-increasing flow of information with telegraphic expansions. In their efforts, the board drew together Chicago's economic future with national politics. The men of the board lobbied Washington for land grants to complete the Illinois Railroad. They were so successful that senators Stephen A. Douglas and General James Shields, both from Illinois, sent special congratulations (Cronon 1991: 583). In 1850, when sandbars blocked the Illinois River, hampering commerce, the board again sent representatives to Washington to lobby for making the port more navigable.

Passable and well-maintained roads, networks of rails, and a navigable harbor brought trade to the city's center. Under the direction of the CBOT, the space of the city became a centralized space of trade.[3] Agricultural markets fused as Chicago's network of railroads and trading connections linked the western plains with the East Coast. Shuttling pig parts and sacks of oats to their ultimate markets in the cities of the eastern seaboard also required a more intangible commodity: fine-grained information. In addition to the physical technologies of the city, the telegraph was critical in making Chicago a value locus; the centralization of information is key.

The great spokes of railroad lines were made far more powerful with the introduction of telegraph wires. The merchants of Chicago could intensify and multiply their relationships with traders in other cities through the wires. The first telegram arrived in Chicago in the same year that leading local merchants founded the CBOT. On January 15, 1848, at the corner of Lake and Clark Streets, a telegraph in the office of Colonel J.J. Speed tapped out a message from Milwaukee. Soon messages from the east and Chicago's urban kin of the northwest were flowing in. The first greeting sent between Detroit and Chicago read:

> To Milwaukee, Racine, Southport, and Chicago.—We hail you by lightning as fair sisters of West. Time has been annihilated. Let no element of discord divide us. May your prosperity as heretofore be onward. What Morse has devised and Speed joined let no man put asunder.
>
> (Andreas 1884: 263)

The telegraph led to a coordination of commerce, prices, transportation, and politics among these regional centers that had been impossible before.

Cities maintained different standards for measuring weight and quality, a technical problem that slowed trade, hampered distribution, and divided commercial regions. Accurate information and interconnection were not enough to establish the regularities that allow Chicago's value to circulate. A certain set of standard measures now had to be imposed on provisions in a market where products could move easily across geographical boundaries. The CBOT had created a series of exact standards for the inspection, warehousing, and shipping of grain to make traffic between cities and regions possible, but their adoption in Chicago was not sufficient. They had to be taken up throughout the country to create the fluidity necessary for national agricultural commerce. The CBOT standardized measures and coordinated commerce. To spread their influence and enhance the importance of their city as a commercial node, the CBOT organized boards of trade from Milwaukee to Buffalo to develop standards of reliable commercial news and price quotations. Such tools of transparency and fungibility ultimately benefited Chicago as a value locus.

Standardization was effective in creating a network of market centers that could work together across geography to coordinate their markets. However, the reliability and accuracy of market information remained a keen interest to Chicago traders. As an outgrowth of Chicago's dominance in agricultural trade, the CBOT established contracts and trading pits to deal in the future reapings of America's heartland. The trade in these agreements for future delivery valued goods that would flow through Chicago's commercial hub within a year's time. Prices of winter's wheat and summer's corn streamed out of the pits, allowing farmers and manufacturers to budget and plan for their businesses. These producers relied on the CBOT to provide sound information, and the CBOT recognized that quality of information available in their markets would provide an edge among urban marketplaces.

At its third annual meeting, in 1851, the CBOT adopted a rule forbidding members to give 'untruthful or bogus reports of their transactions, on pain of expulsion' (Andreas 1884: 583). This move toward truthfulness and transparency in commerce was necessary for making Chicago a center of nationwide commerce. In a national arena, where the reputations of individual traders were not known, commercial agents could rely only on the reputation of the organization. It was imperative for the CBOT to police both the conduct of its members and the information that flowed into and out of its pits, thus establishing fair prices for commodities and disseminating accurate information.

The combination of the physical centralization of America's farm trade, the informational centralization of agricultural prices, and the development of local expertise in market building did more than establish the city as a commercial center; it made Chicago a value locus.

EXPERT MARKETS

Creating the conditions for the highest-quality information sits at the center of Chicago's mission. As in the nineteenth century, building a physical infrastructure of market information was central to another signal moment in the CBOT's history: the construction of their signature skyscraper at the end of

LaSalle Street. The design of the trading floor, the exchange's heart and the center of its value locus, was particularly crucial and, therefore, contentious. The process offers a view of how centralizing space was put to work in the pursuit of creating reliable prices. From the design of the trading pits to the placement of the telephones and the material of floorboards, the architects, board officials, and members debated their vision of optimal arrangements with intensity. The physical placement of each conduit of information – whether technological, like a telephone, or embodied, like a competitor – was critical to the production of Chicago's value engine.

To create prices that could be transported across locations as true worth, the trading floor had to be arranged to adhere closely to market ideals – or at least as closely as a market in flesh, steel, and wire can ever approximate a model. The managers, architects, and other experts who designed the market sought to build a space that would maximize competition among buyers and sellers by ensuring that the room provided each participant with equal access to the information that converged on the trading floor. To do this, these market professionals embedded their expertise into the trading floor's structure. As they aspired to ideals of competition, they repeated and intensified key qualities of the modern city now built into the trading floor: centrality, density, and open circulation.

Creating prices that could move across space as value meant designing a specific kind of trading place. The construction and layout of the dealing floor guided the daily paths of the traders and configured whom they could see and hear, their access to information, and what communications technologies they could use instantly and which they stretched to procure. To create a maximally competitive environment, all traders had to have equal access to the market and its sources of information. The conduits which channeled the board's prices to the outside world also had to be efficiently arranged. Each of these problems demanded close attention to the construction of physical space. But which kinds of communications technologies to include? Where to place them? How to set up the pits to optimize their operation? Each question provided fodder for a range of experts and interests to weigh in.

Competition arose between the interests of some powerful firms at the CBOT that believed they would profit more if they had protected access to market information, and the experts – architects, sound designers, and market managers – who wanted to build a space modeled on open competition. Henry Rumsey, the board's chair of the New Building Committee, was convinced that creating an exclusive exchange would undermine the future of Chicago's marketplace, because it would reduce the number of players who would engage in economic activity. A valid price would emerge only if Chicago could design a market that brought large numbers of individual traders together in time and space. Without this kind of structured competition, bids and offers would be unreliable and the city's market would suffer. Rumsey's victory in this dispute resulted in the design of a trading floor (or 'board room', as it was ultimately called) where the market came to life through the hands and voices of a dense multitude of traders.

To bring the marketplace more closely in line with the ideals of commerce, the trading pits and the room that housed them needed to enhance the market principles of individual competition and smooth circulation. The pursuit of these abstract principles required a vast open space at the building's core. Six huge trusses, each weighing 227 tons, held up the skyscraper over the enormous hall, eliminating the need for support columns that would block the movement and view of the traders. The wide-open arena allowed traders to circulate easily between pits, the telegraph and telephone operators, their offices, and the smoking room where traders met clients. This space, at 165 feet long by 130 feet wide, with a 60-foot ceiling, gave all traders equal access to the markets and to the information they needed to trade.

Floor trading was taxing physical labor. To take full advantage of the markets, a trader stood for hours a day among throngs of competitors jostling each other for advantage. For the trading floor to work, it had to provide for the traders' bodily comfort and allow them to hear and see the sounds and gestures of their trading partners and the market as a whole. In particular, the acoustics of the hall were crucial to the operation of the market and to profits. Rumsey conveyed this to the architects, Holabird & Root, in a request to use a flooring material for the trading room that would absorb excess sound and be 'easy on the feet'. The architects dismissed wood, rubber tile, cork, and linoleum as options. The softer materials would quickly give way under the floor traffic and would 'in a short time present a dilapidated appearance'.[4]

Unhappy with the architects' aesthetic intransigence, the board turned to a scientist for help. Rumsey hired Professor F.R. Watson, a physicist at the University of Illinois, to analyze the acoustics on the trading floor. There was apparently much room for improvement. Traders in the corn pit were especially upset. They complained that they couldn't make out one bid from another and that they and their competitors had begun to shout more and more loudly to break through the noise, raising the cacophony to levels their eardrums could not sustain, and that their throats were raw with the effort.

Worse than the toll it took on the feet, vocal cords, and minds of the traders, the noise threatened the accuracy of the information coming out of the pits. The director of the CBOT's price reporting recounted that his clerks were having difficulty in recording 'quotations correctly and the traders themselves are unable to hear properly across the pit'.[5] If the quotations were incorrect, false price information would flow from the trading floor by way of the telephones, telegraphs, and pneumatic tubes that connected locations as near as offices in the building and as far away as England and Argentina.

Demand for information from the CBOT was growing rapidly. How to arrange the informational conduits of the trading floor was a hot button issue. The New Building Committee had to mediate arguments between traders, telegraph companies, and the architects over where the telegraph and telephone stations were to be located on the floor, and how to allot them to Western Union and other telegraph companies. The Holabird & Root plan provided for 16 telephone booths in sight of the quotation board. Traders suggested other configurations for the telegraphs and telephones. For instance, one firm suggested that stadium-style telephone banks would help make sure their quotes to customers were valid, since each clerk at his desk could have a 'very clear view of the whole floor'.[6] Although Rumsey did not adopt the plan, the suggestion turned out to be prescient. The stadium-style phone banks were installed in the electronically augmented trading floor of the 1990s, once again linking the basic problem of physical design and the uninhibited flow of information.

Under the guidance of the committee, the architects were primarily concerned to make market information available to all participants. In their report on the standing building, they noted that the older trading floor had become a jumble. Desks and bodies even obscured the quotation boards. A fair market where skill and speed would determine profit, and competition would determine price, required equal access to information. The firm set out to construct a board room that would give no inherent advantage to place. However, not all participants were willing to give up their privileges. Some member firms tried to manipulate access to telephones, aiming to gain advantage in the market by influencing the arrangement of space and technology on the trading floor. They pressured the board of directors to secure extra telephone lines that would support their own business. The president, in turn, requested that Rumsey accommodate them.

Rumsey responded by asserting the authority of experts. He would not acquiesce to the installation of more private telephones. Instead, he insisted that 'the best architectural and engineering talent have counseled us in determining the best possible arrangement for our floor facilities'. The two years of study and thought in the design furthered the interests of 'the entire active membership rather than the few', a distribution of advantage that also supported the quality of the CBOT's price information.[7]

Private telephones threatened equal access to information in another way as well. The building committee objected because the fixed location of the telephones had spawned another new informational practice. Some firms had begun 'flashing' their orders to the pit from the telephone lines, relying on the rapid hand signals that would become an integral and identifying part of financial pits. The committee saw this as introducing informational disorder, threatening the centrality of the trading pit, both in time and in space. Flashing blurred the boundary between the pit and the outside market. The hand signals made customers' orders visible to attentive traders, who could see them before they reached the open market. Rather than acting only with information available within the market borders of the pit, these traders could act with information from outside the physical and temporal boundaries of the market. Adding new telephones would therefore not only favor a select group of trading houses but also allow the market to spill over out of the trading pit.

The building committee pushed the experts ahead of interests to support the quality of information over the pursuit of any trader's or firm's profits. Rumsey and his committee had worked to create a board room that would draw traders, firms, and information into the pit. The design of the floor anchored the market inside the pit and created lines of communication that stretched away from it. They had assembled the expert opinions that showed them how to shape the space of the trading floor and, in the process, engineer information. By creating open sightlines, good sound conditions, and even-handed distribution of information technologies, designers engineered a physical field that maximized competition and created a setting that participants believed created the best prices. From this carefully crafted space of economic contest, value emerged.

VIRTUAL URBAN MARKETS

In the most contemporary marketplaces, experts virtualize urban qualities to create a new kind of value locus. Beginning in the mid-1990s futures markets both in Chicago and around the world began to move online. Chicago's traders staked out contending positions around the new devices. Some Chicago firms moved quickly onto these new market 'platforms', while others resisted the rise of technologies that would challenge the trading pits as the preeminent arrangement for futures markets. Those firms and exchanges that developed and embraced screen-based technologies wrote a new chapter in the mutual development of markets and cities. At first glance, the relationship might seem conflictual: that online technologies obliterate the need for urban centralization. However, a closer examination of electronic trading systems tells a different tale. Like their historical counterparts, online markets validate value through geographic arrangements. Once again, the principles of centralization and atomization animate these systems.

Each of Chicago's influential futures exchanges – the CBOT and the Chicago Mercantile Exchange (CME) – reacted to the rise of online technologies in significantly different ways. The CBOT struggled particularly hard around the question of how to virtualize their urban market center. The CBOT's first foray into online trading took a very literal approach to moving the trading pits online. The failure of Project A showed the limitations of this approach.

Launched in 1998, Project A extended the reach of the trading pits both in the design of its screen and in the times of its use. Drawn as a bird's-eye view of a trading floor, traders encountered each other online as they did in more traditional trading: among the octagonal rings of trading pits. The screen included each of the CBOT's pits, arranged as they are on the floor, their size drawn to reflect the heft of the market they contained. The interface design brought traders together inside these virtual pits, mimicking the spatial concentration that the physical trading floor was designed to afford. When traders logged onto the system, their identifying badges – three-letter markers tied to their accounts – were shown 'standing' in the pit where they made their trades. Just as neighbors could track a trader's deals when standing near him in the pit, Project A linked each trade with its initiator. On Project A, as in the trading pits, centralization was achieved through a coordination of sightlines and traders' mutual monitoring of price.

Project A extended the reach of trading pits in time, too. At first, the CBOT opened Project A only in the hours after the trading pits had closed. The board's traders shunted the new technology off to the temporal edges of their markets, concerned that Project A would offer an alternative and threaten their livelihoods. The floor traders knew very well what they were doing: an after-hours market would not possess the liquidity, or ability to transact easily, of the pits. There simply would be so many fewer traders available in the marketplace that executing large orders would be difficult, creating conditions undesirable to the banks that would use the service. Fewer participants meant less reliable prices and more volatility; in other words, less dependable prices were created. By decentering Project A temporally, the floor traders retained control over both the time and the place of the primary market. A weak online platform would not possess the qualities of spatial and temporal centrality that undergird sites of value creation.

However, with competing exchanges deepening their commitment to online trading, CBOT management began to challenge the floor traders. Within a year from Project A's introduction, the CBOT began to offer 'side-by-side' trading, allowing traders to deal either online or in the pit during the regular hours of the trading floor. The electronic market then continued after the pits closed, allowing ambitious traders to continue their dealing from screens in their offices both in the building and around the world.

Project A offered online dealing as a virtual trading pit. However, traders adopted the new technology only reluctantly. As electronic trading began to flourish on other, more flexible systems, the CBOT abandoned Project A. The screen design and the restrictions on time mimicked and extended a particular place, the CBOT's storied trading floor. But, as a trading platform, Project A failed to establish the spatial qualities that the pits organized so effectively. Project A's more effective competitors instead virtualized the spatial qualities that the trading pits embodied. These systems used their screens and connections to organize traders in space and time in new formations. The virtualization of the urban qualities that had anchored Chicago's markets since the mid-nineteenth century depended on new physical instantiations of centrality and density.

The trading technologies and electronic interfaces that came to dominate futures markets bypassed the trading pits entirely. First, novel and stream-lined images of the market rendered electronic trading central and atomized. Second, these markets offered a new density centered on time. In the trading pit, each bid or offer came through the voice and body of an individual trader. The market was fragmented, a piece residing in each of the bodies and voices which competed to make a single deal. The trading screens that most traders use today create a value locus by aggregating all the bids and offers available at a single time.

Online, each trader sits behind a bulwark of screens, confronting the market as an aggregate number. The bolded typeface of bids and offers in front of traders' eyes distill the financial evaluations and intentions of their competitors into a single set of bid and offer. Electronic trading systems establish a centralized space of the market by aggregating the intentions of physically distant competitors. The market's new image, as an aggregate number, replaces the physical centrality of the market. These numbers come to stand for a market that exists in the network.

Creating a visually centralized market space shapes competition in online trading. In electronic markets, traders compete as individuals against the market as a whole. The anonymity of competitors and their aggregation renders the market visible as an object beyond the individual intentions of its participants. Traders across the globe can apprehend the market through bid and offer numbers. The atomizing design forwards the project that Henry Rumsey's experts worked into the steel and cables of the physical trading floor.

The designers of online markets extended this foundational project of the CBOT: to maximize competition and, therefore, to create a center for establishing valid prices. An aggregate bid/ask number now stands in for the thousands of individual traders who inhabit the market space. At the same time, the individual traders inhabit their own decision-making space. They no longer confront other traders and their idiosyncratic strategies; rather, they must compete against the market as a whole. Once an abstract notion, the market is now made observable in the fluctuations of the bid/ask number, which changes second by second on futures trading screens. The design embodies a particularly urban paradox: futures traders, now exquisitely connected, act alone.

Before online technologies, the centrality of time and the centrality of space came together neatly in Chicago's trading pits. The network technologies of online trading seem to pry them apart. No matter where traders are located across the globe, online technologies offer real-time access to the markets of their choice. However, time zones still assert an organizing force over financial geography and the spatial organization of markets. The importance of traders operating in time-zone proximity reasserts the relationship between geography and time, but one stretched over different territory.

Online value loci rely on temporal centralization indexed to their place of origin. Markets, even when available to traders across the globe, remain centered in particular cities. For instance, in the late 1990s, as the CBOT struggled to compete with online markets, London's pits were in danger too. The London International Financial Futures Exchange (LIFFE) had long been the site for trading German Treasury futures. Based in trading pits, these markets were some of the most profitable for Britain's traders. The founding of the all-electronic Eurex exchange changed the European geography of futures trading. As soon as it opened, the German–Swiss exchange quickly attracted the business of the German banks that primarily traded their country's debt. Although traders around the world could now deal directly on the exchange, they had to do so during Frankfurt hours. The exchange never closed; however, the hours of greatest participation produced the most reliable values. Temporal density made the market.

The new temporality of online trading also shaped competition. Online the market's composition shifted as the trading day moved from morning to afternoon. The bid/ask numbers of 8.00 a.m. London time included a different cohort of traders than those after lunch. In the Eurex markets, London traders who worked on German markets arrived an hour before those who traded on the British markets. Even those who had formerly traded Bund contracts on the LIFFE floor now had to deal on Frankfurt time, one hour ahead. Although no names or numbers identified their competitors, the London traders could reasonably assume that they and their German counterparts dominated the market in the morning hours. At 1.00 p.m. GMT, however, the market changed. Chicago's traders, having finished their morning preparations, entered the market at 7 a.m. local time. Because of their city's preeminence in futures markets, Chicago's traders have a reputation for an aggressive and skillful style of speculation. Some traders claimed that the afternoon hours gave the best opportunities for competition because Chicago brought larger volumes and more expert trading to the market.

Online, markets confer value through temporal density. Because participation varies as traders rise, log on, take lunch breaks, and hit quit for the evening, the market itself contracts and expands as the day goes by – first in Frankfurt, then in London, and finally in Chicago. Time zone rather than physical proximity organizes the spatiality of electronic markets.

In their virtualization of urban qualities, the designers of today's online futures markets animate a 'price discovery' model based in maximum competition. Project A failed, in part, because its screens and times of operation literally replicated the trading pits and extended them to electronic space. The more successful market designs instead centralized a new, entirely electronic space and atomized traders to maximize competition. These expert-designed systems created the possibility for trade at any moment; however, in practice, the traders themselves restored another urban quality to electronic trading: density. Concentration in time is an essential element of the new electronic market. A great number of traders, gathered at once in front of their screens, animates the virtual space. As they buy and sell, they bring the urban marketplace, now in digital pulses, to life.

URBAN EXPERTS, VIRTUAL MARKETS

In moments of city building, Chicago has constructed markets out of urban materials. Such infrastructure, architecture, and digital projects do more than just bring markets to life. They establish the city as a value locus, a place where physical structures, expertise, and dense trading networks combine to make prices that act as general symbols of worth. Values, stamped with the legitimacy of such an urban nexus, can travel well beyond the site of their production.

The urban value locus brings together place and human capacities in powerful ways. By nature, spaces of value resolution concentrate expertise. Such cities confer legitimacy on their value specialists. Places equipped with proficiency in establishing worth also create an exportable commodity in design skills. For Chicago, this expertise lies in assembling markets.

As the businessmen of the CBOT built Chicago to be America's market city, they also established it as a location of market know-how. Today, exchanges across the world rely on Chicagoans' talents for making markets. In the 1980s, the London International Financial Futures Exchange brought in Chicago consultants to design their exchange and Chicago traders to seed their pits. Today, one of the most prominent market technology companies, Chicago's Trading Technologies, equips traders around the world with the software that connects them to futures markets. Chicago's market expertise has become virtual as it structures online environments that reach around the world.

The city's trading prowess has also fueled another trend that again virtualizes the urban attribute of concentration. Gaining speed in the 1990s and rolling through today, exchanges have been merging to create a new kind of financial geography. Chicago, in particular, has created a new network of connection.

In 2006, the CBOT's cross-town rival, the Chicago Mercantile Exchange, bought its more storied sibling. Today, the CME Group also owns the New York Mercantile Exchange (NYMEX) and has extended its reach through strategic partnerships in Korea, Brazil, and Dubai. Also in 2006, the Chicago trading firm Archipelago merged with the famous New York Stock Exchange, which had struggled, with bleak results, to modernize its markets. Archipelago brought its cutting-edge electronic trading platform and Chicago skills in market building to the Big Board.

In each instance, Chicago firms concentrated and virtualized urban market arrangements, combining organizations across boundaries of both state and nation to assemble more centralized markets. These new, virtual urban agglomerations concentrate traders, create new temporal geographies and spatial landscapes, and coordinate flows of information, just as the CBOT's founders did on the swampy shores of Lake Michigan. Although contemporary global networks are often considered to be without discrete centers, certain urban nodes continue to exert outsize influence. In the world of assembling urban exchange, nowhere is as powerful as Chicago.

NOTES

1 For a full account see Zaloom (2006).
2 William Cronon's ironically titled *Nature's Metropolis* (1991) details many of the ways in which the landscape of Chicago was transformed to make it amenable to commerce.
3 James Carrier (1998) notes that separating marketplaces from the general life of the city is a 'practical abstraction' supporting the notion that markets operate with their own sphere and with their own laws. However, in Chicago, the whole city was material for the practical abstraction of the market. City and market rose together and supported the idea that economic arrangements undergird the social life of the city.
4 Holabird & Root to Rumsey, October 4, 1928, CBOT Archive.
5 Ibid.
6 Arthur Lindley to John Bunnell, January 5, 1926; Rumsey to Lindley, January 18, 1926, CBOT Archive.
7 Rumsey to the President of the Board of Directors, April 5, 1930, CBOT Archive.

REFERENCES

Andreas, A.T. (1884) *History of Chicago: From the Earliest Period to the Present Time*, Chicago: A.T. Andreas.
Carrier, J.G. (1998) 'Abstraction in Western Economic Practice', in J.G. Carrier and D. Miller (eds), *Virtualism: A New Political Economy*, New York: Berg.
Carruthers, B. (1996) *City of Capital*, Princeton, NJ: Princeton University Press.
Castells, M. (1996) *The Rise of the Network Society*, New York: Blackwell.
Cronon, W. (1991) *Nature's Metropolis: Chicago and the Great West*, New York: W.W. Norton.
Kynaston, D. (1988) *The City of London*, London: Pimlico.
Leyshon, Andrew and Thrift, Nigel (1997) *Money/Space: Geographies of Monetary Transformation*, New York: Routledge.
Miller, D.L. (1996) *City of the Century: The Epic of Chicago and the Making of America*, New York: Touchstone.
Sassen, S. (2001) *The Global City*, 2nd edition, Princeton, NJ: Princeton University Press.
Zaloom, C. (2006) *Out of the Pits: Traders and Technology from Chicago to London*, Chicago: University of Chicago Press.

12 Second empire, second nature, secondary world

Verne and Baudelaire in the capital of the nineteenth century

Rosalind Williams

TWO WRITERS AND A CITY

In November 1848 Jules Verne set out for Paris from his home city of Nantes, telling his family he would continue his law studies but in his own mind determined to make his mark in literature. The times were unsettled. In France the Second Republic had just been proclaimed, while other governments across the Continent were trembling and falling. Verne was eager to get to Paris in time to see the Second Republic inaugurated in a solemn ceremony on the Place de la Concorde. Along with a friend of his age, he took a stagecoach to Tours – the railroad did not yet extend all the way to Nantes – intending to take the train the rest of the way. At Tours, however, they discovered that all the railway cars were reserved to transport National Guardsmen to the capital. Verne and his friend tried to sneak on board but were detected by a gendarme who noticed they had no uniforms, swords, or papers. They had to continue by stagecoach. By the time they arrived in Paris and made their way to the Place de la Concorde, the ceremony was over and the crowd had dispersed, leaving only litter and tattered decorations (Allott 1940: 13–14; Allotte de la Fuÿe 1956: 35–6; Butcher 2006: 71–2).

Jules Verne moved out of Paris in 1871, at another revolutionary moment: the defeat of France by Prussia and the subsequent bloodbath of the Paris Commune. Verne had indeed made his mark in literature, but only after a long period of financial and artistic struggle. His first success finally came in 1863, 15 years after his arrival, with the publication of *Five Weeks in a Balloon*. This novel led to a contract with publisher Jules Hetzel committing Verne to produce three (later two) novels a year – a treadmill of literary production that was financially rewarding but personally exhausting. The novels written while Verne lived in Paris include many of his best and best-known works: *The Adventures of Captain Hatteras* (1864–65), *Journey to the Center of the Earth* (1864), *From the Earth to the Moon* (1865), and *Twenty Thousand Leagues under the Seas* (1869–70).[1] Verne and Hetzel gave the series the collective title of 'Extraordinary Journeys', *Voyages extraordinaires*, with the subtitle *Mondes connus et inconnus*, 'Known and Unknown Worlds'.

By the time *Twenty Thousand Leagues* appeared, Verne was in the process of a slow but steady withdrawal from Paris, which he found distracting, expensive, and landlocked. In 1857 he had married a widow from Amiens, a provincial city 70 miles north of Paris (connected to the capital by a direct railway line) and 40 miles inland from the North Sea on the river Somme. Beginning in 1864 Verne spent summers and later whole years in Le Crotoy, a fishing village on the estuary of the Somme. There he bought a small fishing boat, named it the *Saint-Michel*, converted the cabin into a floating office, and sailed it for weeks at a time on writing voyages to North Sea ports and beyond.

During a voyage up the Seine to Paris in the summer of 1870, Verne was almost caught behind the Prussian lines when war broke out. After months of turmoil – four Prussian soldiers were billeted with the family, Hetzel fled Paris, one of Verne's cousins died in Paris during the Commune – he decided to move to Amiens (Allott 1940: 116; Butcher 2006: 213–15). For him it was a middle landscape between city and sea, between the civilization he needed to publish and the escape from civilization he needed to write. In a letter to a friend, Verne explained:

> On the desire of my wife, I am settling in Amiens, a wise city, safe, even-tempered, where the society is cordial and lettered. One is close enough to Paris to catch its reflection, without the insupportable noise and the sterile agitation. And, after all, my *Saint-Michel* remains anchored at Le Crotoy.
>
> (Béal 1985: 14, my translation)

Verne settled in Amiens, buying a house overlooking the train tracks to Paris. In Amiens he wrote scores of novels, researching them using the resources of the city's 'cordial and lettered' society, and taking frequent sailing trips as long as health permitted. He died there in 1905, in a second-floor bedroom of the house he had bought in 1871.

Amiens enabled Jules Verne to write the 'Extraordinary Journeys' – but the series was defined and launched during his years in Paris. Between the revolutionary bookends of 1848 and 1871, Verne lived and worked in Paris when it seemed more than a city, more than the capital of France, when it was *la capitale du dix-neuvième siècle*, in the memorable phrase of Walter Benjamin. City, art, capitalism: this is the tripod supporting the now-standard interpretation of the significance of Paris under the Second Empire, as both crucible and emblem of modernity. According to this interpretation, the raw energy of mid-nineteenth-century capitalism, manifested in the material reconstruction of Paris under Baron Haussmann, played a key role in generating artistic modernism, epitomized in the poetry of Charles Baudelaire.[2]

Jules Verne arrived in Paris a few years after Baudelaire settled there and left a few years after Baudelaire's death in 1867.[3] Unlike Baudelaire, however, Verne plays no role in the standard interpretation of Paris as capital of the

nineteenth century. This may have been understandable in earlier generations, when critical opinion treated Verne as a writer of adventure stories
for children. It is less understandable since an impressive crew of literary
and cultural critics in the late twentieth and early twenty-first centuries
has explored the literary and intellectual richness of the 'Extraordinary
Journeys'.[4]

The most obvious reason for the neglect of Verne's connections with Second
Empire Paris is that he rarely writes about the city. Instead, he takes readers
to the moon, to comets, to the depths of the oceans, to the center of the earth,
to the poles, to regions of the earth far away from Europe. Verne himself
declared that his artistic mission is 'to paint [*peindre*] the entire earth, the
entire world, under the form of the novel, in imagining adventures special to
each country, in creating characters special to the environments where they
act'.[5] In practice, these environments were almost anywhere but the heartland
of Western Europe. In his imaginative writing (as opposed to geographical
atlases Verne wrote for Hetzel primarily as money-makers), Paris, Western
Europe, and the British Isles are typically places to leave from to go exploring
and to return to with stories of exciting adventures of faraway places.[6] The
story happens elsewhere.

The one novel that Verne set in 'the capital of the nineteenth century' was
his first: *Paris in the Twentieth Century*, written about 1860, before he started
working on *Five Weeks in a Balloon*. It was written at a time when Verne was
discouraged and embittered after a dozen years of mostly fruitless efforts to
succeed as a writer. *Paris in the Twentieth Century* is a thinly disguised por-
trait of the capital of the nineteenth century: it shows the city as the capital
of a cruel, greedy, utilitarian capitalism which leaves little room for artistic
imagination or even simple human kindness. It exacts some measure of
revenge through nasty portrayals of the captains of finance and seeks pity for
the fate of the poor starving artist. The novel careens between angry realism
and limp romanticism until, at its end, it veers into its most gripping episode,
a fantastic but futile journey to the graveyard.

After the success of *Five Weeks*, Verne submitted the already completed
Paris manuscript to Hetzel, who rejected it brusquely: 'It is a hundred feet
below *Five Weeks in a Balloon*', Hetzel wrote to Verne. Hetzel went on to
assail it as poorly written, boring, and populated by unpleasant characters,
notably the hero, 'an idiot'. Publishing it would be a disaster for Verne's
career, Hetzel concluded.

Verne was already writing the book that would become *Captain Hatteras*.
Hetzel had read some of that manuscript, loved it, and published it in 1864–
65 as the second installment in the 'Extraordinary Journeys' series. *Hatteras*
too was a hit. As for *Paris in the Twentieth Century*, the draft manuscript was
locked in a metal safe. It remained there until 1989, when one of Verne's
descendants moved the safe upon sale of a family home. Blasted open with a
blowtorch, the safe yielded the manuscript placed under a pile of linen. After
authentication the novel was published in French in 1994 (selling over

200,000 copies in the first year) and was thereafter translated into 30 languages, including an English edition in 1996 (Weber 1996: xxiv–xxv).[7]

It is understandable why Verne is rarely associated with Paris of the Second Empire. He hardly ever writes about the city and, when he does, he mainly expresses bitterness about it and desire to escape from it. For both Verne and Baudelaire, living in Paris of the Second Empire was not a happy life experience, but it was a defining one and critical in shaping their art. How does it change our understanding of the relations between material and literary history if we examine Verne as another representative artist of the capital of the nineteenth century?

MATERIAL AND CULTURAL CHANGE

Leading writers on Paris at this transformative time – Walter Benjamin, T.J. Clark, David Harvey, and Marshall Berman, among others – are too sophisticated to rely upon a simple model of substructural material changes resulting in superstructural artistic responses. A more subtle version of this model may be implied, however, when Paris of the Second Empire is analyzed as an interactive network of individuals, social organizations, technological systems, cultural products, and non-human nature – an ensemble linking, in endless loops of co-evolution, material, social, and artistic change.

T.J. Clark points out the circularity in this argument: modernism has its circumstances in modernity. 'It is not enough to say, as we all do now, that the terms of modernism and the facts of Parisian life are somehow linked' (Clark 1984: 14). Attributing historical change to a co-evolving network highlights connections that might otherwise be ignored, but it also blurs analysis of causality and priority. Above all, it blurs a fundamental distinction among the elements of the network: of all of them, only individual human beings possess self-consciousness. Self-conscious individuals interact with the world (including the collective products of human action) in uniquely reflexive ways. They are stimulated by their immediate environment, they express it, and they also use their imagination to go beyond it.

In Verne's case, in two important respects the 'facts of Parisian life' in a straightforward material sense – that is, the rebuilding of the city during the years he lived there – were not the key ones stimulating his invention of the 'Extraordinary Journeys' as a form of literary modernism. In the first place, the material facts of life that most influenced him were those of Nantes, the provincial city where he was born in 1828 and lived until his departure in 1848. As a once thriving and then declining port, the estuary, docks, and vessels of Nantes were the source of Verne's strongest and most persistent imaginative obsession: that of leaving the constraints and corruptions of human-settled land for the unsettled sea, or some other mobile medium, preferably in the company of a few other men in a technologically advanced vehicle that allows experience without entanglement. Because of the nature of the shipping trade of Nantes – for centuries it was the main slaving port of

France – Verne's obsession is complicated by his awareness that even the sea is the site of submission (never more evident than in the 'captivity narrative' of *Twenty Thousand Leagues under the Seas*).

In the second place, once Verne moved to Paris, it was not the material but the cultural facts of Parisian life that had the most profound effect on his evolution as a writer. When he arrived in 1848, he headed immediately for the literary and especially the theatre crowd, still swept up in Romanticism, which had dominated French letters in the last generation. Verne's instincts for individuality, creativity, and freedom were aligned with those of writers such as de Vigny, de Musset, Hugo, and Dumas father and son. For most of his time in Paris, Verne concentrated on writing romantic comedies: Dumas *fils* helped bring one of his first plays to the stage.

Like so many other artistic immigrants to European capitals, Verne found (in the words of Raymond Williams, speaking of early-twentieth-century London) 'a community of the medium; of their own practices':

> To the immigrants especially ... language was more evident as a medium – a medium that could be shaped and reshaped – than as a social custom ... Over a wide and diverse range of practice this emphasis on the medium, and on what can be done in the medium, become dominant.
>
> (Williams 1985: 21–2)

Verne, unusually, was part of two very different media communities. Almost from the start, in addition to writing for the theatre, he also wrote short non-fiction or lightly fictionalized articles for popular magazines on exploration and discoveries. During his years in Paris, he became increasingly involved with Saint-Simonians and enthusiasts for new inventions such as Jacques Arago, Henri Garcet, and Nadar (Félix Tournachon), who was active as a photographer and also as a promoter of heavier-than-air flight. At the same time that he was involved with Romantic dramatists, Verne also moved in circles of learned societies and popular journalism. Some of his earliest published articles were imaginative non-fiction, or a precursor of what would now be called science journalism. During his 20-plus years in Paris, through many moves from one cheap apartment to another, Verne took with him a desk that had two separate drawers: one for science, one for comedies (Allotte de la Fuÿe 1956: 89).[8]

Eventually, it would be Verne's non-literary friendships that helped him more than salon society in shaping his destiny as a writer. His great literary invention, launched in *Five Weeks in a Balloon*, was the geographic romance. Combining comedy and science, he invented a new kind of novel, uniting the crackling dialogue and fast-paced events of the stage with exposition of new geographical knowledge and settings. He combined obsessions brought with him from Nantes and cultural experiences found in Paris to create extraordinary journeys carrying him far from the city.

Verne's first, long-forgotten novel about Paris did not succeed, either as art or as commerce. Much of the narrative is that of a realist, even naturalistic, melodrama: the grim story of an artist, trapped in an uncaring mercenary world, who seeks to build a romantic refuge within the cruel imperatives of bourgeois economic and political power. But neither the gritty details of realism nor the imaginative escapes of romanticism were adequate in expressing the social, economic, and material powers at work in mid-nineteenth-century Paris. When the haven collapses, another Paris emerges: a city transformed by mysterious processes of climate change into an unearthly terrain, where realism and romanticism alike give way before an unknown world dislocated in both time and space. *Paris in the Twentieth Century* is a strange and unstable brew of realism, romanticism, and fantasy.

ACCURSED PARIS OF THE TWENTIETH CENTURY

When *Paris in the Twentieth Century* was first published in the 1990s, critical and popular attention focused on the supposedly prophetic technologies of the fictional future city: an extended subway system with driverless trains, power from subterranean atmospheric systems, fax-like machines, and 'gaz-cars' running silently on the surface. If these systems seem prophetic, it was because Verne, as was his habit, based them on current prototypes, reports, and speculations. (Two years after the book appeared, there were some 500 gas-powered carriages on the streets of Paris.) The inventions were intended to be imaginative, but not imaginary. In the text Verne names some of the sources of these devices, and his descriptions are thickly larded with names of contemporary inventors (Unwin 2000a: 130, 2005: 36–7).

The theme of the novel, however, is not the organization of the material world but that of human consciousness. On the opening page the reader is transported, without any explanation of how or why, to August 13, 1960. A crowd gathers on the former Champs du Mars to attend the prize ceremony for the Academic Credit Union (*Société générale de crédit instructionnel*). Michel Dufrénoy – 16 years old, an orphan from Brittany, an aspiring writer – is introduced as winner of the Latin prize for his centralized public high school. When the audience mocks him for such a useless accomplishment, he throws the prize book, the latest factory manual, on the ground – the first of many defiant but futile gestures he makes in the face of this brutally utilitarian society. Michel is oppressed by a machinery of instruction that favors mathematical, descriptive, mechanical, physical, chemical, and astronomical subject matter for the purpose of training the young in industry, commerce, and finance.

This is not fantasy; it is not even the future. *Paris of the Twentieth Century* exists in a temporal twilight zone, where the imagined future is an exaggerated version of the Second Empire. The instructional credit union running Michel's school is transparently an educational version of the Crédit Mobilier established in 1852 to provide financial backing for the rebuilding of Paris:

'Now, construction and instruction are one and the same for businessmen, education being merely a somewhat less solid form of edification'[9] (Verne 1996: 5).

There is nominal liberty to do what you will, in the future city, even if it means studying Latin. However, investments and institutions are all organized to reward commercial pursuits only. The unidentified narrator bemoans the lack of interest in and support for literature and the arts, especially the French language. It is 'an age when everything was centralized, thought as well as mechanical power' (Verne 1996: 173).

Even more oppressive than the organization of education is the organization of labor. Michel, needing a job, approaches his banker-uncle Boutardin, one of the inwardly mechanized men of the time:

> he moved quite regularly, with the least possible friction, like a piston in a perfectly reamed cylinder; he transmitted his uniform movements to his wife, to his son, to his employees and his servants, all veritable tool-machines, from which he, the motor force, derived the maximum possible profit.
>
> (Verne 1996: 30)

Michel becomes part of this machine, put to work in the evil uncle's bank, where he is hitched to an array of office machines: a calculating machine, with sensitive keyboards resembling a vast piano; electrical telegraphs sending printed messages; and the *grand livre*, 20 feet high, a book-keeping system with a mechanism that allows it to be pointed in various directions like a telescope, and with a system of bridges or gangways (Verne uses the nautical term *passerelles*) that allows it to be raised or lowered according to the needs of the scribe. Michel is assigned to read information aloud to another employee, the copyist Quinsonnas, who enters it into the *grand livre*, using different colors of ink to code his entries. He works for months before he has a single day off.

Michel gamely tries to make a human world for himself in the midst of this iron cage of capitalism. He strikes up a friendship with Quinsonnas, an aspiring musician, visiting his walk-up apartment, where they are joined by a third *déclassé* friend, Jacques. These 'three mouths useless to society' (the title of Chapter 7) spend a lively evening together conversing about piano-playing, eating, and drinking: 'Since that memorable evening, the three young men had become close friends; they constituted a little world of their own in the vast capital of France' (Verne 1996: 99).

On Michel's one vacation day after an entire winter of work, he visits another uncle, Huguenin – his good uncle, an artistic soul who lives in a modest home on the outskirts of Paris. Michel enjoys a long talk with this uncle about French literature as they sit in a small room overflowing with books and furnished with well-worn easy chairs. There is even a token of non-human nature: once a year, a ray of sunlight penetrates into the room from the high walls of the surrounding courtyard. They are later joined by some of Huguenin's friends, a professor and his granddaughter. Michel is smitten by the granddaughter, lovely Lucy. The little group keeps talking, exchanging Latin phrases, and enjoy an excellent dinner.

To round off this fine day, they take a walk to the most astonishing public works project of Paris of the twentieth century: a canal connecting the plain of Grenelle with Rouen, so that Paris has been transformed into a port. With the precision typical of Verne's voice when he gets excited about the possibilities of human achievement, the narrator reports that the canal is 140 kilometers long, 70 meters wide, and 20 meters deep, culminating in a long series of drydocks and wetdocks that can accommodate a thousand deep-draft vessels. The Rouen–Paris canal had been dug with a railway network running alongside to provide towing for the vessels, and the network of docks is connected by drawbridges operated by compressed-air machines. The entrance to this astounding port of Grenelle is proclaimed by an electric lighthouse, 152 meters high, the highest monument in the twentieth-century world, visible from the tower of the cathedral of Rouen.[10]

This scene of Sunday sightseeing at the port of Paris sharply alters the tone of the novel. Up to that point, the text has been dominated by black humor attacking the oppressive commercial values around which everything, 'thought as well as mechanical power', has been organized in twentieth-century Paris. With the description of the port, another voice emerges, that of technological desire: 'In this herculean task, industry seemed to have achieved the extreme limits of the possible' (Verne 1996: 131). This is not the banal, oppressive industry of the bank office. Instead, the port manifests the liberating possibility of large technological systems. Sightseers stroll along the splendid granite quays, admiring the spectacle of steamers flying the flags of all nations. Now Paris – like Nantes in its heyday, but with more benign cargo – is connected with the larger world. The urban prison begins to open; escape seems possible; new vitality enters the text and the city:

> Certainly it was a magnificent spectacle, these steamers of all sizes and all nationalities whose flags spread their thousand colors on the breeze; huge wharves, enormous warehouses protected the merchandise which was unloaded by means of the most ingenious machines . . . ships towed by locomotives slid along the granite walls . . . all the products of the four quarters of the world were heaped up in towering mountains of commerce; many-colored panels announced the ships departing for every point on the globe, and all the languages of the earth were spoken in this Port de Grenelle, the busiest in the world.
>
> (Verne 1996: 135–6)

The tour ends with a description of one of Verne's fantasy vehicles: the great ocean liner *Leviathan IV*, with 30 masts and 15 chimneys, 61 meters wide, so powerful it could cross the North Atlantic in three days, with railroads connecting the decks, some of them large enough to hold public gardens with shade trees and bridle paths. 'This ship was a world' (Verne 1996: 137).[11]

Alas, this happy day is soon over, as Michel and Quinsonnas go back to the workaday grind. Back in the office, back to work on the *grand livre*, they get into a heated conversation about women and love, during which Quinsonnas makes a sweeping gesture to make a point. He spills bottles of colored ink all over the book. The banker enters and sees the disaster: 'That veritable atlas which contained an entire world, contaminated! ruined! spattered! lost!' (Verne 1996: 151).

Michel and Quinsonnas are fired. At first they revel in their escape from the prison of employment, which gives them the opportunity at least to get out in fresh air. But there are the first ominous signs that non-human nature is faltering in this industrialized world. Quinsonnas warns Michel that, metaphorically speaking, it may be a new day, 'but physically, it is growing dark; night has fallen; now we don't want to sleep by starlight – in fact, there is no starlight. Our astronomers are interested only in stars we cannot see' (Verne 1996: 153–4). They go to Uncle Huguenin to tell him what has happened, suggesting that they celebrate their new freedom by spending the day in the country; the good uncle responds, 'But there *is* no country, Michel!', going on to explain:

> For me, the country, even before trees, before fields, before streams, is above all fresh air; now, for ten leagues around Paris, there is no longer any such thing! . . . by means of ten thousand factory chimneys, the manufacture of certain chemical products – of artificial fertilizers, of coal smoke, of deleterious gases, and industrial miasma – we have made ourselves an air which is quite the equal of the United Kingdom's.
> (Verne 1996: 157)

Michel tries another job, at a centralized state-run theatre warehouse, but after 'five long months of disappointments and disgust' (Verne 1996: 187) decides he can stand it no longer. After he quits this second job, it becomes evident that the atmospheric change is even more serious than the loss of fresh air. Northern Europe begins to chill with the advent of a new ice age:

> The winter of 1961–62 was particularly harsh; worse than those of 1789, of 1813, and 1829 for its rigor and length. In Paris, the cold set in on November 15, and the freeze continued uninterrupted until February 28; the snow reached a depth of seventy-five centimeters; and the ice in ponds and on several rivers a thickness of seventy centimeters; for fifteen days the thermometer fell to twenty-three degrees below freezing. The Seine froze over for forty-two days, and shipping was entirely interrupted.
> (Verne 1996: 195)

Grapevines, olive trees, and chestnut trees die; wheat and oat harvests are lost; cold and snow bring railroads to a halt; their engineers and ordinary people in the streets perish from the cold. Fertility and nature itself seem to retreat. This new ice age is the pivot of the story. Paris of the twentieth century has been a future but familiar world, described in the language of social realism, even naturalism, with romantic interludes and one burst of technological enthusiasm. But in the last three (of 17) chapters, when glacial ice and polar snow invade the city, it is transmuted into a truly strange world, depicted in the language and images of fantasy.

Michel's 'little world' of friendship and art collapses. Quinsonnas leaves Paris for Germany. Lucy and her grandfather fall on hard times. Without money, Michel falls into bad habits. He becomes ashamed of himself, rarely visiting his good uncle or Lucy and her grandfather, and is reduced to eating 'coal bread'. The progress of science ensures that no one would die of hunger, 'But how did he live?' With his last pennies, he decides to buy Lucy a bouquet. 'Like a madman' (Verne 1996: 198), in temperatures 20 degrees below freezing, he runs into the streets, in the snow, in the dark, heading northeast on foot, only to discover that Lucy and her grandfather have been evicted from their apartment. They are all homeless now, and he does not know where to find her.

The last chapters of *Paris in the Twentieth Century* are a *voyage extraordinaire* within the capital. A twilight unreality settles over the city. Dialogue and humor fade from the text as the voice of an imagined observer takes control, indicating that the now nearly speechless protagonist has entered an unfamiliar world (Unwin 2000a: 53). The new narrative voice, like that of a shaman, takes Michel and the reader beyond everyday reality and beyond waking consciousness. The background sounds of ringing clocks give way to eerie silence (Platten 2000: 81, 87; Unwin 2000a: 53). Verne was already working on *Captain Hatteras*, and the sources he was reading as background on arctic explorations surely stimulated his imagination of Paris as an arctic waste. Michel, like Captain Hatteras, makes an obsessive, doomed trek to a terminal point on the earth – but Michel does this in the middle of Paris. He too goes mad as he becomes obsessed with finding a point of meaning: not the north pole, as for the captain, but the whereabouts of Lucy and her grandfather.

As he continues his journey to the unknown, the world becomes transformed not only by cold but by 'the demon of electricity' (the title of Chapter 16). There are many descriptions, in the nineteenth century, of the wonders of 'the electrical fairy', including ones by Verne himself, most notably in a chapter of *Twenty Thousand Leagues under the Seas* titled 'Everything by Electricity'. In *Paris of the Twentieth Century*, however, the electrified landscape is one of nightmare and horror. As Michel walks along the Seine, he notes that 'over his head the sky was cluttered with electric wires passing from one bank to the other and extending like a huge spiderweb' (Verne 1996: 206). He passes a morgue, where electric apparatus is used to restore life to bodies 'still harboring some spark of existence' (Verne 1996: 208). He arrives at Notre Dame, just as mass is ending: the altar shines with electric light, as does the monstrance raised in the priest's hand. 'More electricity', Michel sighs, 'even here!' (Verne 1996: 207). He is disgusted by glaring lampposts, electrical signs displaying advertisements, and an electric concert (200 pianos wired together!). Pursued by the demon of electricity to a scaffold, Michel sees preparations for an execution by electrocution. All these images and sounds mingle and amplify each other into a terrifying synesthesia of death.

In his frozen trance, Michel finds himself back in a village, a refuge from 'this deadly capital, this accursed Paris' (Verne 1996: 211) – but it turns out to be a village of the dead, the cemetery of Père-Lachaise, its tombs grouped like tiny cities. There, from the top of a cemetery hill, he once again sees Paris, from Montmartre to Notre Dame to the lighthouse of the frozen port. He looks at the sky: water may freeze, but the sky seems to offer him a way out, if he could fly like a bird. But no: even the sky has been weaponized: there the government has floated armored balloons designed to protect the city from lightning storms. There is no escape by land, water, or air. Trapped, Michel wishes that he could cut the ropes of the balloons to clear the sky so Paris can be destroyed in a rain of fire. Instead, he swoons on the snow. The reader has no reason to think he will ever regain consciousness.

ART AS CONSTRUCTION OF A SECONDARY WORLD

Bitterness at bourgeois civilization, desire to escape from it, trapped in a city covered as if with a lid, burdened by consciousness: Verne and Baudelaire share these themes, as well as other significant affinities in their artistic goals and practices. Both seek to create language adequate for new experiences found not just in Paris of the Second Empire but more generally in a world being transformed with unprecedented scope, scale, and pace. The world they confront is not just a new urban environment, but a new environment for human life, one dominated by structures and processes created by humans themselves: a critical tipping point in the long evolution of human-generated 'second nature' (Marx 2008: 8–21; Williams 2008: 257–74).[12]

At this threshold moment, and in response to it, Verne and Baudelaire redefine the very role of literature. For them its traditional purposes – to entertain, soothe, create beauty, give pleasure, impart experience and even wisdom – are inadequate. They assert that the mission of art is nothing less than the construction of a 'secondary world', a goal as ambitious and audacious in its realm as the creation of second nature in the material realm (Clute and Grant 1997: 847).[13]

Baudelaire's great literary innovation is symbolism in poetry; Verne's, the scientific romance (or, to use the term that Verne preferred, the geographic romance) in prose. Symbolism and romance are typically treated as separate chapters of literary history, but they share this crucial affinity: the purpose of the work of art is to transport writer and reader from the familiar world to an unfamiliar one, with the boundary between the *mondes connus* and *inconnus* as uncertain as that between first and second nature. Their aesthetic goal is more than evasion, escape, transport, voyage, or journey, no matter how extraordinary. Their desire is to go beyond surrounding reality by means of the potentialities within it (Chesneaux 1972: 196–7). Thus Baudelaire in symbolism and Verne in romance both create a destination, a hybrid of the known and the unknown, a derivative but independent world, a secondary world constructed from language.

For Baudelaire, the fundamental unit of construction is the symbol. He builds poetic worlds from sensory images extracted from the visible world, rearranging them to make an aesthetic system, a web of universal analogy woven by the creative imagination. The poet establishes connections among the sensuous phenomena of the visible world – whether natural in the usual sense of non-human nature, or the 'transformed, fabricated, unrecognizable nature' of the cityscape – and the supra-sensible world behind them, as they are linked through signs, symbols, and images. Everywhere the poet finds stores of analogies and images to build systems of connections between the soul and the world (Raymond 1970: 13). The symbol makes a transfer, a connection, in a technical sense, so that the poem is an intricately designed system of switches (just as *correspondances* are points on the Paris Métro system where travelers change trains).

Verne uses texts as his primary building blocks. He sifts through reports, articles, and other texts to assemble these fragmentary signs of the contemporary world into a comprehensive and coherent realm of art. Verne's library was full of practical works of information about geography, history, and languages of faraway places, gleaned from newspapers and magazines. He never claimed to have any expertise in science or engineering but was, by his own account, 'a great reader', always with a pencil and notebook in hand. Each day in his Amiens club he read 15 newspapers, as well as other periodicals and publications, especially those dedicated to science and exploration (particular favorites were Camille Flammarion on astronomy and Élisée Reclus on geography). In an 1894 interview Verne explained: 'I have thus amassed many thousands of notes on all subjects, and to date, at home, have at least twenty thousand notes which can be turned to advantage in my work, as yet unused' (Evans 2001: xiv).[14]

As Timothy Unwin (and Daniel Compère before him) has shown, Verne writes through grafting, collage, and assembly, 'in which the author's readings are exploited and recycled'.[15] Unwin calls these practices 'reflexive realism', a highly self-conscious form of realism in which the author creates his fictional world not out of direct experience but out of symbolic representations. Verne said in a newspaper interview shortly before his death, 'I think that an attentive reading of the most documented works on any new subject is worth more than concrete experience, at least when it comes to writing novels' (Unwin 2000a: 58).[16] Like his instrument-wielding explorer-heroes – and later symbolists – Verne leaves actual life to servants (Wilson 2004: esp. 211–13).

Verne organizes these realistic points, laid down through dogged research, into a narrative arc that transports readers to worlds known and unknown. The arc may come from classical myths, medieval romances, or contemporary stage plots, or a mix of them all. Verne's adventurous stories about exploring the universe combine some of the oldest Western story-telling traditions with the emerging habits of the information age; they demonstrate that hyper-realistic attention to information is compatible with an obsessive desire to escape from reality or, to put it a little differently, with world alienation (Arendt 1998: esp. 6, 248–57). Both in Verne's novels and in Baudelaire's poems, the selection and arrangement of symbols and texts are carefully designed to convey both writer and reader to another world, 'Anywhere so long as it is out of this world!' (as the poet's soul cries in the prose poem by that title).[17]

This is a desire not only to flee the oppressive world but to replace it with another world, one constructed by the artist rather than the bourgeoisie. This understanding of the goal of art may spring from disdain for the bourgeoisie, but it expresses a culture of nature similar to that motivating the rebuilding of Paris, France, Europe, and the European-controlled world at the time. The drive to create a secondary world of art is similar to the drive to create second nature in this respect: both assume the availability of non-human nature for human exploitation. In his essay on 'The Question Concerning Technology', Martin Heidegger (1993) describes the essence of 'technology' as nothing material but rather as a human attitude towards the non-human world which regards it as a 'standing reserve' available for exploitation. The essence of technology, Heidegger proposes, is a culture of nature in which nature is regarded not primarily as something living of its own accord, nor as something to be nurtured, nor as something fragile, but as a resource for humanity.

Baudelaire and Verne share this assumption. Baudelaire, for his part, repeatedly describes nature as a storehouse of images. In an article on Victor Hugo, he writes that for 'the best poets . . . comparisons, metaphors, and epithets are drawn from the inexhaustible store of *universal analogy*' (Raymond 1970: 11n). Critic Marcel Raymond comments:

> Baudelaire assumes an extremely significant attitude toward outward nature. He regards it not as a reality existing in itself and for itself, but as an immense reservoir of analogies and as a kind of stimulant for the imagination. 'The whole visible universe', he writes, 'is only a store of images and signs to which the imagination accords a relative place and value; it is a kind of fodder that the imagination must digest and transform.'
>
> (Raymond 1970: 11)

Nowhere is this attitude expressed more famously than in Baudelaire's iconic poem 'Correspondances':

Nature is a temple where living pillars
At times allow confused words to come forth;
There man passes through forests of symbols
Which observe him with familiar eyes.[18]

(Fowlie 1992: 27)

The 'forest of symbols' is a standing reserve from which the poet extracts poetic material in the form of 'comparisons, metaphors, and epithets'. These resources allow the poet to decipher the meaning of the visible universe and to design and build an aesthetic one. In this forest the poet 'proceeds to cut himself a path in the world of analogies and to arrange and order the material supplied by nature' (Raymond 1970: 11).

Verne shares the attitude that non-human nature is a standing reserve for human exploitation. In particular, he expresses the Saint-Simonian belief that the exploitation of non-human nature can remain separate from the exploitation of humankind (Chesneaux 1972: esp. 82–3). This belief is evident in *Paris in the Twentieth Century*, in which the narrator attacks white-collar office slavery, as well as all the inner and outer mechanizations associated with the profit system of capitalism, while also celebrating the conquest of non-human nature through systems that turn Paris into a port and through fantasy vehicles such as *Leviathan IV* that provide a controllable 'little world' as a refuge from the cruelty and stupidity of the larger one.

As daily experience became increasingly that of a human-built world, literary experience became increasingly conscious of its self-construction. If the realist metaphor for art is the mirror (reflecting the world), and the romantic metaphor is the lamp (illuminating the world), for Verne and Baudelaire the appropriate metaphor for literature is a construction site: a vast project built by the writer from the raw materials of images and texts (Abrams 1953).

Building such a secondary world is a grandly ambitious enterprise. Baudelaire constantly expanded and refined *Fleurs du mal*, with careful attention to the sequence and juxtaposition of this ever-expanding poetic system. Verne was constantly writing more *Voyages extraordinaires*, keeping track of geographical regions he still needed to cover in order to create a prose map of the entire world: 'a mad, colossal, monstrous literary ambition' (Unwin 2005: 26). These aesthetic empire-builders proposed a new and encyclopedic mission for writing, one 'that draws attention to itself as substituting art and its creations for the once-possible synthesis of the world empires' (Said 1994: 189).

This literary mission is congruent with the size and swagger of the creation of second nature in mid-nineteenth-century Europe. The rebuilding of Paris is only part of the colossal rebuilding of the world in France at that time:

Everyone knows that the nineteenth century was an age of change, but for many people of the time, roads, railways, education and sanitation were trivial innovations compared to the complete and irreversible transformation of their physical world . . .

... By the mid-nineteenth-century, huge tracts of land were being reclaimed at a rate of several thousand acres a day. Half the moorland in Brittany disappeared in half a century.

Large parts of Mediterranean France were transformed within a generation.

(Robb 2007: 268–9)

The material events that reshaped the lands and waters of mid-nineteenth-century France depended upon the 'idea of management (*aménagement*, which may also be translated as 'development') of territories and cities on the scale of the entire earth ... the progressive shrinking of the planet which primes itself [*qui s'amorce*]' to become 'the dwelling place of man'. This goal, based on an ideology of circulation, was expressed in fractal-like patterns of technological systems lacing cities, regions, continents, and eventually the whole globe (Williams 1993). It is certainly 'the triumph of the networks' (Picon and Robert 1999: 192–4), but the qualitatively unique component of all these networks is the motivating idea behind them, the very idea of global management. As Karl Marx noted, among all the animals only the human species consciously designs: only the human architect 'raises his structure in imagination before he erects it in reality' (Marx 1867).[19]

These material events of the mid- to late-nineteenth century – the rebuilding of Paris, the 'complete and irreversible transformation' of France, the beginnings of the networking of the planet (Mattelart 2000) – depend upon and generate *events of consciousness*: in this case, events of imagining that humanity can and should remake the world in this way. To be sure, as long as we human beings have dwelled on earth, we have been actively creating second nature. What is new in the mid-nineteenth century is a self-conscious project of doing this in a comprehensive and totalizing manner. In Paris of the Second Empire, in France, Europe, and European overseas holdings, the critical event of consciousness is the conviction that humans can and should ransack the standing reserves of the planet to create a comprehensive second nature organized around our needs and wishes.

All the material evidence in the nineteenth-century world – bridges and railways built, sewers and canals dug – cannot demonstrate when and how human beings came to think about their relationship to their earthly home in this way. Imaginative writers like Verne and Baudelaire, who are acutely attuned to the new culture of nature and unusually skilled at probing human motivations with the most flexible tool of all, that of language, provide archives of this event of consciousness. They not only reflect what is happening around them, not only express new goals for art that stimulate what would later be called modernism, but they also engage in projects to construct secondary worlds of art as ambitious and innovative as contemporary efforts to construct second nature. In literary and practical arts, in Paris of the Second Empire and in the larger human empire of the same period, the supreme achievement of humankind is defined as the creation of artifice.

NOTES

1   See the complete list of titles, composition, and publication dates in Butcher (2006: 313–14).
2   'The capital of the nineteenth century' is the title of Benjamin's (1892–1940) sketch or précis of his massive Arcades Project (*Das Passagen-Werk*), written in 1935. For the text, in the context of the larger project, see Benjamin (1999: 3–26). For recent commentaries, see Jennings (2006). Key texts, besides those of Benjamin, are Berman (1988: 131–64); Clark (1984, 1999), for discussion of Baudelaire's politics; Harvey (1985: 36–62) and a more extended treatment in Harvey (2003); and Higonnet (2002).
3   There is no evidence that they met, or that Baudelaire knew of Verne, but Verne certainly knew of Baudelaire, primarily through reading his translations of Edgar Allan Poe's tales, published between 1852 and 1860 under the collective title *Histoires extraordinaires*. Verne's copy of Baudelaire's translation of Poe was dated 1862 (Unwin 2005: 208 and note). On some of Verne's references to Baudelaire in the *Voyages extraordinaires*, see note by Pierre-André Touttain in Verne (1989: 8n), with special reference to Marcel Moré (1963: 78–9). More generally, there are fascinating similarities between Verne's years in Paris and Baudelaire's: the constant background of illness, depression, sexual frustration, and frequent moves from one cheap apartment to another; having their portraits done by Nadar; both eventually publishing with Hetzel; a basically apolitical stance combined with interest in utopian socialism of various hues; and a deep, broad anger toward bourgeois civilization as a whole. See Butcher (2006: 73, 212–13) and Platten (2000: 85).
4   For summaries of Vernian studies, see Evans (2000: 34), who emphasizes English-language studies and organizations in the 1990s. Also see Unwin (2005: 3–5) (including notes) and Butcher (2006: 307–12). A recent update of Vernian studies, with special attention to machinery and technology in his work, was held on the occasion of the centenary of his death: the colloquium on 'Jules Verne: Les Machines et la Science', held in Nantes in October 2005, was organized by l'École Centrale de Nantes and the Université de Nantes (Mustière and Fabre 2005).
5   Unwin (2005: 27); his translation of the 1890 autobiographical text (Verne 1989: 3–8).
6   Butcher (2006: 224) contends that, because of Hetzel, Verne's works were 'not allowed to visit France', so that his corpus of geographical novels covers the 'known and unknown world' except for his own country. It is certainly true that Hetzel acted not just as a publisher and editor, in the conventional sense, but also as a sort of patent attorney. Verne had come up with a new invention, the geographical romance, and Hetzel was determined to stick with the winning formula. Still, Verne did not seem to complain about Hetzel's proscription of local or regional topics, as he

often did with other counsels from his publisher, and there are innumerable independent statements from Verne about his fascination with distant places. See the helpful geographic index to Verne's works in Jean Chesneaux (1972: 8–10), who notes that the most vital zones of Europe in his imaginative writings are the exotic fringes of Europe such as Greece, Bulgaria, Transvylvania, Hungary, Russia, Scotland, and Ireland.

7   See summary of Hetzel's response in Piero Gondolo della Riva, preface to Verne (1994: 15–17).

8   Allotte de la Fuÿe (1956: 89). Butcher warns us that she is not a reliable biographer, though there is no particular reason to challenge this particular detail.

9   'Or, construire ou instruire, c'est tout un pour des hommes d'affaires, l'instruction n'étant, à vrai dire, qu'un genre de construction, un peu moins solide' (Verne 1994: 30).

10 Many canals and related riverworks were constructed in late-nineteenth-century northwest Europe to improve connections between cities and the sea (Vernon-Harcourt 1886: 576; 1896). At the time Verne was living in Paris, major projects were being undertaken around Rouen to improve the navigation on the mouth of the estuary of the Seine. He would also have been aware of similar projects on the estuary of the Loire, intended to revive Nantes as a port at a time when it was rapidly losing trade to coastal cities better able to accommodate ever-larger ocean-going vessels. Finally, at that time work was starting on the 88-mile-long Suez Canal, 'one of the most remarkable engineering works of modern times', with most of the capital and engineering coming from France (Stevenson 1878: 789).

11 The name of the vessel alludes to the largest steamship of the day, the *Great Eastern*, launched in 1857 under the name *Leviathan*; the narrator comments that the nineteenth-century ship would not even serve as a tender to the imagined twentieth-century *Leviathan IV*. The *Great Eastern* was built to carry emigrants but was soon converted for laying transatlantic telegraph cables. Jules Verne and his brother Paul made an Atlantic crossing on the *Great Eastern* in 1867, the first sailing after a French company bought and refitted the vessel primarily to bring Americans to France for the Paris Universal Exposition (Butcher 2006: 175; see also Butcher 2006: 176–81 for more discussion of Verne's actual and imaginary ocean liners).

12 The concept of 'second nature' was used by Aristotle, Kant, and Nietzsche in a very general sense to describe acquired rather than innate 'human nature'. Hegel and Marx used the term in the context of evolutionary biology to describe the material and cultural environment that human beings have superimposed on the given nature of the planet over many millennia. In this latter sense, the concept of second nature was appropriated by early-twentieth-century critics such as Georg Lukács,

Theodor Adorno, and Walter Benjamain (Buck-Morss 1977: 52–7; Lukács 1971; see references in Dimendberg 1995: 112 and note). More recently William Cronon has proposed reviving the first nature–second nature distinction (Cronon 1991: xvii–xvix), as have other environmental thinkers such as Janet Biehl (1991: 117–18); see references in Marx (2008: 20–1).

13 The term was coined by J.R.R. Tolkien in his essay 'On Fairy Tales' (1939), subsequently published in his collections *Essays for Charles Williams* (1947) and in *Tree and Leaf* (1964).

14 The quotation comes from Robert H. Sherard, 'Jules Verne at Home', *McClure's Magazine* (January 1894): 120–1. Verne reportedly accumulated 20,000 notecards from his reading (Butcher 2006: 270, 285).

15 Unwin argues that the paradox of nineteenth-century realism is that its claims to objectivity triumph at the same moment as intense self-consciousness about the act of writing itself. This 'reflexive realism' arises from 'a schism between the represented world and the medium of representation', this schism becoming a 'paradoxical and unexpected' source of artistic riches. 'Reflexive texts offer a vision of the world while reflecting on their own standing' (Unwin 2000b: 6–7). See also Unwin (2005: 57) and Compère (2000: 40–5).

16 Unwin is quoting Compère (1991: 45), who in turn is quoting the Chicago *Evening Express*, March 25, 1905. Unwin (2005: 52) praises Compère, 'who more than any other critic has emphasized the polyphonic, composite nature of the Vernian text'. See also Compère (2000: 40–5).

17 'N'importe où hors du monde', in *Le Spleen de Paris* (1869).

18 'La Nature est un temple où de vivants piliers/Laissent parfois sortir de confuses paroles;/L'homme y passe à travers des forêts de symboles/Qui l'observent avec des regards familiers.' For other English translations of 'Correspondances', see Clark and Sykes (1997: 14–20) and, most recently, Waldrop (2006) ('Nature is a temple whose columns are alive and sometimes issue disjointed messages. We thread our way through a forest of symbols that peer out, as if recognizing us').

19 'A spider conducts operations that resemble those of a weaver, and a bee puts to shame many an architect in the construction of her cells. But what distinguishes the worst architect from the best of bees is this, that the architect raises his structure in imagination before he erects it in reality. At the end of every labour-process, we get a result that already existed in the imagination of the labourer at its commencement. He not only effects a change of form in the material on which he works, but he also realises a purpose of his own that gives the law to his modus operandi, and to which he must subordinate his will. And this subordination is no mere momentary act' (Karl Marx 1867, http://www.marxists.org/archive/marx/works/1867-c1/ch07.htm).

REFERENCES

Abrams, M.H. (1953) *The Mirror and the Lamp: Romantic Theory and the Critical Tradition*, New York: W.W. Norton.

Allott, K. (1940) *Jules Verne*, London: Cresset Press.

Allotte de la Fuÿe, M. (1956) *Jules Verne*, trans. Erik de Mauny, New York: Coward-McCann; French edition, 1953, Paris: Hachette.

Arendt, H. ([1958] 1998) *The Human Condition*, 2nd edition, Chicago: University of Chicago Press.

Béal, J. (1985) 'Un Amiénois visionnaire', in B. Roux (ed.), *Le Printemps technologique de Jules Verne*, Amiens: Édition Courrier Picard et Crédit Agricole de la France.

Benjamin, W. (1999) *The Arcades Project*, trans. Howard Eiland and Kevin McLaughlin, Cambridge, MA: Belknap Press of Harvard University Press.

Berman, M. ([1982] 1988) *All that Is Solid Melts into Air: The Experience of Modernity*, New York: Penguin Books.

Biehl, J. (1991) *Rethinking Ecofeminist Politics*, Boston, MA: South End Press.

Buck-Morss, S. (1977) *The Origin of Negative Dialectics*, New York: Free Press.

Butcher, W. (2006) *Jules Verne: The Definitive Biography*, New York: Thunder Mouth Press.

Chesneaux, J. (1972) *The Political and Social Ideas of Jules Verne*, trans. Thomas Wikeley, London: Thames and Hudson; French edition, 1971, *Jules Verne: une lecture politique de Jules Verne*, Paris: François Maspero.

Clark, C. and Sykes, R. (eds) (1997) *Baudelaire in English*, London: Penguin.

Clark, T.J. (1984) *The Painting of Modern Life: Paris in the Art of Manet and his Followers*, Princeton, NJ: Princeton University Press.

Clark, T.J. ([1973] 1999) *The Absolute Bourgeois: Artists and Politics in France 1848–1851*, Berkeley: University of California Press.

Clute, J. and Grant, J. (eds) (1997) *The Encyclopedia of Fantasy*, New York: St. Martin's Griffin.

Compère, D. (1991) *Jules Verne écrivain*, Geneva: Droz.

Compère, D. (2000) 'Jules Verne and the Limitations of Literature', in E.J. Smyth (ed.), *Jules Verne: Narratives of Modernity*, Liverpool: Liverpool University Press.

Cronon, W. (1991) *Nature's Metropolis: Chicago and the Great West*, New York: W.W. Norton.

Dimendberg, E. (1995) 'The Will to Motorization: Cinema, Highways, and Modernity', *October*, 73, Summer.

Evans, A. (2000) 'Jules Verne and the French Literary Canon', in E.J. Smyth (ed.), *Jules Verne: Narratives of Modernity*, Liverpool: Liverpool University Press.

Evans, A. (2001) 'Introduction' to Jules Verne, *Invasion of the Sea*, trans. Edward Baxter, ed. Arthur Evans, Middletown, CT: Wesleyan University Press.

Fowlie, W. (ed. and trans.) (1992) *Flowers of Evil and Other Works/Les Fleurs du Mal et oeuvres choisies by Charles Baudelaire*, New York: Dover Publications.

Harvey, D. (1985) *Consciousness and the Urban Experience: Studies in the History and Theory of Capitalist Urbanization*, Baltimore, MD: Johns Hopkins University Press.

Harvey, D. (2003) *Paris, Capital of Modernity*, New York: Routledge.

Heidegger, M. (1993) 'The Question Concerning Technology', in *Basic Writings*, ed. David Krell, New York: HarperCollins; originally published 1954 in M. Heidegger, *Vorträge und Aufsätze*.

Higonnet, P. (2002) *Paris: Capital of the World*, trans. Arthur Goldhammer, Cambridge, MA: Belknap Press of Harvard University Press.

Jennings, M.W. (ed.) (2006) *The Writer of Modern Life: Essays on Charles Baudelaire*, trans. Howard Eiland *et al.*, Cambridge, MA: Belknap Press of Harvard University Press.

Lukács, G. ([1914–15] 1971) *The Theory of the Novel*, trans. Anna Bostock, Cambridge, MA: MIT Press.

Marx, K. (1867) *Capital*, Vol. 1, Part III: 'The Production of Absolute Surplus-Value', Chapter 7: 'The Labor-Process and the Process of Producing Surplus-Value', Section 1: 'The Labor-Process and the Production of Use Values', Online, available at: http://www.marxists.org/archive/marx/works/1867-c1/ch07.htm.

Marx, L. (2008) 'The Idea of Nature in America', *Daedalus*, 137(2): 8–21.

Mattelart, A. (2000) *Networking the World, 1794–2000*, trans. Liz Carey-Libbrecht and James A. Cohen, Minneapolis: University of Minnesota Press.

Moré, M. (1963) *Nouvelles explorations de Jules Verne*, Paris: Gallimard.

Mustière, P. and Fabre, M. (eds) (2005) *Jules Verne: Les Machines et la Science*, Nantes: Coiffard.

Picon, A. and Robert, J.-P. (1999) *Les dessus des cartes: un atlas parisien*, Paris: Pavillon de l'Arsenal, and Paris: Picard.

Platten, D. (2000) 'A Hitchhiker's Guide to Paris: Paris au XXe siècle', in E.J. Smyth (ed.), *Jules Verne: Narratives of Modernity*, Liverpool: Liverpool University Press.

Raymond, M. (1970) *From Baudelaire to Surrealism*, London: Methuen; French edition, *De Baudelaire au Surréalisme*, 1933.

Robb, G. (2007) *The Discovery of France: A Historical Geography from the Revolution to the First World War*, New York: W.W. Norton.

Said, E.W. ([1993] 1994) *Culture and Imperialism*, New York: Vintage Books, Random House.

Stevenson, D. (1878) 'Canal', *Encyclopaedia Britannica*, 9th edition, Vol. 4, New York: Charles Scribner's Sons.

Unwin, T. (2000a) 'The Fiction of Science, or the Science of Fiction', in E.J. Smyth (ed.), *Jules Verne: Narratives of Modernity*, Liverpool: Liverpool University Press.

Unwin, T. (2000b) *Textes réfléchissants: réalisme et réflexivité au dix-neuvième siècle*, Oxford: P. Lang.

Unwin, T. (2005) *Jules Verne: Journeys in Writing*, Liverpool: Liverpool University Press.

Verne, J. (1989) 'Souvenirs d'enfance and de jeunesse', *Bulletin de la Société Jules Verne*, No. 89, 1er trimèstre.

Verne, J. (1994) *Paris au XXe Siècle*, Paris: Hachette.

Verne, J. (1996) *Paris in the Twentieth Century*, New York: Random House.

Vernon-Harcourt, L.F. (1886) 'River Engineering', *Encyclopaedia Britannica*, 9th edition, Vol. 20, New York: Charles Scribner's Sons.

Vernon-Harcourt, L.F. ([1882] 1896) *Rivers and Canals*, 2nd edition, Oxford: Clarendon Press.

Waldrop, K. (2006) *Flowers of Evil: Charles Baudelaire*, Middletown, CT: Wesleyan University Press.

Weber, E. (1996) 'Introduction' to Jules Verne, *Paris in the Twentieth Century*, New York: Random House.

Williams, R. (1985) 'Metropolis and Modernity', in E. Timms and D. Kelley (eds),

Unreal City: Urban Experience in Modern European Literature and Art, New York: St. Martin's Press.

Williams, R. (1993) 'Cultural Origins and Environmental Implications of Large Technological Systems', *Science in Context*, 6(2): 377–403.

Williams, R. (2008) 'Afterword', *Notes on the Underground: An Essay on Technology, Society, and the Imagination*, new edition, Cambridge, MA: MIT Press.

Wilson, E. ([1931] 2004) *Axel's Castle: A Study of the Imaginative Literature of 1870–1930*, New York: Farrar, Straus and Giroux.

Interview with Rob Shields

Ignacio Farías

Ignacio Farías: I would like to start talking about the urban logic of contemporary capitalism and maybe begin with the issue of gentrification. What it started as a say mid-range theory of neighbourhood regeneration has become a now integral part of larger frameworks to make sense of cities and their economic performance: global city hypothesis, the notion of the creative class, etc. Are these global theories the adequate framework to think about what's going on in neighbourhoods?

Rob Shields: Gentrification generalizes complex changes in family life and housing which are not only economic and sociological, but also cultural. The term further generalizes across city-regions and states. But one should ask: have we really understood both the similarities and the differences between cities which are glossed in this term? I am not that sure. Gentrification mirrors studies of globalization. Both represent a moment of global theorizing whose historicity has not been examined, and both share some qualities with discourses of modernity such as progress. They stand in for discredited meta-narratives by providing a generalized story of the present which is both anchored in empirical studies (even if they sometimes contradict the overall thesis) and answers in spirit to a desire for a framework that facilitates judgements and prescriptions for the directions of new social projects. Gentrification is not only 'seconded' as a meta-narrative but, like globalization, finds its roots in a theory vacuum created by the destruction of positivist social science theory by postmodern critiques in the 1980s and 1990s.

How this displacement happened is worth its own sociological or anthropological investigation. In retrospect, there is much more to be said than that there was a simple clash of paradigms. This was not a Kuhnian scientific revolution internal to the social sciences, a bit like a palace coup. Rather, the broad power of narratives such as 'progress' and the vocation of social science as a source of nomothetic prescriptions (i.e. policy) came into disrepute and was de-legitimized. As a result 'policy' goes its own way as 'opinion polling', 'business studies' and 'social policy'. It continues as a pragmatic response to the need to extrapolate short-term trends from data and to create narratives from case studies.

IF: Nowadays gentrification has become such a prominent public discourse, not just in social science, that it is difficult to imagine gentrification processes occurring in the subtle and almost automatic ways in which they were first thought: artists and creative scenes moving first into a neighbourhood, followed by bars and restaurants, refurnishing of buildings, young middle classes, etc. Around the corner of my place there is a sign that says 'please gentrify this', and the irony of the graffiti is that it can be read as a quite legitimate political demand, for city governments are explicitly aiming at the gentrification of city areas. It is as though it had become a reflexive process. Maybe one should speak of a 'reflexive gentrification'?

RS: I also suspect that the financial flows are quite different now. If you think about how these buildings were bought and renovated, often by young people who rented out rooms and then slowly expanded their own quarters. They used the rent to pay the mortgage or they got money through family networks. I'm not sure what the banking arrangements are now like. What is the 'trick' that allows some people to be in a position where they can afford to gentrify? I suspect it's much more developer-led now. It has become a professional activity: you buy the whole building, you do it up once and you sell it back on the market. With the loft market now it is the same way, at least in North America. It's become professionalized, not just reflexive, but professionalized, and that means governed by building codes standards. Whereas if you would do the work yourself in your own house, it's much less transparent or visible to the authorities: they can see if you ask for a building permit, but otherwise it's harder for them to see how you change it. So it's possible for people to change things without following the standardized procedures. This is not to say that they're sub-standard but simply that they can experiment with the standards. Whereas for an architect in commercial construction, the foyer, the stairs, everything is almost already designed, because it's in the building codes: the doors open certain ways and so on.

But there is a further issue in the notion of 'reflexive gentrification', or the becoming-reflexive of gentrification and gentrifiers. I don't know whether this phrase is a product of the translations we are making between Spanish, German and English languages and theoretical contexts but it will be a new term for English speakers. Is this reflexive in the German sense of a reflexive modernity, a modernization of modernity? No. Reflexive gentrification denotes the professionalization and even industrialization of gentrification or, to use other terms, neighbourhood renewal or urban revitalization. Not only is this a matter of a home renovation industry but services, planners, policy makers and all those arbiters of taste who set the urban ideals of a domestic 'good life' that the 'everyperson' of contemporary urban society might aspire to.

I have an image in mind too – a small poster glued up along an American street proclaiming 'The Next Hot Neighbourhood' on which has been graffitied 'Yeah Buy me next.' This is self-conscious gentrification as a lifestyle consumption opportunity: a product on offer. An image of a street signpost with an added no-right-turn sign forms the left margin of this poster and suggests planning practices such as one-way traffic and other typical 'traffic-calming' measures for desirable North American neighbourhoods. These often adjoin thoroughfares which offer a wealth of boutiques and the most trendy consumption opportunities. 'Buy me next.' This is reflexive gentrification in a tidy aphorism. Gentrified neighbourhoods are not only systematic results of urban financial cycles but are professionally scouted and produced by speculators as new areas for the cultural industries and houses tidied up or 'flipped' for purchase by the 'creative classes'. Gentrification has reached its terminus in a real estate agent's formulaic practice.

IF: ... what you say reminds me much more of the notion of a cultural industry in the singular than that of a creative class ...

RS: This is the paradox of the creative city and the creative class, because in a way for a creative class you have to have a somewhat disorganized city. Berlin, not Bonn. Detroit or post-Katrina New Orleans, not LA nor New York. The truly creative class may be a kind of bohemian creative class. A different creative class could be professional engineers or doctors launching a new gaming venture. Not inventors, but innovators implementing something commercially. But together these don't make up a coherent socioeconomic class fraction in the analytical sense. If you want a creative industry then you're talking about something very different.

I think the words 'creative' and 'class' have been hijacked. We have to be very careful, very critical, because these are really important words. We are not in the realm of analysis but of metaphor and what Dewey called 'scene words'. The same has occurred with 'knowledge', which in the discourse of the knowledge economy was redefined as information. Knowledge is not information; it's much more emergent than that: computers have information; human beings have knowledge; animals have knowledge. There's a big difference, because knowledge is a virtuality, a processual sense-making; information is an object. But if one wants to commodify knowledge, one has to make it appear as an information-like object rather than an intangible, performative understanding. Without saying more than that, what if the same thing is happening with creativity? So they take the word 'creativity' and they use it to talk about a kind of entrepreneurial class that you might get in Barcelona or Manchester. Creativity is presumed to be an economic activity statistically regularized amongst a specific group. I'm leery about accepting the self-description of marketing entrepreneurs as 'creatives'. 'Creative' describes a performative activity, not the quality of a socioeconomic group. What of the shared cultural heritage of elites and other citizens of these cities? Both have a powerful and parochial identity, accent, language and regional history where one suspects that local entrepreneurs are strongly rooted in place. How would one 'break in' to culture industry cliques in these capitals of their own culture? The question of redevelopments to attract this creative class is, 'If you build it, will they come?' Is it really a question of attracting globally mobile entrepreneurs or is it a process of providing a stage for local talent? Monica Degen's work on Manchester is very interesting in this regard: old industrial areas that got turned into bars, theatres and squares function like a kind of backdrop or 'props' where entrepreneurs who are doing things like publishing or music production can meet, have coffee and make a deal and be seen as being 'trendy'.

However, these kinds of places provide a backdrop for small and medium-sized enterprises where a large part of what they're selling – be it shoes, greeting cards, clothes, whatever – is a kind of effervescence or style. It is not the production, something which is going to happen in China. This is not a creative class, but a styling class, a posing class, posers. So it's not the Blues; it's Britney Spears. They're not selling music; they're selling the merchandising, the dance video; they're selling the game, the DVD, the earrings, whatever. There is something about the creative city that is very disciplined, very moralizing, about taste, very normative, to be bourgeois and beautiful, which actually means talking fewer risks. So this creative city seems to want not the artists, but these 'distributors' who are in a controlling position in the supply chain, or value chain, where they are able to broker between what consumers are brought to desire through advertising, and what artists or the street produces. So you don't want authentic ghetto designers; you want Tommy Hilfiger. And you want it in your city, because this is the point where maximum wealth is accumulated by playing on relations of production and consumption, not content.

Whether one considers the sites of the 1990s or future ones, these are the locales that capture the spirit of the current modes of production and consumption and which contribute to a social encounter or to a business deal. They do this by contributing a frame for a specific type of entrepreneurial social action, that is, an expressive context, an environmental metonym of the values and 'projects', as Boltansky and Thevenot would conceive it, of a time and place. In as much as such an environmental frame is a type of ethic-moral framework, it references an entire network of sites, a spatialization in which certain places or realities are included and others are deftly obscured, or pushed into the background, or erased altogether.

IF: Which perspectives are in your view more adequate to theorize these kinds of urban transformations?

RS: Anthony King has warned that in focusing urban studies on economic analysis and then using this as a basis for policy recommendations enormous damage has been done. We have neglected the social and cultural. There are other, very interesting lines of theoretical enquiry, particularly around this idea of the new spirit of capitalism. Luc Boltanski, Eve Chiapello and Laurent Thevenot, for example, have been working on the notion of regimes of justification; this is about the way in which decisions are made in cities, decisions based not necessarily in monetary criteria. Style could become very important, and hence the creative city becomes a stylish city. It's not the kind of Fordist, low-cost-production city, but this means that you have to have new skills and new criteria for making judgements. So this process of decision making and judgements changes, but also the basis of social solidarity changes because people are no longer anchored to a specific position in industrial class structure but rather they're, they argue, formed into temporary teams and projects where their tenure at jobs is much shorter. This is a continual networking and, hence, you need to use bars to get contacts. In order to get on to the next contract, you need those contacts, because those contacts are distributing the opportunities for employment amongst themselves. They call this the project city. Others such as Toscano would call this the precarious city.

In that context I think it's interesting to ask in what way do place images start to become important in these decision-making moments? City governments and administrations need to manipulate place images, because they emerge as a new basis of inter-urban competition, and a new basis of judgement and justification or legitimacy. Terms such as 'legitimacy' and the discussion of public spaces as stage-sets for action also remind us of Habermas, who needs to be reread, and then there is Serres, who provides the philosophical foundations glossed as Actor-Network Theory, but whose work is barely translated into English.

IF: This is interesting, for, at least in the way I read it, your book *Places on the Margin* was much more about images as representations that are socially and historically constructed, while in your recent work you seem to emphasize more on the performative effects of images . . .

RS: I have been stressing more the performative quality of spatializations, but in *Places on the Margin* it's all there. There I always talk about what people do, not just about pictures and ideas. In fact, most of what I do is not about representation, but about activities: people that go to these places, meet different people, follow the rules or don't follow the rules. Mostly the perspective is historical succession: in Brighton, the 1700s working beach, the early 1800s medical beach, followed through the 1900s by the leisure beach of the British seaside holiday. In this sense, the chapters are complementary. You see a series of displacements, but they're not exhaustive. The Brighton chapter is one of historical succession, but you don't see, for example, Gay Brighton. Nor did I research the way in which Brighton is a kind of a musical landmark. In the Canadian North chapter I didn't research the native images of the Far North, as counter-discourses to the southern metropolitan images which were my focus. But in the case of a destination such as Niagara Falls I talk about how there is a real disjunction in the early 1990s between what tourists want and what hotel operators provide. They provide a kind of accommodation based on North American expectations, and an aesthetics of exception and desire, a type of Hollywood vision of pleasure, sex and desire. But Japanese tourists want to see natural beauty. So there's a real struggle over how the Falls are understood as a site – of natural exception or of libidinal explosion. Despite this case, the overall thrust is that, rather than simply a set of social constructions, places are spatialized as 'places for this and places for that', as sites of action and the loci which concretize and thus stabilize cultural values.

IF: I understand that when you talk about place image it is not about icono-graphic images. Still, my sense is that there is a major change in your recent work, for when you talk about the virtual you take a more ontological stance. So, under that light, the question about place images seems a more epistemological question.

RS: Maybe the terminology is confusing and it's possible to organize it in a number of ways, because of course images are virtual. What you get in a mirror is a virtual image. The problem with images is the way we understand them, in a representational way, as paintings. Myths, on the other hand, are more clearly representational discourses. They are much more represen-tational than the images. This is my sense of it. However, around this axis, it is true that I've been more focused recently on the ontological, on the 'it-ness' of places and of place images, that is, of spatializations.

But, looking at virtuality, the process of spacialization is virtual, because it's relational, linking different places. Spatialization is not about the place in itself; it's about its difference from other places. That's why the advice for urban placements is always to look at other places and explicitly to think of the city not as an island but in its relationship to other places. How does it fit to other regions and spaces, not just cities even?

IF: I would like to come back also to this distinction between the 'city' and the 'urban', in order to understand to what extent when we talk about the virtual we talk about virtual entities or we talk about some kind of a virtual medium.

RS: It is a medium and refers to processes, not to identities. Strictly speak-ing, identities only exist in a present moment. Then they age, develop; they change. This chair, for example, its identity only holds for a very short time, for chair becomes: it may be scratched a bit; maybe it's painted; maybe it can lose its back and still maintain its capacity to function as a chair. So with virtuality we are not talking about things; we are talking about real capabil-ities and processes. These introduce time and change, thus a degree of fluidity, to identities, singularities, which permits their ongoing identification even as they change. The urban is also more than a purely intangible thing, which would be an idea, an abstraction, not a virtuality.

IF: The Austrian psychologist Franz Heider proposed in the 1920s this distinction between medium and form, which I think might be helpful. A medium contains elements coupled in a loose way, which when tightly coupled constitute forms which emerge conditioned by the medium. If the urban is a medium, cities might be understood as emergent temporary forms or, as you say, identities. *Places on the Margin* could be thought of as focusing on certain historical forms, and *The Virtual* as referring to this medium. Does it make any sense?

RS: In working on virtualities as an ontological category I came up with the idea that the urban is virtual or, to put it better, if urbanity is the virtual then a city at a fixed point in time is actuality; it is the actualization of that urbanity. The urban is not a thing; it is a capacity or process. It is the capacity to be a city. A village, for example, doesn't have the same concentration of capacities that makes an agglomeration recognizable as 'city'. It doesn't have the capacity to do or to host the things that you associate with an actual city.

IF: So despite what the Chicago School would have said urbanity is not a human way of life. It's not about humans, not about their habitus, about their possible urban habitus, but it's a capacity ascribed to cities.

RS: Yes, but also to persons and to communities . . . The difference with the Chicago School is that here things are players as well. You can say that this is what actor-network says, but it's older than that. It goes back to the 1930s, to what Gibson says about affordances, environmental affordances. Affordances are the kind of interactions you can engage in conjunction with a given site or element. For pavement, you can walk on it; you can sit on it; you can drive on it . . . And what de Certeau says about the material city is that you have to actualize it as this or that. What will it be? It is your choice at any given time. So, in the actualization of things, people play essential roles. But one should not underestimate the materials: their hardness, their softness, their ability to maintain a shape. All this makes the material a player in a way that is significant, causative, not causal. Urban-ness is causative as well, not just an outcome or a representation of a set of ecological interactions as the Chicago School might figure it. While Bruno Latour and also Michel Serres give us a sense in which objects can be causative, what I'm trying to do is to show the way in which virtualities, such as myths, are also causative, to show that there's really some reference to a reality, not just semantics, when, for example, a criminal says 'Society made me do it'.

Cities offer affordances that are not just physical, but also social and cultural affordances as well. This is important, even though the higher the order of agglomeration the more diverse are the understandings of, and relations to, the urban environment. For cities it is indeed harder and harder to achieve a public univocal statement of identity or problem definition. Nonetheless virtualities as place myths or regional character need to be taken seriously, for they are stakes in struggles over the environment. Judgements and affordances on the basis of regional ascription, such as you come from the South or the North, are significant in people's life chances. Yet if you take the apparatus of modern social science it sounds like you're talking about something trivial. I am trying to validate the importance of these virtualities, not identities or identifications. They work as a medium through which things pass, through which one may have certain entitlements, connections and so on. This is not an apolitical, philosophical game.

IF: This might also be understood as an implicit critique of ANT and its possible contribution to urban studies. What can urban studies learn from ANT? And which could be also the limitations and problems of a fully ANT-inian take on the city?

RS: It could contribute to studies of institutional interaction by foregrounding the importance of relations, as Serres suggests. In an actor-network approach, a stress is generally placed on describing people's or groups' ongoing mobilization of other actors and intermediaries to secure objectives or maintain a situation by 'enrolling' them into a network which involves objects, texts, technologies and cultural values, as well as human subjects. And it is also ethnographic and hence descriptive and attentive to the anomalous rather than suggesting laws governing the outcomes of, say, a certain arrangement.

There is however a diversity of approaches, some of which risk a *post hoc ergo procter hoc* fallacy ('after the fact therefore because of the fact'). Especially for early actor-network, critiques have focused on the residual structuralism of this post-humanist approach, and its tendency to focus on ongoing processes which produce outcomes, rather than the traditional social science strategy of revealing social forces or relations of domination or appropriation which are not discernible to common sense or are implicit in everyday activities or judgements. There has also been a risk of hypostatizing networks as objects in and of themselves rather than the relationality of the settlement of diverse elements into a constellation that is able to produce or impose broad outcomes on a societal scale. Thus, ANT does less well at integrating the complex totalities studied by Todorov, or Lefebvre's work on the city as an isotopia, a fundamentally disaggregated and non-coherent but contiguous environment.

Overall the key contribution of actor-network approaches is to bring process and relation to the centre stage of social science research. It shows how to account for the contribution made to outcomes by process. This is an important improvement to fairer governance, to better city management and to planning. Why? Well, because it alerts players to the decisive impact that a chosen process can have by creating points along the way which are vulnerable to contingent events – a project misses its deadline, materials are not available or only at a higher price, taxpayers refuse to pay increased taxes, the supporters in the network are replaced and so on, all events which are familiar to anyone trying to get something done in a municipality or a region.

Cities are also ideally suited for relational approaches, because of the importance of sunk costs in an urban environment which tends to stabilize a certain way of life or material culture and the significance of effects caused by, for example, densities of people and institutions. Rising out of ethnographies of single labs and scientific teams, actor-network approaches have tended not to theorize beyond the boundaries of their own networks, which are like Aristotelian spaces or spheres. There has thus been an attempt to move toward a stronger emphasis on relationality and process rather than overarching networks. Because they are totalizing, a network emphasis would not easily respond to the question of whether or not one could, for example, reach beyond the network; all new elements are simply enrolled in the firmament that is the network.

For all their flaws that I've discussed elsewhere, we might think of world city networks as resembling actor-networks in as much as they are outcomes of complex interrelations between humans which involve natural resources, distance from other places and cultural and political opportunities. The weakness is that actor-network approaches don't specify the scale or type of actor that will be the stars of the research. How do we recognize what elements are actors, or what Serres originally deemed *quasi-actants*, in a given situation? Not only at the human scale, but many possible players at the bacteriological scale, for example, are muted or neglected, until a virus such as caused SARS comes along. Actor-network describes what happens next – maybe even a pandemic in which humans exit the network. But of course it doesn't address how to detect the next significant actor or quasi-actor that will enter the network. So, in Latour's terms, ANT allows us to better describe the dynamic 'settlement' that has been made between environmental affordances and social interaction, and may provide a warning of fragility given that these are never perfect equilibria, but it doesn't respond easily to threats, latent risk and emergent capacities (virtualities), nor new actors.

IF: I have the sense that there is a great divide in urban studies between the sort of cultural urban studies and those focused more on political economic issues. How do you think that the concept of 'the virtual' helps to overcome this gap?

RS: Political economy works with a notion of the market as an actual reality, but markets are virtualities in the sense of being 'intangible but real'. This comes originally from Carrier and Miller. What they say is that, when you treat actual people as rational economic actors, you make what is an abstract idea into a self-fulfilling prophecy which they refer to as 'virtualism'. But they don't use the word strictly. They use the words 'virtual', 'virtuality' idiomatically, but still they fall more or less on the target. We can talk about neoconservative economics, in particular, as being a virtual economics. Other progressive approaches, like the Canadian School of political economy, which try to integrate community rely on virtualities too. In this sense, 'the virtual' lends more precision to their discussion, because it allows them to distinguish between actual possibilities like risk, and ideal but real entities such as brands, goodwill and community.

If the virtual is a medium, not a thing, but a kind of a set of capacities that are actualized continuously, it provides the bare bones to theorize the new, even the radically new. Correspondingly the critique of my work is that I should talk about the radically new, because what the virtual provides us is a basis for discussing the unexpected. The fall of the Berlin Wall can be discussed as an inexplicable event – as Lefebvre says, 'Events overturn theory.' However, it could also be theorized from the point of view of the set of affordances, capacities and tendencies which are actualized in the event. One can then provide a discussion of not just the magical realization of abstract ideas but the actualization of intangible but real entities (virtualities) – old, overarching intangibles such as nation, for example, reasserted and exercised on the urban landscape. This opens a much more nuanced approach than just saying it's a surprise.

IF: If I read it well, your theory of the virtual has come out of a rereading of poststructuralist texts, particularly Deleuze. But isn't such rereading problematic when based only on one concept?

RS: What I'm saying is different from Deleuze, who revives the concept of the virtual out of Bergson's reading of Proust and his conception of the nature of memories. However, you cannot operationalize Deleuze's theory of the virtual, which has no space for the concrete or the material. A close reading of Deleuze's many comments on the virtual shows that he understands the material as the totality of the actual, whereas social science since Quetelet has understood probability as also belonging to the realm of the actual or, in my formation, the 'actually possible'. I can't help but think visually and architectonically when it comes to ontology. Because Deleuze drops the term 'material' from his discourse, a characteristic of his writing is a certain weightlessness and ideal tone. What is the way in which the virtual is related to the material? That's my question. Spinoza, for example, wants to know about the virtual and its connection to ethics, to judgement. I want to know about the virtual in connection to action. My contention is that we have done a very good job of talking about the material, the actually real. We have a mathematical language for describing other actualities – the probable – and social science statistics does that very well. What one needs to do is to show how these approaches work together rather than having some kind of separatism. What's interesting is the way the virtual is always in play and in a process of exchange with other registers, the way urban politics involve virtualities – ideal but real objects such as community – as well as discussions about abstract ideas and political promises, about the probability that something might happen, in concrete situations.

Berlin, 2005; Edmonton, 2009

Postscript

Reassembling the city: networks and urban imaginaries

Thomas Bender

Thirty years ago, before ANT had become a part of the tool kit of social inquiry, I wrote a history of community in America from 1600 to 1900 (Bender 1978). As much as the book was a history of community, it was also a history and critique of the way anthropologists, sociologists, and historians conceptualized community and implicated it in explanations of social change. The established definitions of community at the time were derived from the Chicago School, which had early been deeply influenced by the important German sociologists Georg Simmel and Ferdinand Tönnies. Later, by the time I turned to my study, this lineage had been inflected by Parsonian modernization theory. With slight variation in different studies, it was assumed that spatial propinquity flavored by ethnicity constituted community. Built into that model was a temporal logic of dispersal and decline. These ideas derived from nostalgic images of the nineteenth-century German 'home town' (Walker 1971) and colonial New England towns of fond memory. These notions ignored both the internal complexity and the translocal connections of those towns. And its temporal logic denied community to contemporary metropolitan life.

Looking for an alternative framework of analysis that was less determined, I discovered the scholarship on African cities by J. Clyde Mitchell (1969) and other British anthropologists.[1] They had recognized that the Chicago School model did not match up to their field work findings, and they turned to a network analysis as well as setting aside the logic of modernization theory. Their work focused on social networks that transcended the rural–urban division, and identified spatially dispersed communities that included both city and village. With that work on African cities in mind, I was able to develop a quite elementary form of network analysis of community in American cities over time that escaped the teleology of modernization theory and the spatial imperative of the Chicago School. Rather than focusing on space, I emphasized the quality of connections – the level of emotional content – to define community, past and present.

In retrospect, I see that I was trying to undercut the solidity and boundedness characteristic of sociological categories for aggregations. But the network theory I used then was too limited for the task. The networks I tracked were defined only as the connective tissue of social life. The Actor-Network Theory (ANT) is a vastly richer idea. ANT goes well beyond identifying chains of relations; ANT redefines aggregates, but aggregates with open borders, capable of continual transformation. The actor-network is generative; it makes things happen. This capacity of ANT to reveal the interconnections of active, continually transforming networks seems to recommend it as a way of exploring urban life. Although practitioners of ANT were not focusing their work on cities, deploying ANT in urban analysis seems to be a natural extension of it, and a significant move forward in urban studies (Bender 2006).

The core idea of heterogeneous networks that included both human and non-human actors was especially fitting for the study of metropolitan life.[2] It will help us overcome a longstanding tradition in the United States that separates city and nature in social thinking going all the way back to a line from the eighteenth-century English poet William Cowper that was constantly repeated in the nineteenth century: 'God made the country, and man made the town.' God's handiwork, not man's, was where value resided. The point in this instance is not the different valuations but rather the absolute binary. In fact, the city is a complex combination of nature (water, for example) and the work of men and women, which would include the infrastructure that makes water available, but also other aspects of the material city. Indeed, as Stephen Graham has recently written – in the wake of Hurricane Katrina in New Orleans – one can no longer imagine city without nature (Graham 2006).

There was another problem haunting the way historians and most social scientists thought about the city. It was assumed to be a social whole – bounded, organic, and solid – which implied a kind of homogeneous unity, an implication that contradicts all modern urban experience and ignores all its fissures and fractures. The image of the city as a filled space misses the multitudinousness contained within that whole, the prevalence of interstices, and even patches of the urban equivalent of deserts. The city is not a whole, but a composite entity.[3] We surely err if we start with an assumption that the city is some kind of whole, a totality, represented as a bounded or at least an identifiable territorial space that gives shape to social relations.

While I was worrying over these issues, it was my good fortune that Michel Callon came to NYU as a visiting scholar, where he participated in a seminar on 'the social' that I also attended. The discussion often turned to the Actor-Network Theory (ANT) that he and Bruno Latour, working at the Ecole Nationale Supérieure des Mines in Paris, and John Law, working at Lancaster University in England, had developed within the small but lively field or emerging discipline of science and technology studies (STS).[4] Thus I was introduced to ANT.

I was skeptical, but gradually its promise (perhaps with some adaptation) as an approach to urban studies began to excite me. Through Ignacio Farías, I became aware of scholars looking at cities through this lens. These studies, like those in STS, tended to be highly focused, but they were also bringing fresh thinking to urban studies. Conveniently, just as I was beginning to seriously explore the possible use of ANT for urban studies, Bruno Latour, with whom ANT is most commonly associated, published a primer, *Reassembling the Social: An Introduction to Actor-Network Theory* (2005b).

ANT as a theory provides a novel way of explaining and connecting the multiple agents implicated in stabilizing and destabilizing social aggregations. 'The project of ANT', Latour writes, 'is simply to extend the list and modify the shapes and figures of those assembled as participants and to design a way of make them act as a durable whole' (Latour 2005b: 72). It brings to the study of cities a fitting assumption: complexity and interaction, which seem to be a nice match to metropolitan life. The strong empiricist commitment to what I would call ethnographies of causation avoids the formalism of pre-established social categories of social action so common in the social science literature.

Yet I was still hesitant. ANT's primary virtue, it seemed, was perhaps a significant weakness. Its seemingly indiscriminate absorption of elements into the actor-network had the effect of leveling the significance of all actors. It seemed to imply only collective responsibility – with non-human as well as human responsibles. Even if, as anyone would accept, Hurricane Katrina was a key non-human cause of the destruction in New Orleans, how much of the playing out of the disaster was the result of human action or inaction in a variety of relevant networks? Does the causal logic of ANT evacuate responsibility? Latour acknowledges that human actors are different from non-human ones; intention (and thus responsibility?) is presumably part of that difference. What is the relation of causality and responsibility? If the whole actor-network produces the action, whether a war or a new life-saving medical intervention, is there individual responsibility (or credit) anywhere? According to the core idea of ANT, no one entity in the chain can be a sufficient explanation for any action. The action is the result not of the action of any one individual person or thing, but rather of the dynamic actor-network. Is every action overdetermined? This apparent leveling of responsibility in an actor-network analysis of causation worries me, for it seems to remove ethics and politics from social analysis.

There are at least two counter-arguments one might make. If an action requires the whole actor-network, non-action would be the result of a refusal (an intentional act, thus human) by any actor in the network. Thus moral objection by a human actor in the chain or actor-network of causation could prevent a particular action. But it would also prompt a transformation in the networks, the outcome of which would be unpredictable, thus raising the possibility of a worsening of the original moral issue. The other counter-argument points away from the one actor who may refuse to the many actors who might act. If the actor-network is an extension or multiplication of the number of actors, there is also an increase in the number of contingencies and points of potential intervention, thus increasing opportunities for responsible action. Does this take care of my problem? Not quite, but it contains possibilities. It could open up relevant political spaces, including the creation of a public of a particular, contingent, and pragmatic sort. I will return to this way of thinking about the politics of ANT at the end of this postscript. In what follows I will examine the city as a conjuncture of multiple networks, and then I will address the city as an assemblage of assemblages.

THE CITY AS A MULTIPLICITY OF NETWORKS

The central question of this postscript concerns scale, both as *space* or *extensity*, since cities are larger than laboratories, and as *intensity*.[5] The most notable work in ANT by Latour and Callon has been developed primarily in studies of science and technology, where it has been very tightly focused, often on a particular innovation, laboratory, or specific market action.[6] The preceding chapters provide excellent examples of specific, small-scale things, collectivities, or actions – from a building renovation, to a new mass transit system, to the development of a cultural center, to an urban commodities market that establishes value, to an experimental music scene, to legal services work-nets, to a tourist bus. Could a method that is so effective in micro-analysis be scaled upward to a metropolitan scale of both geometric extension and metropolitan intensiveness? Ignacio Farías and Nigel Thrift address this issue in this volume. Farías asks whether there will be ANT-driven studies of typical urban issues: urban poverty, urban development, urban governance, and the like. Thrift seems to think not:

> I think ANT works best in strongly defined situations and I think that's difficult to deny, truth to tell. You talked about the laboratory, where in a sense ANT started out. It's moved into the trading room; it's moved into other milieux where you can be very sure of what you are getting . . . I think it is more difficult for it to work when you are looking at, if you like, everyday life as a whole, or even when you are looking at political movements . . . which . . . don't have bounded spaces.

In speaking of what has been done, Thrift is surely correct. Yet is not one of the virtues of ANT that it does not acknowledge bounded totalities? What are we to make of a social science that refuses boundaries yet is able to do its work only in 'strongly defined situations' and 'bounded spaces'?

Latour and others have pointed to work much like ANT that has addressed histories on a much larger scale. Most often mentioned are the brilliant histories by Thomas P. Hughes on the electrical grid and by William Cronon on Chicago. Hughes's *Networks of Power: Electrification in Western Society, 1880–1930* (1983) compares the introduction of modern electrical grids in Berlin, London, and Chicago. This is urban history, or so it seems, and as its title indicates it addresses networks. The study traces various networks – whether of technologies or political forces, of people or things – in the different cities as they produced grids. His account links heterogeneous entities, whether human or non-human, in a description and explanation of the development of these elaborate technological and commercial urban systems. Yet he does not actually make the city an object of inquiry. He examines the particular and locally distinctive combinations of human and non-human resources gathered together to produce the electrical grid in each place. The grid is his object of inquiry, and it is, to repeat Thrift's words, a 'strongly defined situation'. Though he shows contingency that produces difference among the cities, his focus on a formal system may not make the case for his method were it exploring a less systemic aspect of urban life. Hughes's work is at least doubtful as a general model, though there are many technical systems in the city, and it would be fruitful to examine them as Hughes did electricity – whether one looks at school systems or health care systems or transit systems or fire protection systems.

William Cronon's study of Chicago, *Nature's Metropolis: Chicago and the Great West* (1991), has been praised and rightly so by Latour as a 'masterpiece' of ANT (Latour 2005b: 11). Surely this looks like ANT, even if the author became aware of ANT after, not before, writing it. 'Nature', which Cronon uses to designate the non-human surround of land and the lumber, crops, and livestock it sustained, is given the status of a cause, perhaps even *the* cause, of Chicago's development. And it seems clear that Cronon's intention was to bring nature into history. A professor of history and environmental studies at the University of Wisconsin, Madison, Cronon indicated in his preface that he was more interested in the history of the environment than society. Humans were far from the center of his account:

> Indeed, I have little to say about individual men and women. The few who do show up in these pages are mainly merchants, who enter my narratives less because they are significant in their own right than because they exemplify so well the broader city–country connections I wish to trace.
>
> (Cronon 1991: xv)

He may overstate his point here, but he certainly does de-center humans in his narrative.

Humans are part of an extended series of 'connections', as he calls them, that produced innovations, moved and processed wheat, lumber, and meat, and resulted in making Chicago the center of the Midwest commodity trade. Chicago, for Cronon, was embedded in and the product of networks that extended to the limits of North America. These networks – most importantly railroad lines and credit lines – pointed in two directions. One, which was a dominating one, spread mostly westward into the hinterland; the other extended eastward to New York, and Chicago was subordinated to it. Yet it connected the Midwestern city to a vital oceanic network that provided access to global markets. This account, which fully acknowledges hierarchies of power and authority, challenges Latour's insistence on keeping the social flat.

The human actors in his account are vital to the story he tells, even if more often than not they are anonymous. Yet they were no more important or central to the narrative than rivers, fertile land, and, especially, the vital rail connections to its hinterland and to the American metropole. All elements of the heterogeneous actor-network were collaborators.

Actor-networks, according to Latour, are rich in contingency. Cronon, however, underplays the contingency of his story; his narrative is pretty tight and seems prefigured, if not actually determined, by natural advantage. The older historiography of the development of the commodity trade of the Midwest stressed entrepreneurs, singular bold actors. In that tale Chicago had smarter and bolder businessmen and politicians than St. Louis. Cronon's address to this question is interestingly asymmetrical: he emphasized human agency in St. Louis's failure and nature in Chicago's success: St. Louis entrepreneurs made bad policy choices that took them out of the race, while Chicago had natural advantages – the Great Lakes and a vast hinterland (Cronon 1991: 295–309).[7]

Unlike the books by Cronon and Hughes, which were colonized by ANT advocates, *Splintering Urbanism: Networked Infrastructures, Technological Mobilities, and the Urban Condition* (2001) by Stephen Graham and Simon Marvin was undertaken with ANT firmly in mind, and it does provide an example of ANT in action. With a focus on infrastructure, this book brings together a whole range of networks, processes, and actors in order to show the inextricable connections of networks in multiple realms, from economy to politics to engineering, to various aspects of the natural world, to the built environment, and more. Again, this book is grounded upon a system, the infrastructural system, something planned as a system, but it reaches toward an important conception and critique of the contemporary neoliberal metropolis. It provides an example showing that grounding an ANT study of the city on basic infrastructural systems enables a metropolitan-wide and intensive examination of the metropolis, and it mounts a powerful political critique. The book would have been different and surely not as well integrated without the deployment of ANT thinking.

Clearly as the chapters in this collection show, ANT produces very detailed descriptions of the actor-networks in relation to a variety of specific urban phenomena. And all ANT advocates concur that there is no limit to the number of networks and connections of them, nor is extension of the networks, whether spatially or intensively, limited. That seems to make for a fit with the metropolis. ANT does not trap the analyst in a bounded space of city; in theory it welcomes a notion, such as that of Michael Peter Smith, of 'translocal' as a description of the city in our global age, but evident in much earlier centuries. There is no such thing as a city without translocal aspects, and in today's city, as Smith argues, it is even difficult, perhaps impossible, to distinguish in any clear way the 'inside' and 'outside' (Smith 1999). Sometimes, as in the case study in this volume by Slater and Ariztía, the most spatially extended network can be of particular importance.

Although much of the work by Latour and his colleagues has been highly particular in focus, his ambition for ANT is much larger. As he presents ANT in *Reassembling the Social*, his recent guide to it, ANT operates in two ways: it is an ethnography of causation and it is a way of describing social life. The two are obviously related, but they are distinct, something Latour and others do not always recognize. The distinction is important, even if they are two sides of the same coin. Much as in the discipline of history, in ANT the making of an empirically grounded narrative description that incorporates heterogeneous actor-networks becomes explanation (Callon 1991: 54).

In Latour's account and his work so far, ANT is most importantly a theory of distributing agency. It answers the question: How do things happen? And, in so doing, ANT provides an alternative to orthodox social analysis. Rather than looking to social-structural explanations or transformations of the whole, Latour turns to actor-networks, complex chains of causation that include non-human as well as human actors. In doing this, he undercuts social or structural explanations of change and dissolves social wholes, whether society or the city. Both are accumulations of networks and assemblages of interconnected networks. The empiricism of this approach also challenges a priori social categories of either description or causal explanation. He prefers close empirical observation and description to concepts and categories. He challenges the causal use of 'society' and, for that matter, 'city' as useful categories of analysis because, while they can be conceptualized or imagined, they cannot be empirically demonstrated. Put differently, he rejects any putatively solid, uniform or homogeneous, bounded whole called 'society' or the analytic, 'the social'. He does not make a point of it, but this logic would also preclude 'class' as an explanation, while a particular description of a worker-and-employer conflict elaborated in terms of actor-networks would be an ANT explanation.

These heterogeneous associations or actor-networks are indeed active. They make things happen or, as we see in Anique Hommels's chapter, they prevent them from happening. Her account is not a unique one. Although ANT emphasizes innovation and contingency, 'ordering and obduracy' are recognized and important (Law 2001). If actor-networks are strongly convergent and irreversible, they become, in the language of ANT, a 'technology', or a 'black box', as Callon (1991: 132) describes it. Thus fixed, it lacks the capacity of constant transformation that characterizes the active actor-network. However, as Michael Guggenheim's chapter shows, there is some middle ground of mutability that theorists of ANT seem to avoid and which he usefully identifies and calls a 'quasi-technology'. Mimi Sheller, building on the work of Harrison White (1992, 1995), in a recent article pushes farther, where she usefully softens the notion of networks as 'clean nodes and lines' by introducing the notion of 'gel'. Gel, as she describes it, makes room for 'softer, more blurred boundaries of social interactions', as one might find in public life (Sheller 2004). One sees the value of this concept in Manuel Tironi's description in this volume of the experimental music scene in Santiago (Sheller 2004).

For the study of cities, time is as important as space. Cities are a precipitate of history. Or, as Henri Lefebvre put it, urban sites represent the 'inscription of time in the world' (Lefebvre 1996: 16). History is not just background or the stage upon which urban life plays out. It is an actant, a participant in the networks. Zaloom incorporates Chicago's temporality both as history and as transactional time, as well as the trading space, into her explanation of how the market establishes value, and Hommels demonstrates that history incorporated into a network can contribute to obduracy. A stabilized network or 'black box' (the existing highway) is made historically, but once made it is also a potential actant in a subsequent phase of urban development. One sees the temporality of city life in both the text and images of *Paris: ville invisible* (1998), a book of texts and images by Latour and Emilie Hermant. Consideration of the temporal transformation of certain parts of the natural and built environment of cities is central to any understanding of them, on par in importance with their political and social history. Such temporality is a form of mutability. For ANT to be adequate to the continual transformations of city life, which includes persistence (but not necessarily permanence), a middling, persistent mutability will have to be addressed.

ASSEMBLING THE CITY

Latour's ambition for ANT – constantly evident in his language – is much greater than the work that has so far been undertaken under its aegis. His is a challenge to social theory as it was developed a century ago and that became the foundation for sociology broadly defined. The 'social' or 'society', like 'class', entered into the vocabulary of reformers and intellectuals in the nineteenth century. This new usage of 'social' and 'society' was a recognition that in modern society individual lives were played out in a larger domain than various traditional (and hierarchical) institutions associated with family, guild, church, and village. Thus freed (or forced by poverty) from the land and the village, massive numbers of people migrated to cities. The result, with help from industrialization and infrastructure, was an explosion of urban scale. The migrants were freer individuals by this extension of their terrain of experience – hence the celebration of 'individualism', especially in the Anglo-American world. But at the same time they were more dependent on strangers in an increasingly interdependent society. Sociology was born out of the need to make sense of this new pattern of social relations. Or, as Latour has phrased it, 'sociologists were trained in the alarming discovery of the masses suddenly rushing into towns, with no one knowing what to do with them' (Latour and Hermant 1998: 95).

Various sociological descriptions of this shift were offered, and they remain classic formulations: Henry Maine phrased it as a distinction between 'status' and 'contract', Ferdinand Tönnies as between *Gemeinschaft* and *Gesellschaft*, and Émile Durkheim as between 'mechanical solidarity' and 'organic solidarity'. In the 1950s to the 1970s, American sociologists, under the aegis of modernization theory, phrased this societal shift as one from 'traditional' to 'modern'. The social relations signified by the second term in each of these binaries defined modern society or, more precisely, made the point that these novel social relations played a larger role in the lives of modern individuals, giving shape to more of their social life (Bender 1978: chaps 1–2). No social scientist of that era pressed the notion of 'the social' as a condition and explanation farther than did Émile Durkheim. Generations of sociologists, including Latour, whether trained in France or elsewhere, but especially in France, learned their discipline in the long shadow of Durkheim. It was as an alternative to Durkheim's seemingly controlling 'social fact' that Latour developed his approach to social inquiry (Durkheim 1982: chap. 1).

In making his challenge to Durkheimian sociological orthodoxy, Latour reached back to another pioneer of sociology, Gabriel Tarde (Latour 2002). Tarde was Durkheim's more senior rival, whose ideas were marginalized by Durkheim's success. Durkheim's sociology was built upon the notion of the 'social fact' that sustained social reproduction. He was concerned with stable and persistent social formations, reproduced through institutions, especially law and religion, and he feared that when these mechanisms failed individuals were liable to have feelings of *anomie*, or rootlessness. Tarde, by contrast, rejected a unitary concept of a social whole. He argued that the small group was the proper focal point for social inquiry. Society, for him, was made up of many small interactive groups. He thought of the interaction in psychological terms, as 'inter-mental activity'. He also imagined society in terms of continual invention, driven by the activity of the groups that enacted an open-ended process of imitation and innovation, with affect playing as large a role as cognition. Here we see a hint of the direction taken by Latour: first a group theory of how things happen, and there is also the plurality of associations and a form of active networks emphasized by Latour. Latour has also embraced Tarde's emphasis on the provisional or changing quality of these associations.[8] Even more than was the case with Tarde's social theory, Latour's theory of ANT is heavy in gerunds.

The generative capacity of actor-networks is constant and operates through particular associations, networks, interconnections, and interrelations of potential actants. Moreover, these multiply rapidly and take unexpected directions. The point to be made here is that in neither origin nor prospect is ANT confined to associations or territories of limited scale. As a method ANT is not hostile to the notion of larger social patterns of the scale of a city or beyond, but it challenges a priori deployments of social categories, particularly that of 'Society'. Instead it promotes a research strategy of building the elements of society from the ground up, from its networks and the connections among them that proliferate yet more networks.

Although Latour's work has mostly been on more restricted topics, he is concerned with the role of social science in helping humans navigate this world, which he recognizes as urban. The choice, he says, is accepting that we now live an unavoidably fractured existence, 'prohibited from seeing ourselves as full and complete beings', with feelings of 'nostalgia' and 'impotence'. Or we can change the way we address the world; instead of the fruitless pursuit of the social totality that supposedly will produce a sense of wholeness, we might more usefully examine the many 'temporary places' where people live their lives – 'streets, corridors, squares, words, clichés, common places' – and 'gather themselves into coherent wholes', which Latour implies the city supplies through its multiple networks (Latour and Hermant 1998: 96).

Quite crucial to ANT as an explanatory scheme is the interconnection of networks. ANT assumes a plenitude of networks, and is much richer in prompts to change and realignment of entities than is Tarde's associational theory. Any social assemblage understood in terms of ANT contains many networks, theoretically infinite in number and capable of unending permutations. The linkages of assemblages are for Latour and Callon horizontal, not hierarchical and not historical. But, whatever the case, we ought not, *contra* Latour and Callon, to presume the social to be always flat and ahistorical. That is an empirical question, as the work of Latham and McCormack in this volume suggests. We ought not a priori to foreclose the possibility by failing to inquire.

In practical inquiry, of course, the number of networks is delimited by the questions being asked, and we see those focused limits in the various chapters. But even when delimited, or 'simplified' as Callon would phrase it, a central point is that networks are 'juxtaposed' (Callon 1987: 93–4). The multiplicity of actor-networks and juxtapositions constitutes a potentially destabilizing force, which is why they are *actor*-networks, rather than the passive lineages that were merely connective, as I had earlier understood them in my study of community. In ANT more than mapping is required. The mediators that link networks disrupt and change the configuration, which in turn produces an unpredictable outcome. Again, it is this quality that makes for the fit between ANT and urban analysis. The presence of plentitude and continual transformation is of a piece with the quality of overflowing and unfinished quality of metropolitan life (Bender 2002: ix–xvi). As the narrator of the classic *film noir Naked City* (1948) says in the opening scene, 'The city never sleeps.' The city is an active assemblage of assemblages; this is what makes urban life so lively and stimulating.

Latour has himself addressed the city. *Paris: ville invisible* (1998), produced in collaboration with the photographer Emilie Hermant, is at once an example of a network approach to the study of a city and, it seems, a love letter to Paris. It has been published in book form, but it is also available in a web version, which is interactive, and the images (provided by Hermant) are richer. The text, moreover, is translated into English, Italian, and Spanish in the web version. The presentation of Paris begins with a panorama and a series of maps of the whole, but then the narrative moves increasingly into the small parts of the city, mostly exterior but not wholly so. Some of the vignettes are focused on an individual, but most are not, though humans are an incidental presence in most of them. Some of the images and the textual characterizations are familiar, but most are novel. Few recall tourist photographs, but some do, and that is important for sustaining a shared (if partial) image of the city. The complexity of the role given to virtuality is explored by Farías in his chapter in this volume and in his conversation with Rob Shields. The urban quality that emerges in Latour's Paris is its capacity of enabling an imagined city. It is open enough for Parisians, notwithstanding the particularity of their experiences in the city, to think themselves into that limited but partially shared imaginary or, in his terms, enrolling in it and achieving a sense of affiliation and location, whether brief or continuing.

Paris: ville invisible is an idiosyncratic book, difficult to categorize. The title itself is confusing: a volume of pictures of a city that is declared to be invisible. I turned out of curiosity to the multivolume list of 'subject headings' produced by the Library of Congress (in Washington, D.C.). How would its cataloguers define its subject? These subject headings are accepted by nearly all scholarly libraries in the United States as definitive. In order of importance, using the Library of Congress headings, the cataloguer lists the subject(s) of every book. For *Paris: ville invisible*, the first two headings are 'infrastructure' and 'city and town life', and then 'photography'. Apparently, the bibliographer, who probably had some expertise in urban studies, recognized – rightly – that the significance of ANT for urban studies was most of all its capacity in connecting human and non-human aspects of city life, the infrastructure and the human life it supports.

The book deliberately refuses closure; it breaks the city down into particulars: workplaces, neighborhoods, café romance, walls of advertising, historical structures and plaques, the green market, but also technologies (especially computer screens) of intellectual control of urban data, such as monitoring of infrastructure or security matters. It is full of juxtapositions. The text, which is quite personal and informal, comments, sometimes directly, sometimes obliquely, on Paris and the images. The book is not at all like Durkheim or contemporary urban sociology. What it most resembles, except that it lacks any reference to Marxism, is Walter Benjamin's famous Paris 'Arcades Project' (Benjamin 2002). Benjamin collected and filed, but never found the unifying thread to transform the voluminous heterogeneous material into a text. His project, now published, remains a grand filing system. His commentary was brilliant on the pieces – his study of Baudelaire or his precis for the project 'Paris, Capital of the Nineteenth Century' – but the pursuit of the whole eluded him.

Latour, by contrast, refuses the quest for totality; indeed the whole project of ANT and of this particular book is to undermine that form of sociological thinking. For Aristotle, the ideal *polis* or city was a settlement that could be grasped in a single view. To the extent we seek the same, we are doomed to failure, and Latour recognizes that the modern city, so many times larger than ancient Athens, cannot and, one senses, ought not to be seen 'in a single glance' (Latour and Hermant 1998: 8). The city for him cannot be grasped as a whole. 'Social perspectives', he points out, 'are always intended to totalize but we know only too well that none of these panoramas can really sum up the social: neither the modest Place de Vosges, nor the elegant row of buildings in front of the Odeon theatre'. No, the city 'was molded by an accumulation of series of views, one after the other, juxtaposed but never summed up' (Latour and Hermant 1998: 88). Perhaps the city we know in our daily life is a partial but functional urban imaginary, typically visual in character.

Latour is skeptical of 'social structures' or, again, the a priori deployment of such a concept. He challenges the move to larger, more distant structures as explanations for the small-scale actions that do not seem to explain themselves. Those who resort to structures 'rarely bother to follow the little paths' that constitute social life and change.

> The social structure is a refuge of ignorance, it allows one to do without representation, or scripting, and to scorn the poor actors overwhelmed by their environment. But those actors are never particularly overwhelmed; let's rather say they know they are numerous, populous, mixed, and that they ceaselessly sum up in a single word whatever it is that binds them in action. How can this multiplicity, these overwhelmings, be explained without reverting to structure? Why not talk of subscription . . .? We could then say that at times Parisians subscribe to the partial totalizations that circulate in the city and enable them to give meanings to their lives.
>
> (Latour and Hermant 1998: 90–1)

As this statement suggests, Latour is loath to put power into his account; that points to the flatness of the ontology. This is most obvious in *Paris: ville invisible* by the ridicule he directs at the mayor of the city. Whatever the character of the person, the office carries power, and Latour declines to acknowledge this when he suggests that a working woman has more power than the mayor (1998: 47–59). She may have more dignity, perhaps, or more friends, or more beauty even, but not more power. If Latour continues to explore the city, it is likely that he will be forced to expand and enrich ANT in order to capture institutional power.

There is perhaps a prevision of ANT as urban studies in Michel Callon's brief account of a plan to make environmentally friendly changes in an urban transit system with the introduction of fuel-cell-powered vehicles. Callon begins with the fundamental observation that the 'actor network is reducible neither to an actor alone nor to a network', and he pointed out that it was constituted by 'heterogeneous elements, animate and inanimate that have been linked to one another for a certain period of time', ranging in this case from electrons to social movements to ministries.[9] He warned against thinking of a network as a mere linkage of stable elements. ANT is more dynamic than that, and it can 'at any moment redefine' the entities that compose it.

'An actor network is simultaneously an actor whose activity is networking heterogeneous elements and a network that is able to redefine and transform what it is made of' (Callon 1987: 93).

The challenge to ANT is to describe the actor-networks in larger, less well-defined domains. That is crucial because for ANT – much like historical narrative – description and explanation are collapsed. In each case, an empirically grounded narrative that includes multiple causal elements becomes an explanation. Since in theory the number of heterogeneous elements in associated networks is infinite, one must simplify. This involves informed judgment. Callon explains that 'in practice actors limit their associations to a series of discrete entities whose characteristics . . . are well defined'. But one must take care not to reduce complexity too far. Engineers, he observes, might reduce the town 'to city councils whose task is the development of a transport system that does not increase the level of pollution' (1987: 93–4). But this assumes a lot, namely that the council fairly represents the town and that all residents have the same view of the matter, that questions of urban transit must be related to other systems – existing infrastructure, the geography of residence and jobs, and more. So the number of associated networks might well grow.

Other networks outside of the politics of the city might become pertinent. Some of these networks would surely extend beyond the city. In this case, they might include knowledge of the success or failure of similar efforts in other cities, or disruption in the networks supplying vital materials that might stop deliveries or produce problematic increases in cost or even decreases might enable a more ambitious plan. The impending success of the endeavor could cause the suppliers of raw materials for the hydrogen fuel cell or other source of energy to rapidly enlarge their own networks, which would be interesting but not part of the examination of this local project. The focus of explanation is on the convergence, not the extension, of actor-networks, unless there is some form of circling back. ANT thus contains many contingencies, and one can see how even this limited example could rapidly multiply the number of causes of success or obstacles that would result in failure. ANT works against any monocausal explanation (Callon 1987: 93–4).

In Callon's account, the engineers and the city council are each seeking to stabilize their networks. That requires ensuring the stable support of the influential middle-class electorate, central city advocates, and the users and managers of the public transport system that links the 'urban spread' with the center. But other elements, including climate and topography, might be incorporated into the network as description and explanation. Remove or lose one of these elements or add another network, and the whole structure of possibilities changes. Yet strategies of stabilization are possible, some being material rather than political. The physical design of the Transmilenio bus system in Bogotá described by Andrés Valderrama Pineda, for example, stabilized the project – and prevented political competition from the already existing and potential rival bus companies – by a material strategy, raising the platform and having the bus doors open on the left, instead of the usual right. In Anique Hommels's study of Maastricht we see the combination of an existing highway simply being there sustaining a number of stabilized networks and the political complexity involved in removing it establishing its obduracy.

For all of its emphasis on changeability, ANT is attentive to such strategies or accidents of design or policy that produce durability (Latour 1991). One of the virtues of ANT is that a non-action requires as much explanation as an action. 'Transformation', Callon observes, 'depends on testing the resistance of the different elements that constitute our actor network.' The extension of the number of 'associated entities' in the analysis extends well beyond what is generally accepted in conventional social science (Callon 1987: 96–7). Actor-networks constantly create new combinations of urban entities, as one can see in the chapters by Manuel Tironi on the music scene in Santiago and the study of 'soundspheres' by Israel Rodríguez Giralt, Daniel López Gómez and Noel García López in Barcelona.

The metropolis is a specified collectivity, but somehow not closed. It is made up of networks – human networks, infrastructural networks, architectural networks, security networks; the list could be almost infinite, and they are not confined by a circumferential boundary. The philosopher Manuel DeLanda has explicitly proposed examining cities as assemblages. Like Latour, DeLanda draws upon Gilles Deleuze for his notion of assemblages: 'A multiplicity that is made up of many heterogeneous terms and which establishes liaisons, relations between them across ages, sexes and reigns . . . Thus the assemblage's only unity is that of co-functioning.' And they are 'constantly subject to transformation' (Deleuze and Parnet 2002: 69, 82). Starting from there, DeLanda proceeds to describe cities as 'assemblages of people, networks, organizations, as well as a variety of infrastructural components, from buildings and streets to conduits for matter and energy flows' (DeLanda 2006: 5). Networks agglomerate into assemblages, perhaps a neighborhood, or a crowd at a street festival, or a financial center like Wall Street in New York City. The metropolis, then, is an assemblage of assemblages.

Assemblages are historical constructions. They are made in time and they dissolve in time. They have no essence, nor are they logical outcomes of a fixed theory or process. They are, DeLanda writes, 'contingently obligatory' (DeLanda 2006: 12). All of these assemblages and the relations among them are liable to change, but most them, most of the time, are stabilized, black boxes, like the traffic signals that are just there at intersections and do the same thing endlessly, quite predictably. The metropolis is a combination of stabilized and destabilized elements, and it is constantly in a double process of transformation and destruction, reconstruction and decay. At one and the same time, DeLanda points out, an assemblage 'can have components working to stabilize its identity as well as components forcing it to change, or even transforming it into a different assemblage'. There is also a constant impulse toward the achievement of homogeneity and the establishment of boundaries ('territorializing'), as in the formation of an ethnic community, and an equal and opposite impulse toward weakening or destabilizing boundaries and enabling more internal heterogeneity (or 'de-territorializing') (DeLanda 2006: 12). The territorializing process, which provides the assemblage with a stable identity, is sustained by '*habitual repetition*' (DeLanda 2006: 50, italics in original).

Is a *weak* ANT, used analogically, more useful in urban studies than the orthodox version? Perhaps what ANT offers is an unusually rich heuristic device rather than a formal method for studying cities. It is a metaphoric approach that encourages a highly developed sense of urban complexity, of the unities and disunities, of the stabilities and instabilities, and especially the complex and heterogeneous networks of connection and association out of which the city as a social and as a physical entity is formed and sustained. It is a sensibility that encourages one to think past the outer surface of the urban world, heightening the visibility of the complex array of working parts that sustain and nourish metropolitan life. In effect, *Paris: ville invisible* tells us that the 'real' city is somehow the unseen city, brought partially to light with the photographs, like the physiology charts in a doctor's office.

THE URBAN IMAGINARY

In fact, *Paris: ville invisible* enacts as sociology the urban theory developed a generation ago by the American urban theorist and planner Kevin Lynch. His remarkable book, *The Image of the City* (1960), reports a series of studies of 'cognitive mapping' as done in Boston by ordinary people in their daily lives. He showed by way of the cognitive maps people made that we live in cities of our own making, but one that shares elements with others. In Boston, for example, the Hancock Building and Boston Commons show up in nearly all cognitive maps. Our own imagined city shares key icons with the city of others. A shared city, a public city, does not disappear. I think that is what *Paris: ville invisible* offers. While many of its images are idiosyncratic to Latour and Hermant, others are familiar to anyone who has spent time in Paris. The city is both a common possession and a particular or unique city for all. This is what Lynch discovered in the cognitive maps of Boston that were carried in the heads of a wide variety of Bostonians. He found that, while each participant in the survey noted unique landmarks, across the whole group there were, Lynch wrote, 'sets of images, which more or less overlapped and interrelated' (Lynch 1960: 85). The city is imagined in parts, but everyone has an imagined whole, and the icons of that whole are surprisingly similar. This common map suggests a patterning that enables common city living and a sense of commonality among city dwellers. The urban imaginary is part of the assemblages, including the media, that make up the city. If the effect of ANT is to dissolve the uniform space of the city by revealing the networks and assemblages that organize the city, the city becomes whole again in the imagination that organizes experience from fragments.[10]

Here the literary critic James Donald is useful:

> To put it polemically, there is no such *thing* as the city. Rather, *the city*
> designates the space produced by the interaction of historically and
> geographically specific institutions, social relations of production and
> reproduction, practices of government, forms of media communication,
> and so forth. By calling this diversity 'the city,' we ascribe to it a coher-
> ence or integrity. *The city*, then, is above all a representation. But what
> sort of representation? By analogy with the now familiar idea that the
> nation provides us with an 'imagined community', I would argue that the
> city constitutes an 'imagined community'.
>
> (Donald 1992: 422)

Donald's response is both correct and too easy. The urban imaginary does
not arise arbitrarily or out of nothing. It is not independent of history
and experience; it is built upon the habitual use of the city. The assemblages
that constitute the city must have enough interconnection for the imagination
to be able to make a whole out of them, even though they are not solidly
so. City dwellers can patch together their city. Their imaginations can fill
in the spaces, make the connections, and make it a fit place into which they
can insert themselves. Various means of communication play a role, from
advertising, to literature, to movies, but also less formal means of communication and human interaction, visual awareness, even a sense of *being there*
(and that others were not). Such an imaginary metropolis gives form to what
is in the analysis of the sociologist or historian only partially interconnected
and bounded.

That urban imaginary is subjective, a point well made by Williams in this
volume. But so is it inter-subjective, a point she also makes. Two writers, Jules
Verne and Charles Baudelaire, lived in Paris at the same time. Williams's
description of the network in which Verne was incorporated – 'an interactive
network of individuals, social organizations, technological systems, cultural
products, and non-human nature' – is matched, but not precisely duplicated,
in the case of Baudelaire. Verne's personal history before coming to Paris, as
Williams shows, was a part of the 'endless loops of co-evolution, material,
social, and artistic change'. Baudelaire's experience was both similar and
distinctive. The two writers lived in the same city, and they produced representations of its particular modernity that were clearly of the same city yet
with a different flavor. For both, Paris was 'not a happy life experience', but,
she notes, it was 'a defining one and critical in shaping their art'.

Both produced dystopian representations of the modernity of Paris, but
Verne invoked a nature that was not part of Baudelaire's personal or literary
repertoire. Actor-networks are rich in contingencies, and these two writers
reveal that. Verne's novel of Paris was not published in his lifetime. What if it
had been, Williams asks us? What if critics had engaged it? Critics have made
Baudelaire's vision of Paris the key to literary and historical understandings

of modernity in Paris. That urban imaginary has powerfully shaped all subsequent inquiries. It is indeed part of the assemblages that constitute Paris, past and present. But what if Verne's recently discovered literary production, the product of many of the same networks, had competed with Baudelaire's? Williams asks: 'How does it change our understanding of the relations between material and literary history if we examine Verne as another representative artist of the capital of the nineteenth century?'

Not only writers, but residents and visitors produce for self-use urban imaginaries. That is what Lynch discovered. And in them the disarticulated social practice of urban life and the imagined city are at once in tension with each other and collaborators in making the experience of urban culture seem coherent. If, as Lefebvre argues, the gift of the city is the production of itself as an *'oeuvre'*, that work is on-going, the joint work of every resident, always open to revision (Lefebvre 1996). The imagination serves as metaphorical glue, ordering or grasping the cumulation of assemblages. The issue here is not one of correspondence of the social and mental, but rather whether the imaginative construction of the city sustains a sense of social ordering (Law 1994: 25–6). Put crudely, metropolitan culture is a benign deceit. That said, reification is the danger, for it masks conflicts, differences, and disjunctures as it unifies, thus misleading our understanding of lived experience. It also invites us to presume the 'thingness' of it all, when in fact it is a perpetual process. Thus Latour usefully exposes in *Paris: ville invisible* the play between the experienced city and the imagined city. Experience both sustains the urban imaginary and disrupts it. The urban imaginary, like history, is incorporated into the urban assemblage, but that in turn prompts yet another formation, transformation, and realignment of networks. This mobility opens up space fissures and fractures in our experience that protect us from the dangers of reification.

The tendency in ANT to refuse dualisms, whether human and non-human, cause and explanation, and its openness to heterogeneity, bears a resemblance to the classic American pragmatism of William James and John Dewey, who were also hesitant about wholes, wholly opposed to absolutes, and comfortable with rather undifferentiated chains of causation. Such a notion of society and social change seems to invite a particular approach to politics, one very much in the spirit of pragmatism. ANT points our analysis and politics in a direction that is both humble and ethically sound. It promises many points of interventions of various sorts and supports an inclusive, practical, incremental approach to scholarship and politics that brings together act, causation, and responsibility. Ash Amin and Nigel Thrift make the point that the modern city is 'so continuously in movement' and so bountiful in 'unexpected interactions' that it is rich in 'resources for continuing political intervention' (Amin and Thrift 2004: 232). It is important as well that Callon comments that the heterogeneous narrative that stands as an explanation in ANT will distribute 'as many voices as there are actors' (Callon 1991: 152).

I am struck that, in his defense of the politics of ANT, Latour makes reference to John Dewey's *The Public and its Problems* (1927), Dewey's only work of political theory. Unlike most political theorists who begin with the origin of the state, Dewey, as a pragmatist, is interested in consequences. That is where politics counts. Dewey judged political acts and the truth value of ideas by their consequences. Dewey imagined politics in the form of an active public taking on a specific social formation (an actor-network?) prompted by a danger to social well-being. It would be initiated by the people most affected, or likely to be affected, and it would then build, broadening the network. What Latour calls 'matters of concern' would for Dewey be the prompt for such acts. Dewey imagined them to arise when state or private actions produce harm to a third party or parties. There is no permanent public for Dewey; it is an ad hoc response of victims and their allies who put pressure on democratic governments for remedy. Latour is not far from that understanding when he suggests that parliaments are supplemented by 'other assemblages' or provisional publics (Latour 2005a: 41). ANT would add non-human responsibles (whether hurricanes, environmental degradation, or freeways and other infrastructure that enforces class division and racial segregation) to the capitalists and governments Dewey had in mind. I doubt that Dewey would object to this expansion of causation and political judgment. The human consequences were always of greater concern to him than origins and motives. And I think that is a reasonable platform for both scholarship and politics.

NOTES

1　As I proceeded I became aware of work on urban networks by sociologists at Berkeley and Toronto: Claude Fischer and Barry Wellman, who has since then produced a large body of work, but at that time it was his early work that I knew (Fischer 1977; Wellman and Craven 1973).

2　I had published an essay on the 'new metropolitanism' stressing the importance of the environmental issues of ex-urban and rural regions of metropolitan areas and possible links with the urban environmental justice movement as a means of breaching the political divide between city and suburb or region (Bender 2001).

3　This problem of representation is similar to one that is revealed in the representation of empires on schoolroom maps, which grossly distort the nature of imperial authority. The colonial possessions that a century ago covered the greater part of the earth were color-coded: pink for the British empire and, usually, green for France, for example. That way of representing empire projected a sense of uniform and total control. Historians of empire today, however, point out that in almost all cases power relations though unequal were negotiated, not absolute. Close examination of actual conditions, moreover, often reveals large areas, often frontier areas, where control was weak and incomplete. Of course these monochrome and featureless maps also fail to say anything about terrain or the ecologies of animal, plant, or human social life. That same superficiality is characteristic of scholarly images and linguistic descriptions of the city. We need a finer-grained analysis of the social entities and material stuff, whether natural or made, that constitute cities. We need a theory of the city that recognizes it not as a unitary, or homogeneous, whole but rather as a composite, not organic, that is semi-stable. This usage of 'composite' derived from current use in the historiography of early modern monarchies and empires is now being presented (Elliott 1992).

4   Even before I wrote *Community and Social Change in America* (1978), Granovetter had published an article (1973) that is recognized by ANT theorists as an early move in the direction of the analysis that became ANT.

5   In defining scale Manuel DeLanda (2006: 6–7) draws upon the usage of 'extension' in physics to develop a sense of extension that is not a geometric one, but one with high energy and numerous components.

6   For markets, see the work of Michel Callon (1998; Callon *et al.* 2007).

7   Recently, another study of Chicago focused on the internal industrial development of the city, written by a geographer, also uses a network analysis, though again without any reference to ANT (Lewis 2008).

8   Latour was not the first twentieth-century sociologist to turn to Tarde to challenge orthodox social theory. It was used by Paul Lazarsfeld and Elihu Katz to challenge the 'mass society' theories of the 1940s and 1950s (Katz and Lazarsfeld 1955).

9   Note that Callon, as in this quotation, in contrast to Latour, tends to omit the hyphen in the actor network.

10 For more on urban imaginaries, see Çinar and Bender (2007).

BIBLIOGRAPHY

Amin, A. and Thrift, J. (2004) 'The Emancipated City', in L. Lees (ed.), *The Emancipated City? Paradoxes and Possibilities*, London: Sage.

Bender, T. (1978) *Community and Social Change in America*, New Brunswick, NJ: Rutgers University Press.

Bender, T. (2001) 'The New Metropolitanism and the Plurality of Publics', *Harvard Design Magazine*, Winter/Spring: 6–17.

Bender, T. (2002) *The Unfinished City: New York and the Metropolitan Idea*, New York: New Press.

Bender, T. (2006) 'History, Theory and the Metropolis', CMS Working Papers Series, No. 005-2006, Center for Metropolitan Studies, Berlin, Online, available at: http://www.geschundkunstgesch.tu-berlin.de/uploads/media/005-2006_01.pdf.

Benjamin, W. (2002) *The Arcades Project*, trans. Howard Eiland and Kevin McLaughlin, prepared on the basis of the German volume edited by Rolf Tiedman, Cambridge, MA: Harvard University Press.

Callon, M. (1987) 'Society in the Making: The Study of Technology as a Tool for Sociological Analysis', in W. Bijker, T. Pinch and T. Hughes (eds), *The Social Construction of Technical Systems*, Cambridge, MA: MIT Press.

Callon, M. (1991) 'Techno-economic Networks and Irreversibility', in J. Law (ed.), *A Sociology of Monsters: Essays on Power, Technology, and Dominance*, London: Routledge.

Callon, M. (1998) *The Law of Markets*, Oxford: Blackwell.

Callon, M., Millo, Y. and Muniesa, F. (2007) *Market Devices*, Oxford: Blackwell.

Çinar, A. and Bender, T. (2007) *Urban Imaginaries: Locating the Modern City*, Minneapolis: University of Minnesota Press.

Cronon, W. (1991) *Nature's Metropolis: Chicago and the Great West*, New York: Norton.

DeLanda, M. (2006) *A New Philosophy of Society: Assemblage Theory and Social Complicity*, London: Continuum.

Deleuze, G. and Parnet, C. (2002) *Dialogues II*, trans. Hugh Tomlinson and Barbara Habberjam, revised edition, New York: Columbia University Press.

Dewey, J. (1927) *The Public and its Problems*, New York: H. Holt.

Donald, J. (1992) 'Metropolis: The City as Text', in R. Bocock and K. Thompson (eds), *Social and Cultural Forms of Modernity*, Cambridge, UK: Polity Press.

Durkheim, É. (1982) *The Rules of Sociological Method*, ed. S. Lukes, trans. W.D. Halls, New York: Free Press.

Elliott, J. (1992) 'A Europe of Composite Empires', *Past and Present*, 137: 48–71.

Fischer, C. (1977) *Networks and Places: Social Relations in the Urban Setting*, New York: Free Press.

Graham, S. (2006) 'Cities under Siege: Katrina and the Politics of Metropolitan America', Online, available at: http://understandingkatrina.ssrc.org/Graham/.

Graham, S. and Marvin, S. (2001) *Splintering Urbanism: Networked Infrastructures, Technological Mobilities, and the Urban Condition*, London: Routledge.

Granovetter, M. (1973) 'The Strength of Weak Ties', *American Journal of Sociology*, 78(6): 1360–80.

Hughes, T. (1983) *Networks of Power: Electrification in Western Society, 1880–1930*, Baltimore, MD: Johns Hopkins University Press.

Katz, E. and Lazarsfeld, P. (1955) *Personal Influence: The Part Played by People in the Flow of Mass Communications*, Glencoe, IL: Free Press.

Latour, B. (1991) 'Technology Is Society Made Durable', in J. Law (ed.), *A Sociology of Monsters: Essays on Power, Technology, and Domination*, London: Routledge, pp. 103–31.

Latour, B. (2002) 'Gabriel Tarde and the End of the Social', in P. Joyce (ed.), *The Social in Question: New Bearings on the History of the Social Sciences*, London: Routledge, pp. 117–32.

Latour, B. (2005a) 'From Realpolitik to Dingpolitik, or How to Make Things Public', in B. Latour and P. Weibel (eds), *Making Things Public: Atmospheres of Democracy*, Cambridge, MA: MIT Press.

Latour, B. (2005b) *Reassembling the Social: An Introduction to Actor-Network Theory*, Oxford: Oxford University Press.

Latour, B. and Hermant, E. (1998) *Paris: ville invisible*, Paris: La Découverte. English: http://www.bruno-latour.fr/virtual/PARIS-INVISIBLE-GB.pdf.

Latour, B. and Woolgar, S. (1986) *Laboratory Life: The Construction of Scientific Facts*, Princeton, NJ: Princeton University Press.

Law, J. (1994) *Organizing Modernity*, Oxford: Blackwell.

Law, J. (2001) 'Ordering and Obduracy', Centre for Science Studies, Lancaster University, Online, available at: http://www.lancs.ac.uk/fass/sociology/papers/Law-Ordering-and-Obduracy.pdf.

Lefebvre, H. (1996) *Writings on Cities*, ed. and trans. E. Kotman and E. Lebas, Oxford: Blackwell.

Lewis, R. (2008) *Chicago Made: Factory Networks in the Industrial Metropolis*, Chicago: University of Chicago Press.

Lynch, K. (1960) *The Image of the City*, Cambridge, MA: MIT Press.

Mitchell, J. Clyde (1969) *Social Networks and Urban Situations: Analyses of Personal Relationships in Central African Towns*, Manchester: University of Manchester Press.

Sheller, M. (2004) 'Mobile Publics: Beyond the Network Perspective', *Environment and Planning D: Society and Space*, 22: 39–42.

Smith, M.P. (1999) 'Transnationalism and the City', in R. Beauregard and S. Body-Gendrot (eds), *The Urban Moment*, Thousand Oaks, CA: Sage, pp. 119–39.

Walker, M. (1971) *German Home Towns: Community, State, and General Estate, 1648–1871*, Ithaca, NY: Cornell University Press.

Wellman, Barry and Craven, P. (1973) 'The Network City', *Sociological Inquiry*, 43: 57–88.

White, H. (1992) *Identity and Control: A Structural Theory of Social Action*, Princeton, NJ: Princeton University Press.

White, H. (1995) 'Where Do Languages Come From? I. Switching between Networks II. Times from Reflexive Talk', Center for the Social Sciences, Pre-print series, Columbia University.

Index

Abler, Ron 205
accumulation 30, 42
Actor-Network Theory (ANT) 2–8, 16,
    20, 110, 155n21, 304–20; buildings as
    technologies 164; CCON project
    99–100; centres of calculation 84;
    controversies 176n9; co-production of
    knowledge 124; critiques of 151;
    description of 85; 'flat' space 53, 54;
    innovators 132; Large Technical
    Systems 131; local networks 86;
    objects 8–9; ontology 13, 225, 231–2;
    political issues 204; quasi-
    technologies 166; scale 67, 74–5;
    Shields on 298–9; Simmel's work as
    early form of 229, 230–1; as 'sociology
    of associations' 73; socio-technical
    assemblages 197–8; technology 144,
    175n3, 186; terminology 126; 'the
    social' 232; 'third elements' 240;
    Thrift on 112–13; topology 37;
    'tragedy of culture' 239
actualization 223–4, 225, 226, 297
Adorno, Theodor 285n12
affordances 297–8, 299, 300
agencement 14, 15
agency 15, 245, 308, 309
agglomeration 27, 29, 30, 37, 39
agora 184, 192n8
Aibar, E. 3
Akrich, Madeleine 126, 144
ambage 48
ambiguity 48
Amin, Ash 7, 14, 109–10, 116, 319
anonymity 12, 211, 231
ANT see Actor-Network Theory
Appadurai, Arjun 63
Aramis transport system 132
architecture 4, 260

Ardila-Gómez, A. 133
arenas of development 133–6
Aristotle 285n12, 313
art 32, 279–80, 281, 282, 283
ASEAN see Association of Southeast
    Asian Nations
assemblages 14–15, 205, 249, 316–17;
    blind practices 247, 248; buildings as
    technologies 164; clusters 37; flat
    approach 65; networks 312;
    relationality 85; Simmel 239;
    socio-technical 197–8; 'thinging' 235;
    tourism 225–6
association 3, 187; ANT as 'sociology of
    associations' 73; flat approach 65;
    'forms of' 229, 231; translation 240
Association of Southeast Asian Nations
    (ASEAN) 80
Asturias 18, 91–108
atmosphere 66, 67
ATMs 243, 244–6
Attali, Jacques 179, 180, 182
Augé, Marc 12
Augoyard, J.-F. 180, 191n3
Aviles 91–108

Back, Les 182
banal overflows 210–11, 213
Barcelona: cacerolada 179, 191n1; as
    'cool city' 102–3; entrepreneurial
    class 293; May Day 2005
    demonstrations 184–6, 187–8
Barco, Virgilio 135
Baudelaire, Charles 20, 270, 272,
    279–80, 281–2, 284n3, 318–19
becoming 239–40
Benjamin, Walter 55–6, 270, 272,
    285n12, 313
Bergson, Henri 13–14, 300

Berlin: bus route 100 219–23; Marathon 54, 55, 56, 60, 61–2, 68n1; Videobustour 215–19
Berlin Wall 216, 219, 226n1, 300
Berman, Marshall 272
Bijker, Wiebe 3, 144, 153, 155n21
Bilbao Guggenheim 94, 95
blind people 231, 244–8
bodily practices 55–6
'Body without Organs' (BwO) 222
Bogotá 18, 123–38; collective transport system 128, 129, 131, 133–4, 136; metro 134–5, 136; private cars 135–6; Transmilenio 123, 124–5, 127–33, 135, 136–7, 315–16
Boltanski, Luc 294–5
boosterism 61–2
borders 203
Boston 56, 60, 199, 317
Boudon, Philippe 167
boundaries 118–19, 131–2
bounded cities 10, 12
bourgeoisie 281
Brain, David 4
Brasher, Chris 55
Braudel, Fernand 86
Brenner, Neil 57, 58, 59, 63, 64
Brighton 295
buildings 161–78; change of use 18–19, 161–2, 167–74; gentrification 292; high-rise 142; as 'moving projects' 4; as 'mutable immobiles' 162, 166–7, 169, 174, 175; as quasi-technologies 162, 165–6, 168, 174; as technologies 162, 163–4, 169, 170–1, 174
Bull, Michael 182, 192n7
Burgess, E.W. 9
bus stations 129, 130–1
bus tours 214–23
buses 124–5, 127–31, 133–4, 136, 315–16
Butcher, W. 284n6, 285n8

caceroladas 179, 180, 181, 183, 191
Callon, Michel: Actor-Network Theory 113, 304, 306, 309, 314–15, 319, 321n9; architecture 4; assemblages 312; economics 6; generalized symmetry 3; innovation 132; 'obligatory passage points' 150; overflowing and framing 213–14; political ecology 5; technological development 144; transformation 316
Camagni, R. 39
Canadian North 295–6

canals 276, 285n10
capitalism 10, 61, 86, 282; critical geography 116; embeddedness 157n56; new spirit of 294; scalar analysis 63
Carlson, W. Bernard 144
Carrier, James 267n3, 299–300
cars: Bogotá 135–6; Maastricht 140, 149
cash points (ATMs) 243, 244–6
Castells, Manuel 254
CBOT *see* Chicago Board of Trade
CCON *see* Centre Cultural Oscar Niemeyer
Celan, Paul 229
central place theory 10
Centre Cultural Oscar Niemeyer (CCON) 91, 92–107; cultural leveraging 97–9; cultural policy 93–6; scaling practices of youth 100–7
Charlton, Peter 75
Chiapello, Eve 294–5
Chicago 255–66, 321n7; Cronon's work 5, 307–8; futures markets 253, 254–5, 262–5, 266; Marathon 55, 56; temporality 310
Chicago Board of Trade (CBOT) 20, 253, 255–66
Chicago School 9, 11, 114, 297, 303
Chile 111; *see also* Santiago experimental music scene
China 79, 83
Christaller, Walter 10
cities 9, 12–13; Actor-Network Theory 304–20; affordances 297–8; 'cool' 102–3; determinism 198; distinction from 'the urban' 189, 296–7; as fluid objects 12; Latour's work 113–14; as multiple objects 14; obdurate objects 139, 143–5; relationality 85; scale 57, 74; Simmel 248, 249; technopolitics of sound 188–9, 191; Thrift on 109; urban imaginary 317–19; as 'value loci' 253–4, 258, 265–6; as virtualities 115; *see also* urban studies
'city border philosophy' 147
Clark, T.J. 272
class 309
clustering 17, 27–8, 29, 31; Eurocentric model 49; firm identities 43; 'immutable immobiles' 48; Santiago experimental music scene 32, 33, 36; 'support spaces' 39; topology 37
'cognitive mapping' 317

Colin Ng & Partners (CNP) 77, 78, 79, 81
collectivity 187–8, 190
Collier, S.J. 15
colonialism 320n3
commotion 182, 187–8, 189, 190
communication: blind people 244; sense-making 223–4; tourist 210, 223, 224–5; urban imaginary 318
communication technologies 44, 45–6, 197; *see also* Internet; mobile phones
communities of medium 273
community 303
Compère, D. 286n16
concrescence 14
congestion 135–6, 149
constructivism 85, 127
consumption 39, 40, 49n1, 102, 231, 294
continued existence, privilege of 169–71
continuity 36–7
conviction 182, 187–8, 189
'cool cities' 102–3
'cool globalization' 106
Le Corbusier 167
Coutard, O. 4
Cowper, William 304
Crawford, Margaret 55–6
creative class 31, 293–4
Cronon, William 5, 255, 267n2, 286n12, 306, 307–8
cultural capital 92–3, 97, 98, 105
cultural industries 31, 48–9, 93–4, 95, 293
cultural leveraging 92–3, 97–9
'cultural life' 238–41
cultural production 30, 31, 32–3, 94
culture: CCON project 92, 93, 95–6, 97–8, 104; 'global' 97–8, 99–100, 102, 104, 106; 'tragedy of' 234, 235, 237–8, 239, 241; urban 11–12
'curvature' 186–7
'cyborg urbanization' 4–5

Dakar 49
de Certeau, Michel 11, 12, 210–11, 297
decenteredness 40, 42, 43
Degen, Monica 294
DeLanda, Manuel 5, 15, 316–17, 321n5
Deleuze, Gilles 7, 14, 73, 198; assemblages 316; becoming 239–40; 'Body without Organs' 222; sense-making 223–4; the virtual 300
Delgado, Manuel 11, 12, 211
delocalization 74

democracy 115, 184
demonstrations 184–6, 187–8
DenCities 233
density 232–3
Descartes, René 236
de-scription 126, 127
determinism 4, 18, 197, 198
de-territorialization 190, 209, 210
developing world 49
Dewey, John 293, 319–20
dialogue 107
Dingemans, F.C.J. 142
disability 244, 245–6, 247
disruption 198
'divided city' 11, 13
dominant frames 144, 145, 152–3
Donald, James 318
Drew & Napier (DN) 78, 79, 81
'dual city' 11
Dupuy, Gabriel 204–5
Durand, J.-N.-L. 164
Durkheim, Émile 310, 311

ecological approach 9–10
ecology, urban 114–15
economic development 29–30, 31, 49
economics 6–7, 10–11, 19–20, 111
electricity 278–9, 306–7
Ellis, Cliff 144
embeddedness 7, 116, 144, 150, 151, 157n56
entrepreneurialism 61–2, 293–4, 308
environmental issues: acoustic pollution 180, 191n2; Maastricht highway 140, 147, 149; product obsolescence 200
events of consciousness 283
everyday life 11, 210–11, 213
'exopolis' 10, 13
experimental methodologies 111
ExtenCities 233
extension 232–3, 321n5
'Extraordinary Journeys' (Verne) 269, 270, 271, 272

family 39
finance 253, 254–5, 258–66
financial services 84–5, 88n12
flâneur 5
flatness 17, 53–4, 63, 65–6, 84, 117, 307, 312
Florida, R. 31
flows 66, 118
'fluid spatiality' 46, 47, 48
'fluid technologies' 215

'fold' 186–7
Fotolog 105, 106
Foucault, Michel 6, 187
frames 212–14, 218, 219, 222–3, 224
France 199, 205; *see also* Paris
French, Sally 249
futures markets 253, 254–5, 262–5

gated communities 202
gel 47–8, 309
generalized symmetry 3
gentrification 30, 161, 162, 291–3
geographic romance 273, 279, 284n6
geography: critical 111, 116; economic
    6, 29–30, 116; Marxist 203–4; scale
    53, 57, 58–9, 203
Germany 265; *see also* Berlin
Gertler, M. 29, 30
Gieryn, T. 29
Girard, M. 5
the 'global': CCON project 91, 98–100,
    106; multiplicity 15; scale 60, 62,
    64–5, 75, 86; young people in Aviles
    102–3
'global assemblages' 15
'global cities' 7, 29, 74, 84–5
globalization 29, 53–4, 291; CCON
    project 92, 99; 'cool' 106; cultural
    production 94; LSE professors 98;
    neo-liberal phase 56; networks 75;
    scale 57, 64, 84; Singapore legal
    services 86
glocalization 64
Goffman, Erving 212–13, 244
Gorman, Michael 144
government 61, 74, 84–5
Graham, Stephen 4, 19, 197–205, 304,
    308
Granovetter, M. 116, 321n4
Great Britain 199; *see also* London
Greeks, ancient 184, 186
Grosjean, M. 180
Guattari, Félix 14, 73, 222, 239–40
Guggenheim, Michael 18–19, 161–78,
    309
Guy, S. 4

Habermas, Jürgen 295
Haraway, Donna 4–5
Harris, Marvin 203
Harvey, David 58–9, 157n56, 204,
    272
Haussmann, Baron 270

Hegel, G.W.F. 285n12
Heidegger, Martin 234–5, 281
Held, David 53–4
Helen Yeo & Partners (HYP) 77, 78, 79,
    81
Hermant, Emilie 6, 310, 312, 317
Hetzel, Jules 269, 270, 271, 284n6
high-rise buildings 142
highways 18, 140–51, 152, 153,
    316
hiphop culture 103, 106
history 217–18, 310
Hong Kong 79, 80, 83–4
Honolulu 55, 56
Hopkins, J. 192n6
horizontal linkages 27, 33–6, 45
Houellebecq, M. 73
hubs 42, 74
Hughes, Everett 110
Hughes, Thomas 3, 125, 144, 306–7
Hurricane Katrina 304, 305
hybrid ecologies 114
hybridity 32

identity 43–5, 137
images 295, 296, 317
imagined community 318
immanence 238, 239
'immutable immobiles' 48
'immutable mobiles' 8, 47, 82, 166,
    230–1, 243; *see also* 'mutable
    immobiles'
indifference 12, 231, 233–4
industrial districts 30, 94, 161
infrastructures 8, 308; Chicago 255,
    256–7; military targeting of 201;
    networked 197–8, 200, 202, 205;
    politics of scale 203; study of 203–5;
    technological failures 198–9; 'value
    loci' 254
innovation process 132
innovators 126, 131, 133
in-scription 126
institutions 96
IntenCities 233
intermediaries 240, 247, 248
Internet: online trading 262–5, 266;
    Santiago experimental music scene 33,
    44, 45–6; as scaling device 100; young
    people in Aviles 105, 106
interruption 198, 199
Iraq 191n1, 200–1
Israeli Army 201, 202
itio in partes 236–7, 238, 239, 244

James, William 319
jogging 64, 68n2
joint law ventures (JLVs) 80–1, 82–5, 88n9, 88n10–13, 89n15
Jørgensen, U. 132

Kant, Immanuel 236, 285n12
Katz, Elihu 321n8
keying 213
Khattar Wong & Partners (KWP) 88n7
King, Anthony 294
King, D. 31
Klein, H.K. 153
Kleinman, D.L. 153
Kloosterman, R. 31
knowledge 29–30, 293; co-production of 124; Singapore legal services 82
'knowledge economy' 29, 33
Konvitz, Josef 139

Large Technological Systems 3, 124, 125, 131–2, 155n21, 276
Latour, Bruno 2, 13, 82, 110; Actor-Network Theory 7, 299, 304–5, 306, 309, 310, 311–12, 319–20; Aramis transport system 132; 'assembling' 15; associations 187; capitalism 86; causative nature of objects 297; on Cronon 307; delocalization 74; Durkheim's influence on 311; experienced/imagined city 319; frames 213; 'immutable mobiles' 8, 166; local interactions 65; myopia 248; 'obligatory passage points' 150; Paris 6, 113, 310, 312–13, 314, 317; political ecology 5, 204; scale 66–7; social structures 314; on sociology 229–30; technology 144, 163, 186; things 235; Thrift on 113–14
Law, John 13, 36–7, 42, 46–7, 144, 304
law of buildings 167–74, 176n11
Lazarsfeld, Paul 321n8
Lebow, Fred 54–5, 60–1, 67n1
Lee Kuan Yew 75–6
Lefebvre, Henri 12, 58–9, 298, 300; city/urban distinction 189; everyday life 11, 210; gift of the city 319; time 310
legal services 75–87
Leibniz, G.W. 193n13
liberalization 80, 88n12
literature 279–80, 282

the 'local': CCON project 91, 98–100, 107; economic development 29; scale 62, 64, 75, 86; young people in Aviles 101–2, 104–5
locality: cultural production 31; economic development 29, 30; meaning of 92; young people in Aviles 101, 102
localized economy 28, 30, 33, 43
London: as 'cool city' 102; financial services 85; futures markets 264, 265, 266; law firms 76; Marathon 55, 56, 61–2, 68n1; public investment 74; Underground 131
London School of Economics (LSE) 91, 98, 99
Los Angeles 56
LSE *see* London School of Economics
Luhmann, N. 47, 223–4
Lukács, Georg 285n12
Lynch, Kevin 317, 319
Lynch, Michael 7
Lyotard, J.-F. 75

Maastricht 18, 139–59, 316
Madrid 102–3
Maillat, D. 39
Maine, Henry 310
maintenance 4, 65, 198, 199–200
Manchester 293, 294
marathons 54–7, 64, 67n1; flat approach 65; generation of affect 65–6; scale 60–2, 63
Marcus, G. 14–15
Marcuse, Peter 10, 11
marginalization 32
markets 19–20, 253, 254–5, 256, 257; expert 258–62, 266; virtual 262–5, 266
Markusen, A. 31
Marshall, Alfred 30
Marston, Sallie 66
Marvin, Simon 4, 197, 308
Marx, Karl 86, 283, 285n12, 286n19
Marxism 1, 30, 86, 204
Massey, Doreen 53
Maturana, Humberto 214
May Day demonstrations (2005) 184–6, 187–8
McKenzie, R.D. 9
McMaster, Robert 59
mediation 230–1, 239, 240, 243, 247, 248
mental models 144
meta-narratives 1, 291
methodologies 111

Miller, Daniel 211–12, 299–300
Mitchell, J. Clyde 303
mobile phones 105, 200
modernism 272, 283
modernization theory 303, 310–11
Mol, Annemarie 13, 14, 42, 46–7, 144
molecular sociology 229–30
money 230, 234, 235, 237, 239, 241–8
Morris, M. 211
MSN Messenger 105
multiculturalism 102, 103
multilingualism 103–4
multiplicity 7, 13–14, 19; networks
    306–10; Santiago experimental music
    scene 42; tourist assemblages 225–6
Mumford, Lewis 55–6, 205
Municipal Flat, Maastricht 142, 143
music: 'cool cities' 102–3; hiphop
    culture 103, 106; May Day
    demonstrations 184–5; music culture
    104; personal stereos 192n7; public
    space 192n6; Santiago experimental
    music scene 27–52; soundspheres
    183
music venues 40, 42
'mutable immobiles' 19, 162, 166–7, 169,
    174, 175; *see also* 'immutable mobiles'
'mutable mobiles' 46–7, 48, 168, 174
myopia 229–30, 248–9
MySpace 44, 45–6, 49n4, 106
myths 197, 296, 297

Nantes 269, 272–3, 285n10
narrative arrangements of bus tours 214,
    217–18, 220–1
nature 5, 281–2, 283, 286n18, 304
nearness 235
Neff, Gina 4
neo-conservatism 201–2, 300
neo-liberalism 56, 64, 74, 117, 205
neo-Marxism 6, 63, 86
neo-structuralism 86
Nero, Emperor 184
networks 66, 303–4, 316; Callon
    314–15; capitalism 86; CCON project
    92; definitions of 82; gel concept 47,
    48; global management 283; hubs 42;
    Latour 311–12; multiplicity of
    306–10; Santiago experimental music
    scene 37, 42, 46; scale 67, 75; Shields
    on 298, 299; Singapore legal services
    75, 76, 82–4, 86–7; Thrift on 109;
    see also Actor-Network Theory
Neufert, E. 164, 176n6

New York City: blackouts 198; as 'cool
    city' 102; financial markets 256;
    financial services 85; Marathon 54–5,
    56, 60–2, 68n1; public assembly 5
Niagara Falls 295–6
Niemeyer, Oscar 97
Nietzsche, F.W. 285n12
nodes 35, 37, 42, 47
noise 179–80, 182, 183
non-city 12, 13
non-representational theory 6, 18, 20,
    73, 110

obduracy 18, 139–40, 143–5, 150–3,
    155n23, 199, 310, 316
objects 8–9, 11; causative 297;
    continuity 37; 'culture of' 234, 235–6,
    238; design of 126–7; enactment of
    13, 15, 37; 'in-hereness' 13; money
    241, 243, 247; multiplicity 14;
    quasi-technologies 165–6; social
    relations 242
'obligatory passage points' 150
obsolescence 200
Offner, Jean-Marc 205
Ong, A. 15
online trading 262–5, 266
ontology 13, 225–6, 231–2
overflows 210–12, 213, 218, 222

Paris 317, 318–19; Aramis transport
    system 132; Latour on 6, 113, 310,
    312–13, 314; Marathon 55, 56; Metro
    131; public investment 74; Second
    Empire 269–83
Paris in the Twentieth Century (Verne)
    271–2, 274–9, 282
Park, R.E. 9, 11, 110
Pastrana, Andrés 135
Peñalosa, Enrique 123, 130, 135, 136,
    137
performativity 91, 99, 107; bus tours
    214, 218–19, 221; clusters 37
Perrow, Charles 199
persistent traditions 145, 151–2,
    155n23
Philadelphia 256
Pink, S. 192n6
Poe, Edgar Allan 284n3
poetry 279, 280, 281–2
Polanyi, Michael 29
polis 184, 186, 190, 313
political ecology 5, 204
political economy 203, 299–300

politics: Actor-Network Theory 319–20; Dewey 320; duty of association 187; maintenance and repair 199–200; new technologies 205; scale 59, 63, 203; technopolitics of sound 182, 183–8, 189, 191; Transmilenio 125; urban studies 203–4; virtualities 301
pollution, acoustic 180, 191n2
Popayan 49
post-institutionalism 7
post-structuralism 6, 75, 110, 197–8
power 59, 117–18, 184, 193n13
pre-inscription 126
Prince of Asturias Trust 97
privatization 202, 205
privilege of continued existence 169–71
production 7, 39, 49n1, 294;
see also cultural production
professional worldviews 144
Project A 262–3, 265
project city 295
protest: *cacerolada* 179, 180, 181, 183, 191; May Day demonstrations 184–6, 187–8
Proust, Marcel 300
public space 19, 189–90, 192n6, 202
public transport *see* transport

quasi-technologies 162, 165–6, 168, 174, 309
Quincy, Quatremère de 175n5

radical relationality 3
rail transport: British/French railway systems 199; Chicago 255, 256, 257
Raymond, Marcel 281
realism 274, 278, 280–1, 282, 286n15
reality 7, 13–14
'reflexive gentrification' 292, 293
'reflexive realism' 280–1, 286n15
reflexivity 107
regeneration 91, 93–4, 95, 96, 104, 292
regimes of justification 294–5
regional development 91, 93–4, 97
regionalization 79, 80, 88n8
relationality 3, 84, 85, 205, 299
repair 4, 198, 199–200
resistance 205
Rice, T. 192n6
Riles, A. 82
Robb, G. 282–3
Romanticism 273, 274, 282
Rumsey, Henry 259, 260, 261, 262

Said, E.W. 282
Saka, E. 14–15
Santiago experimental music scene 27–52; cluster characteristics 33–6; computer-mediated 'buzz' 45–6; as gelleable mobile 48, 49; key spaces 37–43; porous identity 43–5
Sartre, Jean Paul 211
Sassen, Saskia 10–11, 56, 110, 254
scalar processes 17, 74, 84
scale 6, 17, 53, 57–62, 85; abstract nature of 66–7; CCON project 91, 92, 96, 99, 100, 107; gel concept 47; globalization 54; Graham on 202–3; problematical nature of 73, 74–5; Singapore legal services 73, 75, 83, 84, 86; Thrift on 116–17; urban marathons 60–2; weakness of scalar analysis 63–5; young people in Aviles 100–7
scaling practices 100–7, 117
Schaeffer, Pierre 191n3
Schafer, Murray 191n3
Schivelbusch, Wolfgang 55–6
science 110
science and technology studies (STS) 3, 7, 139–40, 204, 225, 232, 304
'science wars' 7
SCOT *see* Social Construction of Technology
second nature 279, 281, 283, 285n12
Seigworth, G. 211
self-consciousness 272, 286n15
self-creation 223–4
Sennett, Richard 56, 192n8
sense-making 210, 223–4, 225, 293
sensory practices 230, 231, 241, 243, 245, 247–8
Serres, Michel: Actor-Network Theory 295; *cacerolada* 181; causative nature of objects 297; quasi-actants 299; quasi-objects 165; relations 298
Sheller, Mimi 47, 48, 309
Shields, Robert 5, 20, 291–301
Short, John Rennie 56, 74
sightseeing bus tours 214–23
Simmel, Georg 11, 19, 55–6, 229–52, 303; 'cultural life' 238–41; 'culture of objects' 234, 235–6, 238; extension and density 232–3; money 230, 234, 235, 239, 241–2; sensory practices 241; social relations 242; 'third elements' 236–7; 'tragedy of culture' 234, 235, 237–8, 239, 241

Singapore 17–18, 73–90
skyscrapers 253, 258–62
Sloterdijk, Peter 118, 182, 187, 188, 192n4
Smith, Michael Peter 308
Smith, Neil 58, 59, 63
the 'social' 3, 15, 232, 239, 240–1, 309, 310
social bonds 181, 190, 212
social capital 92
Social Construction of Technology (SCOT) 3, 153, 155n23
social constructivism 18, 155n23
social polarization 11
social power 59
social relations 181, 190, 230, 242, 248, 310, 311
social science 13, 232, 249, 311; Graham on 198; Shields on 291, 298, 300; Thrift on 111, 112
Social Systems Theory 175n3
sociality 47, 211, 234, 240–1, 248
society 239, 309, 310, 311, 319
sociology 229–30, 232, 310–11
'sociotechnical ensemble' 155n21
sociotechnical structures 139, 197–8; Maastricht highway 143; obduracy 144; tourism 223, 224, 225
Soja, Edward 10
Sontag, Susan 85
Sørensen, O. 132
sound 179–95; CBOT trading floor acoustics 260; as social practice 179–80; soundscapes 181, 191n3; soundspheres 181–3, 191, 192n4; technopolitics of 19, 182, 183–8, 189, 191
space 6, 10, 17; boundaries 118–19; commotion 190; extension and density 232–3; 'flat' 53, 54; 'fluid' spatiality 46, 47, 48; multiplicity 14, 42; online trading 263, 264; Santiago experimental music scene 27–8, 31–2, 37–43, 46, 47; 'scalar' 53, 54; sound in urban space 181, 182, 183, 187–8, 189, 191; spatial arrangements of bus tours 214, 215–16, 220; topology 36–7
spatial approaches 9–10
spatialization 294, 296
spillovers 27, 33
Spinoza, Benedict de 300
St. Louis 255, 256, 308
stability 27, 28, 48
standards 257–8

Stark, D. 5
states 74, 75
Stegmeijer, E. 31
structuralism 1, 6, 298
STS *see* science and technology studies
style 294, 295
subjects 234, 235–6, 241, 242, 243, 247
'support spaces' 39
surveillance 203
Sweden 49n2
Swyngedouw, Erik 5, 58, 59, 63, 64, 204
symbolism 279–80, 281

Tarde, Gabriel 109, 110, 113, 229, 311
Taylor, Peter 56, 57, 58
technological determinism 4, 18, 197
technological frames 144, 152–3
technologies: buildings as 162, 163–4, 169, 170–1, 174; communication 197; definition of 163; 'fluid' 215; mobile 114; money 243, 245, 246; network 202; new 105–6, 205; *Paris in the Twentieth Century* 274, 275, 276; sound 184, 186–7, 190–1; virtual markets 262–5
technology: as adjective 186; failures 198–9; Heidegger on 281; obduracy 144, 145, 151, 152; Santiago experimental music scene 45–6; stabilizing processes 175n3; urban transport systems 133
technopolitics of sound 19, 182, 183–8, 189, 191
telegraphs 257, 260–1
telephones 260–1
terrorism 200–1, 202
'thematization thresholds' 224–5
Thevenot, Laurent 294–5
Thibaud, J.P. 180
things 234–5
'third elements' 236–7, 240
Thrift, Nigel 1, 198, 306; concrescence 14; economic role of cities 7; interview with 18, 109–19; political intervention 319
time 6, 67, 310; 'fluid' spatiality 48; gel concept 48; multiplicity 14; online trading 263–4, 265; Santiago experimental music scene 41, 47
Todorov, T. 298
Tokyo 56
Tolkien, J.R.R. 286n13
Tönnies, Ferdinand 303, 310
topology 36–7, 46

Torgue, H. 191n3
totalizing overflows 211–12, 213
tourism 19, 209–28; bus tours 214–23; Canadian North 295–6; CCON project 95, 96; frames 212–14, 218, 219, 222–3, 224; multiple tourist assemblages 225–6; overflows 210–12, 218, 222; 'thematization thresholds' 224–5; virtuality 223–4, 225
trade 257, 258
transcendence 238, 240
translation 132, 133, 240, 243, 248
Transmilenio 123, 124–5, 127–33, 135, 136–7, 315–16
transparency 107
transport: Bogotá 18, 123–38, 315–16; British/French railway systems 199; bus tours 214–23; Chicago 255–6, 257; *Paris in the Twentieth Century* 276–7
tunnels 142, 146–7, 148–9, 150–1, 152

underground 151–2
Unwin, Timothy 280, 286n15, 286n16
the 'urban' 189, 296–7
urban actor-networks 233–4, 235
urban imaginary 317–19
urban planning 115, 125, 131, 139
urban renewal 161, 292
urban studies: ANT's contribution to 2, 8, 16, 73, 298–9, 304–5, 313; cultural/ political divide 299–300; economic analysis 294; gap in 203; Graham on 203–4, 205; multiple assemblages 226; object of 8–9, 113–14; political issues 203–4; Simmel 231; sound 180, 181, 190; Thrift on 112
urbanism 28, 111, 175; critical 203; splintering 202

US Army 201

value 19–20, 253–67
value-added creation 27, 30, 33
Varela, Francisco 214
VBT *see* Videobustour
venues 40, 42
Verne, Jules 20, 269–82, 284n3, 284n6, 285n11, 318–19
vertical linkages 27, 33–5
Videobustour (VBT) 215–19
violence 200–1
Virilio, P. 184
virtuality 300–1, 313; images 296; political economy 299–300; 'the urban' 297; tourism 223–4, 225; virtual markets 254, 262–5, 266
vision 180, 231, 241, 242, 243–4, 248
visual arrangements of bus tours 214, 216–17, 220

war 4, 201, 202
Washington 56
Weber, Max 10, 254
Wells, H.G. 151–2
White, Harrison 47, 48, 309
Williams, Raymond 273
Wirth, Louis 11
Woolgar, S. 8, 108n1
world cities 56
world economy 57

young people 98, 99, 100–7
YouTube 107

Zimbabwe bush pump 46–7
zoning 161, 162, 168–9, 174, 176n11
Zukin, Sharon 110, 175n1
Zürich 161, 168